ENVIRONMENTAL IMPACT STATEMENTS

ENVIRONMENTAL IMPACT STATEMENTS

A Comprehensive Guide to Project and Strategic Planning

Charles H. Eccleston

JOHN WILEY & SONS, INC.

New York • Chichester • Weinheim • Brisbane • Singapore • Toronto

This text is printed on acid-free paper.

Published simultaneously in Canada.

Library of Congress Cataloging-in-Publication Data:
Eccleston, Charles H.
Environmental impact statements : a comprehensive guide to project and strategic planning : a total federal planning strategy integrating modern tools and techniques / Charles H. Eccleston.
p. cm.
Companion volume to the author's : The NEPA planning process.
Includes index.
ISBN 0-471-35868-1 (alk. paper)
1. Environmental impact statements—United States. I. Title.
TD194.55 .E28 2000
333.7'14'0973—dc21 99-042175

Printed in the United States of America.

10 9 8 7 6 5 4 3 2 1

Contents

CHAPTER 2

Preface

There is enough for all. The earth is a generous mother; she will provide in plentiful abundance food for all her children if they will but cultivate her soil in justice and in peace.

—Bourke Coekran

It is not uncommon to find that an environmental impact statement (EIS) process has been conducted in excruciating detail, to comply with the minutest procedural requirement. Such analyses are prepared with a resolve and perseverance verging on vengeance; they may examine every conceivable impact, significant or not, in excruciating detail, yet blatantly ignore the most basic regulatory requirements regarding efficiency. Inevitably, such practice leads to increased costs and delays, and sometimes to poorly planned projects. This book is not about preparing bigger EISs but better EISs. Instead of presenting a document preparation process, this text focuses on the EIS process as a comprehensive framework for planning future programs and projects. As the reader will become aware, this book is unique as it focuses on the EIS process from a planning perspective. State-of-the-art tools, techniques, and approaches are provided to assist federal agencies in meeting their diverse missions.

As depicted in Figure P.1, this book is structured so that the EIS process is described in a logical progression; each chapter is a building block to the subsequent chapters. The book is designed for use as a companion to the author's text *The NEPA Planning Process: A Comprehensive Guide with Emphasis on Efficiency,* which describes the entire National Environmental Policy Act (NEPA) process and its regulatory requirements. While *The NEPA Planning Process* comprehensively details NEPA from the perspective of what must be done, this text provides a practical hands-on guide that focuses on how EIS planning can most effectively be implemented.

To this end, chapter 1 describes a contemporary step-by-step procedure for managing the preliminary or pre-scoping effort, while chapter 2 provides a comprehensive examination of the formal EIS planning process, with

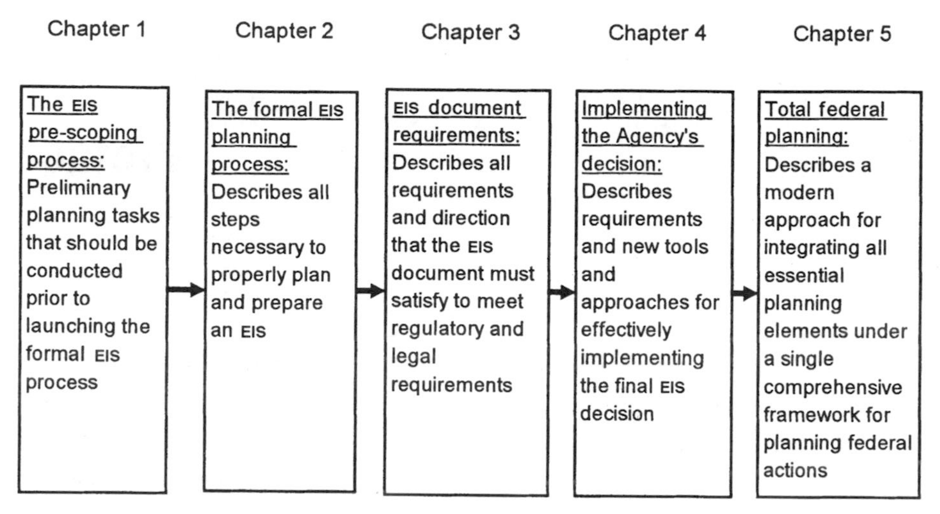

Figure P.1 This book is structured so as to lead the reader through a logical progression whereby each chapter is a building block for the subsequent chapter.

emphasis on application of modern tools, techniques, and methodologies (see Figure P.1, first and second boxes, respectively).

EIS documentation requirements are disseminated throughout NEPA's implementing regulations, executive orders, guidance, and case law. To date, no single text has fully compiled and synthesized all such guidance and requirements into a single source; chapter 3 provides the reader with a single comprehensive source for all significant guidance and regulatory requirements that the final EIS document must satisfy (see third box, Figure P.1). Chapter 4 concludes the EIS planning process by describing the procedure for implementing the agency's final decision once the EIS is completed (see Figure P.1, fourth box). A comprehensive set of checklists for reviewing the adequacy of the EIS and its planning process is provided in the appendixes.

Chapter 5 completes the book by introducing a modern strategy, referred to as *total federal planning* (TFP), that provides a comprehensive approach to unifying large-scale federal planning efforts under a single, systematic, structured, and holistic process (see Figure P.1, fifth box). Under the TFP approach, the EIS planning process provides the unifying glue or framework for integrating all early environmental and nonenvironmental decisionmaking factors into a comprehensive planning process. Modern tools and principles from the disciplines of value engineering, systems engineering, and total quality management (briefly described in the introduction to this book) are applied in enhancing the effectiveness of TFP. The goal of the TFP strategy is a unified and interdisciplinary framework that effectively facilitates decisionmaking while reducing the time and cost necessary to comply with environmental and other planning factors. At a time when Congress is striving to reengineer the governmental framework, TFP offers a systematic approach to integrating what is often a disjointed and convoluted planning cycle. Potentially, this approach has widespread implications in the way agencies approach federal planning.

Objectives of This Book

This book is unique in that it provides the reader with an arsenal of up-to-date approaches, methodologies, and tools for enhancing the effectiveness of federal planning. Emphasis is placed on integrated planning from a holistic perspective, whereby all pertinent decisionmaking factors (e.g., environmental, technical, economic) are incorporated in a single integrated planning process. Important features distinguishing this book include:

- A new strategic approach (TFP) for managing large and complex programs and projects.
- A description of the EIS planning process from the perspective of planning federal actions, as opposed to simply preparing a document. A step-by-step approach is used in comprehensively describing the entire EIS planning process in detail.
- State-of the-art tools, techniques, and methodologies for efficiently managing the EIS process. Emphasis is placed on addressing problems and dilemmas that have traditionally hindered NEPA practice.
- A single compendium that synthesizes and describes all relevant guidance and requirements that the EIS document must satisfy. Lessons from case law are integrated with the relevant requirements.
- Subject matter that is organized to assist the reader in quickly locating issues, problems, and topics of interest.
- The latest ideas, concepts, and direction from the Council on Environmental Quality, professional societies, practitioners, and NEPA experts.

A Word of Caution

I was confronted with two choices in writing this book: (1) introduce the basic elements common to a simple or moderately complex EIS, or (2) provide a comprehensive arsenal of approaches, concepts, tool, and techniques, from basic to advanced, that can be applied to virtually any planning scenario regardless of size or complexity. I chose the later. The intent is neither to instill fear into the novice nor to foster the impression that the EIS process is unduly esoteric or complex. Quite the opposite! Instead, the intent is to arm the reader with modern approaches and techniques from which to choose, based on a given set of circumstances. Not all approaches or elements described in this text are intended to be, or should be, applied to every EIS effort. Thus, for smaller or less complex planning efforts, the reader will probably implement only a subset of the steps and elements presented—those that will contribute most effectively to their particular planning needs. With this in mind, it is left to the reader to exercise professional judgment and discretion in choosing those steps, approaches, and techniques that are most relevant to the individual situation.

Audience

Beginning with fundamental elements and advancing into more complex issues in later chapters, this book is designed for use by beginners and experts alike. While it is primarily aimed at professionals in government, consulting, and the private sector who are involved in NEPA and seek to master it, the book's organization lends itself equally to individuals who desire only an introduction to selected aspects of the EIS planning process. Experienced practitioners may use the book as a resource for quickly reviewing issues or as a comprehensive text describing the entire EIS process. Professionals interested in improving the effectiveness of their planning process will find this text particularly useful.

Individuals, groups, and professionals who will find this book of interest include planners, analysts, scientists, project engineers, regulators, decision-makers, educators, and lawyers, to name just a few. Advocacy or citizen groups involved in EIS compliance issues will also find this book useful as they review and perhaps challenge an agency's EIS compliance process. Finally, the book can be used by those for whom it will be an introduction to the EIS process, including university students in environmental or planning curricula.

In addition to providing a comprehensive overview of the entire EIS process, this book also presents state-of-the-art tools and techniques for enhancing both the efficiency and effectiveness of EIS planning. If you have technical questions or issues, or need assistance, contact me at ecclestonc@aol.com.

Charles H. Eccleston
Chairman, Tools and Techniques (TNT) NEPA Practice Committee
National Association of Environmental Professionals

Acknowledgments

I am indebted to the many people who reviewed and provided comments on this book. While this text does not endorse or necessarily represent the formal view or position of the National Association of Environmental Professionals (NAEP), many respected members of that organization contributed untold hours in reviewing it. Although it is infeasible to mention all individuals by name, I would like to call attention to Mat C. McMillen (Energetics, Inc.), former chairman of the NAEP's NEPA Symposium, performed an in-depth review and contributed important improvements. Particular mention also goes to Judith Lee (Environmental Planning Strategies, Inc., President) and Frederic March (Sandia National Laboratories and former chairman of NAEP's NEPA Working Group), who performed a painstaking review of the entire manuscript and provided numerous suggestions for improvement.

Clifford Duke and Bruce Kemp (The Environmental Company) intensively reviewed the manuscript. In addition to correcting many errors, they contributed important suggestions for improving the content.

Barton Marcy (NEPA office, U.S. Department of Energy's Westinghouse Savannah River Company) and Rhey Solomon (Assistant Director of Ecosystem Management Coordination, U.S. Forest Service) provided an in-depth review and contributed many valuable suggestions for enhancing the organization and content of this book.

Troy Caver (Systems Management and Development Corporation, President) provided many important suggestions and strongly endorsed integrating NEPA with a modified systems engineering approach (chapter 2) for synthesizing EIS alternatives. Bruce L. Lenzer (president-elect of the Society of American Value Engineers [SAVE] International) contributed comments on integrating value engineering and total quality management with the NEPA process.

Special thanks is extended to Barbara D. Williamson (Office of General Counsel, U.S. Department of Energy), who performed a thorough legal review of the manuscript and contributed to various legal concepts. Owen Schmidt (lawyer for the U.S. Department of Agriculture) reviewed and provided comments on the section in chapter 2 describing how the agency's statement of purpose and need can be used to define the range of reasonable alternatives.

Finally, John Gould (NEPA Group, WMH, Inc.) comprehensively reviewed the draft manuscript and contributed numerous suggestions for enhancing readability.

Acronyms

ADREC Administrative Record
AHPA Archaeological and Historic Preservation Act
ANPLAN Analysis Plan
ARPA Archaeological Resource Protection Act
ARTS Administrative Record Tracking System
CD Compact Disc
CERCLA Comprehensive Environmental Response, Compensation, and Liability Act
CEQ Council on Environmental Quality
COMTRACK Comment Tracking
CONSPEC Continuous Spectrum
DBS Decision-Based Scoping
DIT Decision Identification Tree
DLW Dangerous Liquid Waste
DOPAA Description of Proposed Action and Alternatives
EA Environmental Assessment
EIS Environmental Impact Statement
EMS Environmental Management System
EPA U.S. Environmental Protection Agency
ESA Endangered Species Act
FAST Functional Analysis System Technique
FR *Federal Register*
IDT Interdisciplinary Team
IDA Interdisciplinary Analysis
IDP Interdisciplinary Planning
ISO Greek word meaning "equal"; also, International Organization for Standardization
IST Interdisciplinary Steering Team
LCA Life-Cycle Assessment
MAP Management Action Plan

NAAQS National Ambient Air Quality Standards
NAEP National Association of Environmental Professionals
NEPA National Environmental Policy Act
NHPA National Historic Preservation Act
NOA Notice of Availability
NOI Notice of Intent
NPR National Partnership for Reinventing
NRC Nuclear Regulatory Commission
NRDA Natural Resource Damage Assessment
NRHP National Register of Historic Places
OPAA Outline of Parameters and Alternatives
P2 Pollution Prevention
P-EIS Programmatic Environmental Impact Statement
PSD Prevention of Significant Deterioration
ROA Record of Assumption
ROI Region of Influence
ROD Record of Decision
SE Systems Engineering
S-EIS Supplemental Environmental Impact Statement
SEPA State Environmental Policy Act
TFP Total Federal Planning
SHPO State Historic Preservation Office
TNT NAEP's Tools and Techniques (TNT) NEPA Practice Committee
TQM Total Quality Management
VE Value Engineering
U.S.C. United States Code

ENVIRONMENTAL IMPACT STATEMENTS

Introduction

No culture can live, if it attempts to be exclusive.

—Mahatma Gandhi

Frogs and Federal Planning

Before introducing the subject of federal planning, let us stop and consider what Felix the Frog and federal planning have in common.[1]

Once upon a time, there lived a man named Clarence who had a pet frog named Felix. Clarence lived a modestly comfortable existence on what he earned working at the local mall, but he always dreamed of being rich.

"Felix!" he exclaimed one day, "We're going to be rich! I'm going to teach you how to fly!" Felix, of course, was terrified at the prospect. "I can't fly, you idiot! I'm a frog, not a canary!"

Clarence, disappointed at the initial reaction, told Felix, "That negative attitude of yours could be a real problem. I'm sending you to class." So Felix went to a three-day class and learned about problem solving, time management, and effective communication . . . but nothing about flying.

On the first day of flying lessons, Clarence could barely control his excitement (and Felix could barely control his bladder). Clarence explained that their apartment had 15 floors, and each day Felix would jump out of a window, starting with the first floor and eventually getting to the top floor. After each jump, Felix would analyze how well he flew, focus on the most effective flying techniques, and implement the improved process for the next flight. By the time they reached the top floor, Felix would surely be able to fly.

Felix pleaded for his life, but his words fell on deaf ears. "He just doesn't understand how important this is," thought Clarence, "but I won't let naysayers get in my way." So, with that, Clarence opened the window and threw Felix out. He landed with a thud.

Next day (poised for his second flying lesson), Felix again begged not to be thrown out of the window. Clarence opened his pocket guide to *Managing More Effectively* and showed Felix the part about how one must always expect resistance when implementing new programs. Then he threw Felix out the window. Thud!

On the third day (at the third floor), Felix tried a different ploy: stalling, he asked for a delay in the project until better weather made flying conditions more favorable. But Clarence was ready for him. He produced a time line, pointed to the third milestone, and asked, "You don't want to slip the schedule, do you?" From his training, Felix knew that not jumping today would mean that he would have to jump twice tomorrow, so he just said, "OK. Let's go." And out the window he went. Thud!

Now understand that Felix really was trying his best. On the fourth day, he flapped his feet madly in a vain attempt to fly. On the fifth day, he tied a small red cape around his neck and tried to think Superman thoughts. Try as he might, though, Felix couldn't fly.

By the sixth day, Felix (accepting his fate) no longer begged for mercy. He simply looked at Clarence and said, "You know you're killing me, don't you?" Clarence pointed out that Felix's performance so far had been less than exemplary, failing to meet any of the milestone goals set for him.

With that, Felix said quietly, "Shut up and open the window," and he leaped out, aiming carefully at the large jagged rock by the corner of the building. With this jump, Felix went to that great lily pad in the sky.

Clarence was extremely upset, as his project had failed to meet a single goal. Felix had not only failed to fly, he hadn't even learned how to steer his flight as he fell like a sack of cement—nor had he improved his productivity when Clarence told him, "Fall smarter, not harder."

The only thing left for Clarence to do was to analyze the process and try to determine where it had gone wrong. After much thought, Clarence smiled and said, "Next time, I'm getting a smarter frog!"

As in the case of Clarence demanding that Felix the Frog learn to fly, traditional approaches often provide an insufficient mechanism for effectively managing project planning, particularly federal planning. As we enter the new millennium, more is needed than simply *demanding* that federal agencies meet the increasingly complex challenges that lie ahead.

Federal government filing cabinets are bulging with projects that have been delayed, revised, or canceled, to say nothing of overrunning their budgets; project disconnects, endless revisions and redesigns, and uncoordinated planning efforts abound. As diverse as these problems may appear on the surface, they can often be traced back to a small number of fundamental (root) causes.

A subset of these problems can be traced directly to the way agencies chose to implement their National Environmental Policy Act (NEPA) responsibilities; in other cases, such problems were not due to NEPA but resulted instead from a poorly implemented planning process. With respect to problems that can be tied to NEPA, many of their causes can be traced to the simple observation that state-of-the-art tools and approaches, essential in effectively managing environmental impact statement (EIS) planning, were ineffectively used or ignored all together; it must be emphasized that NEPA's mandate and implementation requirements have rarely been the underlying problem. Instead, problems associated with NEPA generally result from the way in which agencies chose to implement their statutory obligations.

As federal funding has tightened, emphasis is increasingly shifting toward doing more with less. As noted by Petronius Arbiter at the beginning of chapter 5, reorganization is often the immediate response when a department or function is not performing to expectations. As will become self-evident, reorganization, by itself, is a practice of questionable merit in terms of providing a framework for surmounting the complexity of challenges that lie ahead in the 21st century. Yes, change is needed, but not simply for its own sake. Specifically, modern approaches and innovative tools designed to address problems that have traditionally hindered federal planning are needed. In large measure, this book is a response to problems that plague NEPA and federal planning in general. While the book places considerable emphasis on the NEPA process, the reader should note that its scope is broader than simply that of NEPA per se. For instance, chapter 5 introduces a new concept for conducting federal planning in which NEPA is but one of several key elements forming a comprehensive system for improving the effectiveness of federal planning.

This text provides the reader with an arsenal of innovative approaches, methodologies, and tools for enhancing the effectiveness of the federal planning. Thus, instead of a smarter frog, the objective is to promote smarter approaches that can effectively address traditional problems. First, though, let us consider some of the traditional problems (reported by practitioners, planners, and decisionmakers). This discussion forms a basis for developing the total federal planning (TFP) strategy (described in chapter 5), which addresses such problems.

Commonly Encountered Planning Problems

There are many examples of federal planning leading to excellent decisions. Yet, all too often, federal planning is fragmented, uncoordinated, narrowly focused, or lacks cohesion sufficient to furnish decisionmakers with a comprehensive picture, essential in reaching an informed decision. Not surprisingly, it is not uncommon to find that a decision has been made in a vacuum where essential consideration of other pertinent planning factors is lacking.

Project proponents, having a vested interest in the outcome of a decision, can exert strong, sometimes overbearing, influence in favor of a particular course of action (e.g., pursuing the proposed action). Once sufficient project momentum is reached, institutional inertia may drive decisionmaking in favor of the proposed action, even in cases where more effective alternatives might exist (in terms of both environmental and nonenvironmental advantages). I refer to this phenomenon as *project* or *institutional inertia.* Table I.1 summarizes problems commonly reported among planners, decisionmakers, and NEPA practitioners.

Traditional NEPA Problems

Compounding the problems described in Table I.1 is the fact that NEPA is often pursued more as a permitting requirement for documenting decisions already made than as a true planning process. Inadequate planning can frequently be traced directly to NEPA not being properly integrated or combined with other federal planning processes nor pursued during the early planning phase, as required by NEPA regulations.[2]

Better alternatives have also been overlooked because NEPA was not con-

Table I.1 Commonly observed problems in NEPA and the federal planning process

1. Preparation of fragmented or uncoordinated plans
2. Failure to view planning from a holistic standpoint
3. Failure to blend pertinent decision and planning factors together (i.e., lack of concinnity)
4. Incomplete information (as a result of an insufficient or an uncoordinated investigation)
5. Failure to integrate NEPA with early project schedules
6. Infrastructure inconsistencies and incompatibilities
7. Failure to identify or rigorously pursue optimum solutions and alternatives
8. Lack of impartiality in analyzing a proposed action versus reasonable alternatives
9. Failure to identify or plan for crosscutting issues
10. Inability to identify long-term inconsistencies with other projects
11. Failure to identify all regulatory and planning requirements
12. Unreliable projections and poor forecasting
13. Miscommunications

ducted using as an open, interdisciplinary, rigorous, or systematic process.[3] Numerous examples have been reported where superior alternatives have either gone unnoticed or fallen by the wayside, primarily because the very requirements that promote rigorous and effective planning have not been complied with. As indicated by the seventh and eighth items in Table I.1, alternatives often are not given consideration equivalent to that of a proposed action. In some cases, alternatives lying outside a particular mindset are afforded only cosmetic treatment.

Not surprisingly, NEPA's proven ability (when properly implemented) to provide a framework for comprehensive long-term planning often goes unexploited. Such practice can lead to uncoordinated and even conflicting plans. As a consequence, project planning and implementation can be a vertiginous process, subject to ceaseless change, modification, and revision. Perhaps the most important attribute of the TFP strategy described in chapter 5 is that it is specifically designed to prevent such problems.

Technical and Cultural Problems

In the past, the EIS planning process has often been conducted in an ad hoc manner with little regard to adopting methods that could enhance efficiency. Moreover, it is not uncommon to encounter pockets of resistance, where EIS planning is conducted much as it was during its early days. In some quarters, practitioners have encountered stiff resistance to new approaches that are key to the twin goals of efficiency and effectiveness.

In the process of preparing an EIS, many practitioners have reported a number of recurring problems:

- All too often, preparation of an EIS is viewed as a document preparation process as opposed to a comprehensive planning process.
- The EIS planning process is often conducted in an ad hoc manner with little regard to use of modern approaches, tools, or methods of professional practice.
- Opposition is often encountered in the acceptance of new or innovative methods, even where they may lead to greater efficiency; the result can be increased costs and delays as some issues are investigated ad nauseam, while others may not even be identified.
- Prescribed methods and regulatory provisions for reducing paperwork, cost, and delays in the NEPA process are often used ineffectively or ignored all together; organizations sometimes fail to apply common sense or the "rule of reason" in preparing the EIS.

The Good, the Bad, and the Ugly

An example illustrating these problems was cited in the companion book, *The NEPA Planning Process: A Comprehensive Guide with Emphasis on Efficiency.*[4] This text began by describing an EIS that was prepared for a relatively modest proposal to stabilize plutonium in a plutonium processing facility located at the Department of Energy's Hanford Site. The degree of detail and analysis expended on this EIS was grossly out of proportion to the complexity of the action or its potential for impact. Mandatory regulatory direction for reducing the effort and cost of preparing EISs was not exercised. Suggested tools and approaches for reducing effort and cost were dismissed out of hand. Little thought was given to comprehensively integrating the EIS with other project or site-wide planning processes, or to plan for future changes that might be encountered during implementation.

In the end, this EIS cost the taxpayers an estimated four to five million dollars and its contribution to the final decision was marginal at best. The story does not end here. As the potential actions were described in such exacting detail, there was correspondingly little flexibility for accommodating later design changes and evolving circumstances: this resulted in increased costs as even more NEPA documentation had to be prepared to accommodate even modest design changes.

As a cost-effective instrument for planning future actions, this EIS has been described as, among other things, a NEPA lemon. Many lessons can be distilled from the plutonium stabilization example, not the least of which is that practitioners need to be trained and competent in the application of modern planning tools and techniques. Management must not only provide close supervision and contractor oversight but also be thoroughly versed in NEPA's purpose (i.e., the purpose of NEPA is *not* to prepare an EIS document) and provisions regarding efficiency. Facility and project proponents need to participate in the EIS preparation but should be limited in terms of their influence and potential for misdirecting the entire planning effort.

In contrast, the example of the New Tank EIS, also described in the companion text, clearly demonstrates how the EIS process can provide decisionmakers with a highly effective tool for planning actions. Properly executed, an EIS can sometimes result in cost savings that run into the hundreds of millions of dollars![5] Yet such successes can be achieved only when the EIS planning process is implemented as it was designed to be used. With renewed congressional emphasis on increasing federal efficiency, agencies can ill afford to repeat such mistakes. In response to such examples, this text views NEPA from the perspective of a comprehensive framework for integrating federal planning—not as a mechanism for describing an action in infinitesimal detail.

Towards Improved Decision Making

Fortunately, the bad examples of NEPA practice are balanced against its innumerable successes. NEPA practice continues to evolve and improve. Since NEPA's enactment, significant progress has been witnessed in terms of minimizing politics while maximizing scientifically based decisionmaking. Consistent with NEPA's fundamental purpose, environmentally questionable projects are often modified during the EIS process. In many cases, the NEPA process was so instrumental in illuminating potential environmental havoc that entire projects were canceled even before a final EIS was issued; in other cases, major proposals were substantially revised based on the environmental results obtained during the EIS process. The effect of the EIS process on shaping American society is immeasurable.

For example, an EIS was completed in 1996 for a highly controversial project involving the disposition of 2,100 metric tons of radioactive spent nuclear fuel at the Department of Energy's Hanford Site. One of the basins in which this fuel is stored was so antiquated that water containing radionuclides had leaked into the soil beneath. The critical priority given to this project was underscored by the fact that a seismic event could result in a fire or explosion having potentially catastrophic implications. The EIS process was instrumental in bringing what might otherwise have been opposing parties together in a unified effort that expeditiously determined a safe disposition for the radioactive fuel.[6] Lacking a comprehensive process that allowed stakeholders to participate in developing an ultimate plan, a project of such controversial nature might otherwise have taken much longer.

In the words of Mr. Eric Gerber, the Spent Nuclear Fuel project manager:*

> A decision was made to involve the stakeholders from the beginning. We discussed pending decisions before they were finalized and actually changed our plans based on stakeholder input. After . . . seeing the impact of their recommendations on our decisions, the Project's credibility became established and stakeholder communications shifted from demands to team participation. An illustration of the success of this effort was the completion of the Project's Environmental Impact Statement in eleven months with few stakeholder comments; previously unheard of for major DOE projects.

Perhaps NEPA's greatest contribution has been to inject a rationally based, scientific planning process throughout all levels of federal decisionmaking. With respect to NEPA's effect on federal decisionmaking, Robert Bartlett writes:

*Personal Communication, 1999.

> NEPA is a great deal more than a mere legal requirement for the preparation of environmental impact statements. NEPA has . . . [compelled] bureaucracies to use science-like analysis as a basis for policies and decisions—an attempt to force greater rationality in government decisionmaking. . . . [I]mplicit in NEPA and underlying its logic as a policy [act] is a distinct form of reasoning—an ecological rationality.

NEPA Provides Agencies with an Unusual Degree of Flexibility

Critics sometimes charge that NEPA is an inefficient process. Inefficiency is not indigenous to NEPA per se. To the contrary, the causes of inefficiencies lie principally with the approaches used and with the lack of or failure to adopt effective tools and techniques for implementing NEPA's requirements.

The Council on Environmental Quality's (CEQ) NEPA regulations establish goals and procedural requirements but leave the question of how such requirements are best implemented largely to the discretion of individual agencies; thus, agencies have been granted an unusually wide degree of latitude and flexibility (i.e., opportunity) in how they choose to discharge their EIS responsibilities. While the CEQ grants agencies an unusually wide degree of latitude and flexibility, many agencies have failed to capitalize on this advantage.

Preparation of an EIS should be welcomed as an opportunity to cultivate the agency's final course of action, integrating all pertinent decisionmaking factors and minimizing cost by preventing disconnects and revisions. For example, the EIS requirement to perform an alternatives analysis can provide a mechanism for identifying more benign courses of action that may minimize or even avoid future permitting procedures, thus reducing costs while expediting project schedules; mitigation measures provide a particularly effective tool for reducing project risks. By incorporating mitigation in early planning, subsequent impact may be avoided or reduced to the point where project implementation and permitting requirements can be curtailed and, in some cases, entirely eliminated. Additional information on enhancing efficiency can be found in *The NEPA Planning Process*.[7]

Applying Management Principles and Tools

This book is designed to equip the reader with a repertoire of modern approaches, tools, and techniques for effectively addressing many of the problems that have traditionally plagued federal planning. Emphasis is placed on cost-effective methods that promote the goal of informed decisionmaking. Some of the modern management tools described in the following sections are integral elements of the TFP strategy, introduced in chapter 5.

The reader may contact either the author or the National Association of Environmental Professional's NEPA Tools and Techniques (TNT) Practice Committee for additional information.

Value Engineering and Value Planning Tools

The concept of value engineering (VE) was pioneered to identify substitutes for scarce materials at the close of World War II. Today, VE provides a practical set of tools for investigating root problems, formulating alternatives, and identifying optimum solutions. Recently, the Society of American Value Engineers (SAVE) revised the VE terminology, redefining it as the *value methodology,* for which new standards have been published.[8]

Ultimately, the results of this process can assist organizations in minimizing unnecessary expenditures, improving planning effectiveness, and enhancing project quality. Specifically, this process can be particularly useful in: (1) reducing the time expended in identifying solutions to problems, (2) determining an optimum direction that best meets all pertinent requirements, and (3) assisting decisionmakers in determining a best-value course of action. Several types of value studies are in common use. Two of the most important are:

1. *Value engineering studies:* usually conducted for engineering, construction, or related activities. It is most commonly advised to perform such a study in the early or intermediate design phase, before management decisions establish a final project direction.
2. *Value planning or value scoping studies:* normally performed in the initial phases of a proposal (i.e., program, project, process, or activity). The purpose of such studies is to determine the mission and direction of the proposal. Such studies are often used to optimize the mission objectives, design, and actions necessary to initiate a project. Scoping, reinvention, criteria and limitations, and similar types of specialized value studies are considered subsets of this type of study.

Bruce Lenzer of SAVE has also proposed an approach integrating total quality management and VE for use in reengineering studies with the aim of improving the effectiveness of operations, processes, and organizations.[9]

The different names ascribed to these value studies denote variations in the general value method process related to the timing, type of activity, or specific application for which the study is intended. The job plan, which is described shortly, remains essentially unaltered for the particular application or variation of the value method used. While value planning studies are actually more closely aligned to the processes described in this book, I have chosen to use the more widely recognized term *value engineering,* even in instances where *value planning* or *value scoping* might more accurately portray the actual process application.

The Value Engineering Process

A preliminary review or pre-study (e.g., of a specific proposal, program, project, process, or design) may be useful in determining if the potential return on investment is sufficient to justify a formal VE study. Emphasis is placed on examining the proposal or activity to determine if potential alternatives show promise for increasing value. If a VE study is deemed to be cost-effective, the pre-study is also used to (1) define the scope, objectives, and expected deliverables, (2) establish logistical elements necessary to ensure a successful result, (3) gather preliminary information, and (4) select the study's membership.

If a decision is made to conduct a VE study, a facilitator is assigned and charged with responsibility for leading the VE team through a number of rigorous procedures designed to identify alternatives or solutions to a problem. An independent, interdisciplinary team is assembled to conduct the VE workshop. Team members are carefully selected to ensure a diverse range of technical expertise and experience. The VE workshop generates, examines, and refines these alternatives. The team challenges preconceived and prejudicial notions in an effort to identify new or alternative concepts that may lead to better solutions.[10]

A standard VE study consists of seven distinct phases. With respect to the NEPA planning process, not all of these are necessarily pertinent or need to be performed. Professional judgment must be exercised, case by case, in determining those phases that are most effectively integrated with NEPA planning:

1. Information phase
2. Functional analysis phase
3. Creativity phase
4. Evaluation and analysis phase
5. Development phase
6. Presentation phase
7. Post-study and implementation phase

INFORMATION PHASE. All information pertinent to the scope of the proposal (or action under investigation) is collected and disseminated to the VE team members. Constraints and limitations that may affect the study results are identified and, if necessary, ranked with an assigned value.

FUNCTIONAL ANALYSIS PHASE. A functional logic diagram (referred to as a *function analysis system technique* [FAST] diagram) is normally generated as part of a standard VE study. The FAST diagram can be used to critically evaluate how versus why critical functions are currently or would be performed. Items that have high potential for added value may be earmarked for more detailed examination. Because of the substantial effort necessary to construct a FAST diagram, its utility (with respect to NEPA planning) should be questioned and carefully examined.

CREATIVITY PHASE. In this step, team participants are encouraged to exercise creativity with the objective of identifying potential solutions for solving a particular problem. Methods such as brainstorming are used to generate innovative ideas for more detailed consideration.

EVALUATION AND ANALYSIS PHASE. Ideas generated during the creativity step are organized into concepts possessing similar features. These can then be solidified into potential alternatives and ranked using one of a variety of techniques—an example is Criteria Weighting Matrix and Evaluation Analysis Ranking. The ranked alternatives are then evaluated with respect to their advantages and disadvantages.

DEVELOPMENT PHASE. Alternatives deemed to have the greatest potential during the analysis step are further evaluated and developed into viable, efficient, and cost-effective options or alternatives.

PRESENTATION PHASE. Options or alternatives that have been fully developed are documented in a report and presented as a study proposal.

POST-STUDY AND IMPLEMENTATION PHASE. The workshop recommendations are considered and evaluated by the team and 'customer' responsible for their implementation. With respect to NEPA, this report might be used as input for the EIS scoping process or for addressing other problems throughout the NEPA process.

> **Opportunities to Apply Total Federal Planning**
>
> If a total federal planning (TFP) strategy is adopted (see chapter 5), an interdisciplinary steering team would review and, if appropriate, adopt the recommendations of the value engineering (VE) workshop. The standard VE procedure may need to be modified and tailored as necessary to meet project and TFP objectives.

The Connection Between VE and NEPA

Because of its proven ability to increase efficiency and reduce cost, the Office of Management and Budget (OMB) has directed federal agencies to apply VE in planning major federal projects exceeding one million dollars.[11] As a result, NEPA and VE are both mandated to be performed on major federal projects. Is this simply another case where overlapping or redundant requirements have been mandated?

This question can best be answered by examining the underlying purpose of VE and NEPA. VE provides a toolbox of problem-solving techniques for analyzing problems and identifying solutions. However, VE is not a planning process. In contrast, NEPA provides a comprehensive planning process, but it lacks the intrinsic tools for effectively implementing its procedural requirements. Thus, VE offers a set of tools for managing various aspects of the NEPA planning process. To date, this connection appears to have gone largely unrecognized. An integrated NEPA-VE approach is advantageous because it provides an efficient means for complying with both CEQ and OMB mandates.

Table I.2 compares principal characteristics and goals of NEPA with those of VE. While NEPA and VE share strikingly similar goals and requirements, these are not redundant; not only are NEPA and VE compatible, they in fact complement one another. The commonality in these goals provides a foundation whereby VE can be used as a tool for increasing the effectiveness of the NEPA planning process. A paper by the author describes an approach in which a facilitated workshop using VE techniques was successfully used to determine the preliminary scope of an EIS.[12]

Systems Engineering Principles in Structured Planning

A system can be defined as a set of components that must work together to perform a particular function. Principles underlying systems engineering (SE) surfaced in the aerospace industry in the mid-1950s, as a means for providing project engineers with a comprehensive and systematic approach to designing and coordinating complex projects. The successful creation of complex systems is increasing reliant on the application of SE concepts.

SE is a term used to describe the general application of engineering skills to the total design and synthesis of a complex system. The difference between SE and other standard engineering disciplines is that SE places greater emphasis on defining requirements, performance goals, and alternative designs; evaluating alternative designs; and coordinating the diverse tasks necessary to synthesis a complex system. The International Council on Systems Engineering defines SE as the "management function which controls the total system development effort for the purpose of achieving an optimum balance of all system elements. It is a process which transforms an operational need into a description of system parameters and integrates those to optimize the overall system effectiveness."[23]

Table I.2 Comparison of the goals and requirements of NEPA with those of VE

NEPA	VE
Requires use of a "public,"[13] unbiased,[14] and "rigorous" process.[15]	Based on an unbiased and rigorous process.
Is predicated on use of a "systematic, interdisciplinary" approach.[16]	Uses a systematic and interdisciplinary process to arrive at a higher-value solution.
"Combines"[17] other federal planning processes during the "early"[18] planning phase.	Encompasses all pertinent planning factors.
"EISs" must be prepared early enough to serve as an important contribution to decisionmaking. An EIS is not to be used to rationalize or justify decisions already made.[19]	VE should be applied early enough to assist in decisionmaking. It is not intended to justify decisions already made.
An EIS analysis must provide a "full and fair discussion" of impacts and reasonable alternatives.[20]	Necessitates a full and fair analysis of alternatives to improve the value of the final product.
An EIS must explore and objectively evaluate all reasonable alternatives.[21] Alternatives form the "heart" of an EIS.[22]	Promotes consideration of all possible alternatives.
Is the only federally mandated planning process applicable to all major federal actions.	The OMB has mandated that VE be applied to the planning of all major projects.
Is a planning process.	Provides management tools useful in planning.
Allows the consideration of cost and other factors in the analysis and decisionmaking process.	Considers all pertinent planning requirements (e.g., cost, schedule, environmental).

Typical System Engineering Process

As a management process, SE is used to control planning and development of complex systems; a disciplined process is used to transform a need into a set of integrated parameters that optimize the effectiveness of a system. A standard SE project typically is composed of eight principal steps, each of which is briefly explained below.

1. Problem statement
2. Requirements analysis
3. Alternatives selection
4. Alternatives generation
5. Alternatives investigation

6. Systems synthesis
7. System construction and operation
8. System monitoring and management

The exact process employed may vary with the application and complexity of the system. A detailed explanation of SE is beyond the scope of this chapter; however, many excellent books have been written on this subject. The reader is referred to the text by Blanchard for additional details.[24]

PROBLEM STATEMENT. A problem statement provides the basis for determining the objectives and the alternatives to be considered. If the problem statement is too narrow, alternatives may be overlooked that might provide better solutions.

REQUIREMENTS ANALYSIS. Once agreement is obtained on the problem statement, the functional requirements of the system and related performance objectives must be defined. These performance objectives are used to develop success criteria for assessing each parameter thus providing a basis for evaluating alternatives during the later phase of the study.

ALTERNATIVES SELECTION. Decision criteria are used to assess the parameters of potential alternatives. A set of alternatives is identified that will be defined in greater detail in the next step.

ALTERNATIVES GENERATION. As a novel approach may lead to significant cost reductions or improvements in effectiveness, a high degree of creativity must be exercised in identifying potential alternatives. In this step, the set of alternatives defined in the last step are defined in greater detail for later investigation.

ALTERNATIVES INVESTIGATION. The alternatives are investigated to predict how they would perform if actually chosen and implemented. Mathematical models are often employed to simulate performance of an alternative.

SYSTEMS SYNTHESIS. This step involves optimizing the entire system by combining selected subsystems and components and by modifying parameters.

SYSTEM CONSTRUCTION AND OPERATIONS. Once the final system design is selected, systems engineers must define and analyze it in sufficient detail to provide instructions for its construction. Engineers must also implement a management system to coordinate implementation and construction activities. Finally, the system begins operational testing.

SYSTEM MONITORING AND MANAGEMENT. The final step involves examining the evolving development and construction so as to prevent or control risk and to optimize system objectives.

NEPA AND SYSTEMS ENGINEERING. There is a subtle yet profound difference between the goals of SE and NEPA. While NEPA provides a comprehensive framework for *planning* federal actions and reaching decisions, SE provides a structured methodology for coordinating the *design and development* of complex systems. These differences aside, many of the structured management and design principles inherent in SE can provide principles essential to efficiently managing a complex planning process. As applied in this book, selected SE principles can be integrated with the NEPA process to provide planners with a systematic methodology for developing optimum alternatives and analyzing decisionmaking factors. A modified SE approach for synthesizing alternatives, as part of the NEPA planning process, is detailed in chapter 2. Selected SE principles are also incorporated into the TFP strategy described in chapter 5.

Total Quality Management Provides Principles for Optimizing Planning

Over the past few years, total quality management (TQM) has ceased to be simply the latest in a long line of management fads and has gained credibility as a component essential to maintaining quality. Essentially, TQM is a structured system for changing organizational culture so as to better satisfy a need through a process of continuous improvement. As originated by W. Edwards Deming, TQM is a cycle for continuously improving a system or process so as to develop products of the highest quality. In essence, TQM involves the following commitments:

- Dedication of an organization to high-quality work processes
- Buy-in from all members of the organization to the principle of improvement
- Training and empowerment of organizational members to recognize and make necessary improvements
- Continual improvement of the organizational system or its processes so that it operates at the highest-quality level

This book considers incorporation of TQM principles into federal planning from two perspectives:

> **Opportunities to Apply Total Federal Planning**
>
> To ensure that decisionmakers have accurate information and optimum alternatives to choose from, selected total quality management (TQM) principles are adopted at critical stages throughout the total federal planning (TFP) strategy described in chapter 5. TQM feedback loops are incorporated at important points in the TFP strategy to improve and optimize planning.

1. To continuously improve performance of an agency's planning process by promoting an organizational structure that is more effectively integrated with essential planning functions
2. As a set of discrete principles and techniques that can applied at various stages to improve the preparation and quality of an EIS planning effort

As described later in this text, NEPA provides the flexibility and framework necessary for incorporating VE, SE, and TQM principles and tools into a single integrated federal planning process.

Modified Deming Cycle

One of the objectives of TQM is to break down unnecessary barriers between groups, organizations, and various disciplines. A modified Deming cycle, consisting of five distinct phases (Plan-Do-Study-Act-Support), can provide a basis for improving an agency's planning process. During the *planning* phase, ideas are generated and developed to improve a system or component. The change is implemented during the *do* phase and then *studied* to evaluate its effectiveness. Actions necessary to adopt the improvement are made during the *act* phase. To be effective, there must be a desire to optimize the process (e.g., developing an optimum course of action). If the results prove less than intended, the process is repeated, returning to the planning phase. The *support* phase captures this requirement by indicating that management must support this process if TQM is to prove effective.

Integrating TQM Principles into Federal Planning

As noted earlier, TQM principles can be applied to improve a plan or proposal, or the agency's actual planning process per se. Planners, engineers, and scientists are brought together from pertinent functions to provide different points of view on how a specific plan or the planning process itself can be improved. With respect to a specific plan or proposal, the focus is on improving the final plan such that it best meets all pertinent requirements, objectives, and resource limitations.

Table I.3 lists basic TQM principles that should be considered for integration into an agency's planning process. Success depends on balancing factors such as politics, technical constraints, and organizational culture with the agency's mission and other functional requirements.

TOP-LEVEL COMMITMENT AND INVOLVEMENT. There must be a commitment and involvement from top management to (1) develop plans of the highest quality and (2) improve the organizational process used for planning actions. Upper management cannot simply pay lip service but must instead

Table I.3 Basic TQM Principles that should be considered and, if appropriate, integrated into an agency's planning process

An agency- or department-wide top-to-bottom institutional approach to quality planning (not just an individual project) must be established.
Both the project advocate and all pertinent planning functions should be involved in defining what constitutes quality in terms of the end planning product.
Project personnel and planners at all levels must be empowered so that they may actively institute or participate in quality improvement.
There must be a commitment to redesigning existing operations so that they function efficiently, unconstrained by unimportant procedures or personal agendas.
A top-level commitment must exist and resources must be provided so that the agency's planning process can be continuously improved.
Planning and support personnel must be committed to optimizing the agency's final plan or design.

demonstrate a genuine commitment to continuous improvement in the planning process. This effort must also seek to incorporate the diverse talents of specialists throughout the organization. Personnel from all pertinent planning functions and levels should be involved in solving problems and improving the agency's proposal or plan.

EDUCATING AND TRAINING. Quality is a function of the skills of the personnel. Agency personnel must receive the necessary training and skills to carry out the planning process in a manner that promotes high-quality planning. Personnel must be trained to perform their planning tasks correctly the first time.

SUPPORT STRUCTURES. Support services should be organized to promote the effectiveness of the agency's entire planning effort. For example, a department serving as an informational resource for the planning function should be organized so that its personnel are both acknowledged by and integrated with the federal planning apparatus.

Additional Considerations

As part of its overall TQM strategy, an agency may also want to integrate factors, such as the principles articulated in Stephen Covey's *The Seven Habits of Highly Effective People,* in their planning process. For example, the fourth habit involves "think win-win." This principle is particularly useful when dealing with circumstances that involve competing resources, objectives, and agendas. Such instances might involve balancing the objective of minimizing development in previously undisturbed areas against a project office's objective of expanding its existing mission. By bringing all essential factors into a single integrated planning process in which the competing merits, requirements, and objectives can be studied, it may be possible to find a win-win alternative that best meets all objectives.

International Principles Governing Impact Assessment

The International Association for Impact Assessment, in cooperation with the Institute of Environmental Assessment, UK, has recently established international principles governing environmental impact assessment.[25] This text includes concrete tools, techniques, and approaches for complying with these principles. The basic principles (Table I.4) apply to all stages of environmental impact assessment. The operating principles (Table I.5) describe how

Table I.4 Basic principles applicable to all stages of environmental impact assessment

Purposive: The process should result in informed decisionmaking and in appropriate levels of environmental protection.

Rigorous: The process should apply methodologies and techniques appropriate to address the problems being investigated.

Practical: The process should result in information and outputs that assist with problem solving.

Relevant: The process should provide sufficient, reliable, and usable information for planning and decisionmaking.

Cost-effective: The process should achieve objectives within the limits of available information, time, resources, and methodology.

Efficient: The process should impose the minimum cost burdens, in terms of time and finance, consistent with meeting requirements and objectives.

Focused: The process should concentrate on significant environmental effects and key issues.

Adaptive: The process should be adjusted to the realities, issues, and circumstances of the proposal under review without compromising the integrity of the process.

Participative: The process should provide appropriate opportunities to inform and involve the interested and affected public.

Interdisciplinary: The process should ensure that the appropriate techniques and experts in relevant disciplines are employed.

Credible: The process should be carried out with professionalism, rigor, fairness, objectivity, and impartiality.

Integrated: The process should address the interrelationships of social, economic, and biophysical aspects of the environment.

Systematic: The process should result in full consideration of all relevant information on the affected environment, of proposed alternatives and their impact, and of the measures necessary to monitor and investigate residual effects (i.e., effects that cannot be mitigated).

Table I.5 Operating principles detail how the basic environmental impact assessment principles should be applied at various steps of the environmental impact assessment process

Scoping: Identify issues and impacts that are likely to be important.

Examination of alternatives: Establish the preferred or most environmentally sound and benign option for achieving proposal objectives.

Impact analysis: Identify and predict the environmental, social, and other related effects of the proposal.

Mitigation and impact management: Establish necessary measures for avoiding, minimizing, or offsetting predicted adverse impacts and, where appropriate, incorporate these into an environmental management plan or system.

Evaluation of significance: Determine the relative importance and acceptability of residual impacts (i.e., effects that cannot be mitigated).

Preparation of EIS or reports: Document clearly and impartially effects of the proposal, measures for mitigation, significance of effects, and the concerns of interest to the public.

Review of the EIS: Determine whether or not the document provides a satisfactory analysis of the proposal(s), and contains the information required for decisionmaking.

Decisionmaking: Approve or reject the proposal and establish terms and conditions for its implementation.

Follow-up: Ensure that the terms and condition of approval are met, including monitoring the effects of development and the effectiveness of mitigation measures; as necessary, undertake an environmental audit and process evaluation to optimize environmental management.

the basic principles should be applied at various steps and stages of the environmental impact assessment process (e.g., screening, scoping, assessment of alternatives).

Assessment Principles: Sliding Scale, Rule of Reason, and Nomenclature

Common sense and professional judgment must be exercised in determining the appropriate scope of an analysis. Together, the following two principles provide planners and analysts alike with powerful tools for reducing cost, delays, and effort expended in the EIS planning process. These principles are drawn on extensively throughout this text.

The Sliding Scale Approach

Some agencies have adopted a sliding scale approach for use in preparing an EIS. This approach recognizes that the amount of effort expended on an analysis is a function of the particular circumstances. Use of a sliding scale approach is justified based on the following regulatory direction:

> Impacts shall be discussed in proportion to their significance. There shall be only brief discussion of other than significant issues. (40 CFR 1502.2[b]) NEPA documents must concentrate on the issues that are truly significant to the action in question, rather than amassing needless detail. (40 CFR 1500.1[b])

Under a sliding scale approach, environmental issues are investigated and other related regulatory requirements are applied with a degree of effort commensurate with their importance. Thus, the amount of effort expended in investigating a specific issue or addressing a particular regulatory requirement tends to vary with the significance of the potential impact and its importance to the decisionmaking process.

Rule of Reason

Experience indicates that a strict or unreasonable application of a regulatory requirement may lead to decisions, courses of action, or levels of effort that are wasteful, ridiculous, or absurd. The rule of reason is a mechanism used by the courts for injecting reason into the EIS process. Common sense must therefore be exercised in determining the scope and detail accorded to issues, alternatives, and effects considered in the analysis.

Nomenclature

Before introducing the EIS process, the nomenclature used in referring to NEPA and its regulations needs to be introduced. Throughout this book, the term *NEPA* is frequently shortened to *Act*. Similarly, the term *CEQ NEPA Regulations* is abbreviated to simply *Regulations*. For brevity, references to a particular section of the CEQ Regulations (40 Code of Federal Regulations [CFR] 1500–1508) are abbreviated so that they simply cite the specific section number in the Regulations where the provision can be found. For example, a reference such as "40 CFR 1500.1" is shortened to the more convenient expression "§1500.1."

The term *proposal* is interpreted to mean the set of reasonable alternatives, including no action and the proposed action (if one is defined).

Overview of the NEPA Compliance Process

Before delving into the EIS process, it is instructive to present a succinct overview of the NEPA compliance process. Emphasis is placed on describing the three levels of NEPA compliance. Variations to this description can and do exist, particularly with respect to the way individual agencies choose to implement their respective processes. The reader is referred to the author's book *The NEPA Planning Process* for details.

The Three Levels of NEPA Compliance

NEPA documents must be prepared early enough so that they will contribute to the decisionmaking process and will not be used to rationalize or justify decisions already made. A NEPA document meets this timing requirement if it is prepared in time for the decision deadline but not so early that it cannot meaningfully contribute to the decisionmaking process (§1502.5).

The NEPA process can be conceptualized as consisting of three distinct levels of environmental review, planning, and compliance. These levels, from the simplest through the most demanding, are:

1. *Categorical Exclusion (CATX):* A potential action is reviewed to determine if it qualifies for a CATX, thus excluding it from further review and NEPA documentation requirements.
2. *Environmental Assessment (EA):* An EA may be prepared to determine if an action qualifies for a Finding of No Significant Impact (FONSI). Actions qualifying for a FONSI are exempted from the requirement to prepare an EIS.
3. *EIS:* With few exceptions, an EIS must be prepared for federal actions that do not qualify for either a CATX or a FONSI.

Initiating the NEPA Process

The NEPA process is typically initiated when a need for taking federal action is identified (see first box, Figure I.1). A limited number of circumstances exist in which a proposal can be exempted from NEPA's requirements. *The NEPA Planning Process* presents a detailed review of circumstances under which an action may be exempted from NEPA requirements.[26]

Reviewing Existing NEPA Documents

No additional NEPA review may be required if a proposal falls within the scope of an existing EIS or EA.

Reviewing Existing EISs

If a proposal is adequately described in an existing EIS (see first decision diamond, Figure I.1), no additional NEPA review is required; the agency may pursue the action. However, as indicated by the box labeled "Prepare supplemental EIS" in Figure I.1, the EIS must be supplemented if:

(i) The agency makes substantial changes in the proposed action that are relevant to environmental concerns; or

(ii) There are significant new circumstances or information relevant to environmental concerns and bearing on the proposed action or its impacts. (§1502.9[c][1])

Reviewing Existing EAs

If the proposal is not described in an existing EIS, a review should also be conducted to determine if it is covered in an existing EA (see second diamond, Figure I.1). If the proposal is adequately described, the agency may proceed with the action.

Categorically Excluding Actions

A CATX is a class of actions that have been previously reviewed and determined not to result in an individual or cumulatively significant impact and is therefore exempt from the requirement to prepare an EA/EIS (§1508.4). Each federal agency is required to prepare a list of CATXs as part of its individual NEPA implementation procedures. CATXs provide one of the most efficient methods for streamlining NEPA compliance.

The proposed action should be reviewed to determine if it falls within an existing CATX (see decision diamond labeled "Categorically excludable?" in Figure I.1). If a action is eligible for a CATX, the NEPA review process is satisfied; no additional review or documentation requirements are necessary and the agency may proceed with the action.

Is the Action Considered Significant?

If, at this point, the action can not be excluded, the proposal is reviewed to determine the likelihood that it may result in a significant impact (see decision diamond labeled "Is action considered significant?" in Figure I.1). The agency's NEPA implementation procedures should be consulted for guidance in determining if the action is one normally requiring preparation of an EA or EIS.

To save time and resources, the agency may elect to prepare an EIS without first preparing an EA. Conversely, an agency may choose to prepare an EA even if it plans to eventually prepare an EIS for the proposal.

The Environmental Assessment

As depicted by the decision diamond labeled "Is action considered significant?" in Figure I.1, an agency may elect to prepare an EA (see box labeled "Prepare EA," Figure I.1) if the impacts are considered nonsignificant, the potential significance is uncertain, or any significant impacts can be mitigated.

The completed EA is reviewed by the decisionmaker to determine if it qualifies for a FONSI (See decision diamond labeled "Significant impact?" in

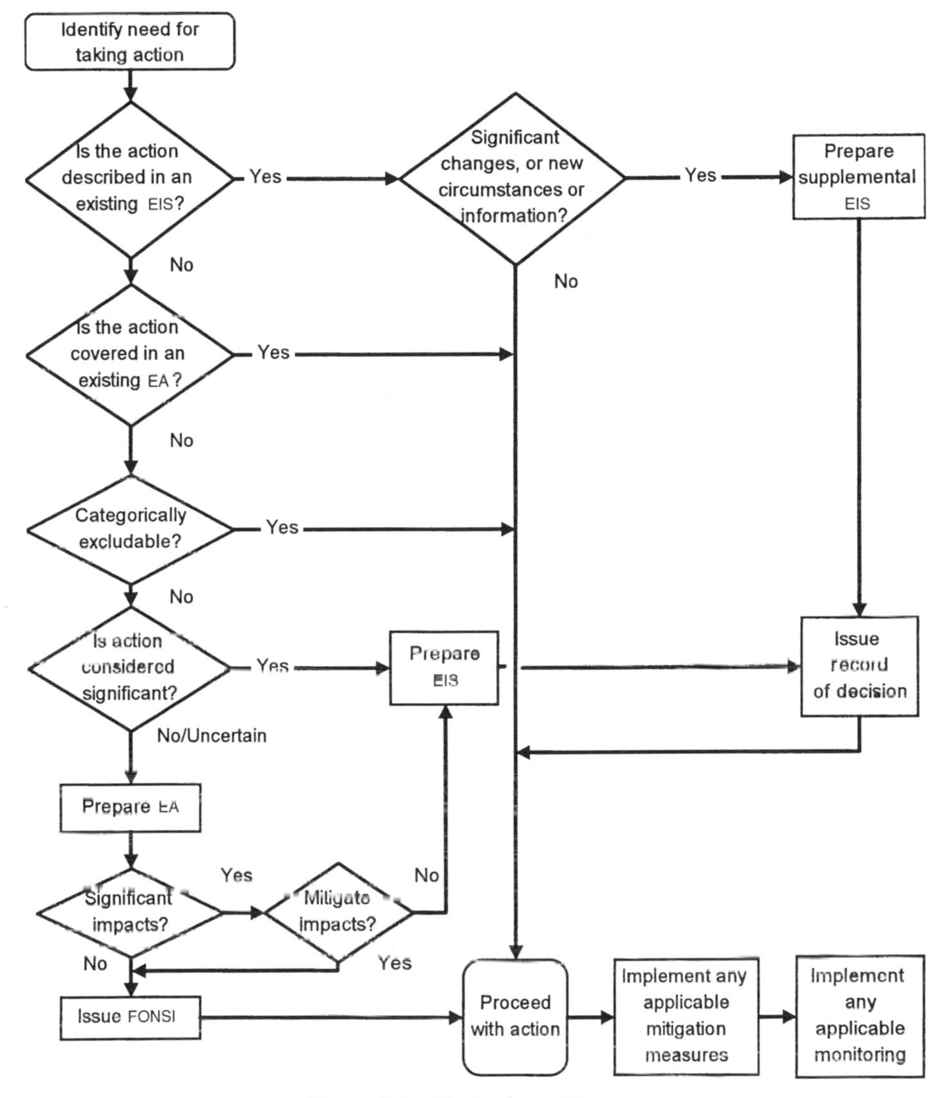

Figure I.1 Typical NEPA Process

Figure I.1). An EIS is not required if the action would not result in a significant impact or if any significant impacts can be mitigated to the point of nonsignificance. If a FONSI is issued, the agency may proceed with the action; any applicable mitigation and monitoring measures are implemented.

Two options exist if the EA does not conclusively demonstrate that the potential impacts are nonsignificant: (1) Mitigate impacts to the point of nonsignificance (see decision diamond labeled "Mitigate impacts?" in Figure

I.1) or (2) Prepare an EIS. For more information on preparing EAs, the reader is referred to the author's upcoming text, *Managing and Writing Effective Environmental Assessments: A Comprehensive Guide to Project Planning and Complying with NEPA's EA Requirements.*

The Environmental Impact Statement

With few exceptions, an EIS must be prepared if a federal action cannot be excluded from NEPA's requirements. The fundamental purpose of an EIS is to analyze potentially significant effects and investigate alternatives for avoiding such impacts.

Prudence should be exercised in conducting and managing the EIS planning process, as experience indicates that a poorly executed process is often characterized by the seven phases depicted in Figure I.2.

The EIS pyramid is depicted in Figure I.3. As indicated in this figure, a well-orchestrated EIS is composed of five quintessential elements that are explained in detail in chapters 1 through 3. An internal and public scoping stage lays the foundation for the integrated planning and alternatives assessment phase that follows. An interdisciplinary analysis of the reasonable alternatives follows and the EIS document is prepared. Review and decisionmaking complete the process. The EIS pyramid is outlined in more detail in the following section.

Typical EIS Process

The principal steps followed in preparing an EIS are outlined in Figure I.4. A preliminary or pre-scoping effort is frequently performed prior to formally initiating the EIS, which occurs with publication of a Notice of Intent (NOI) in the *Federal Register* inviting the public to participate in the public scoping process. The public scoping process is used to determine the scope of analysis that will be investigated in the EIS.

An integrated interdisciplinary planning effort is performed in which a

1. Enthusiasm
2. Disillusionment
3. Panic
4. Search for the guilty
5. Persecution of the innocent
6. Praise for the uninvolved
7. Request for additional funding to revise the EIS

Figure I.2 The seven phases that typically characterize a poorly executed EIS process

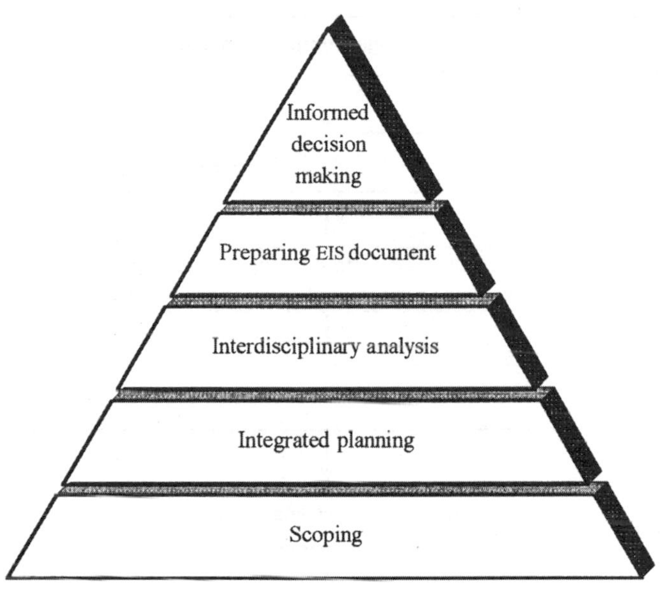

Figure I.3 The EIS triangle

range of alternatives is identified and then defined in detail. The alternatives are then analyzed. The standard EIS process typically consists of two distinct phases: A draft EIS, followed by a final EIS; the term EIS is used to describe a statement in either stage. To promote efficiency, an EIS required to be concise and to the point:

> Agencies shall focus on significant environmental issues and alternatives and shall reduce paperwork and the accumulation of extraneous background data. Statements shall be concise, clear, and to the point, and shall be supported by evidence that the agency has made the necessary environmental analysis. (§1502.1).

The completed draft EIS is circulated for public review and comment, and filed with the Environmental Protection Agency. Once comments are incorporated, the final EIS is circulated and filed. The decisionmaker uses the final EIS, in conjunction with other relevant information, in selecting the course of action to be taken (§1502.1).

The responsible decisionmaker reviews the analysis and chooses the alternative to be pursued. Unlike an EA, the decisionmaker has the latitude to select any alternative that was adequately analyzed in the EIS, even though it may result in a significant impact; A Record of Decision (ROD) is issued recording the agency's final decision; with respect to NEPA's requirements, the agency

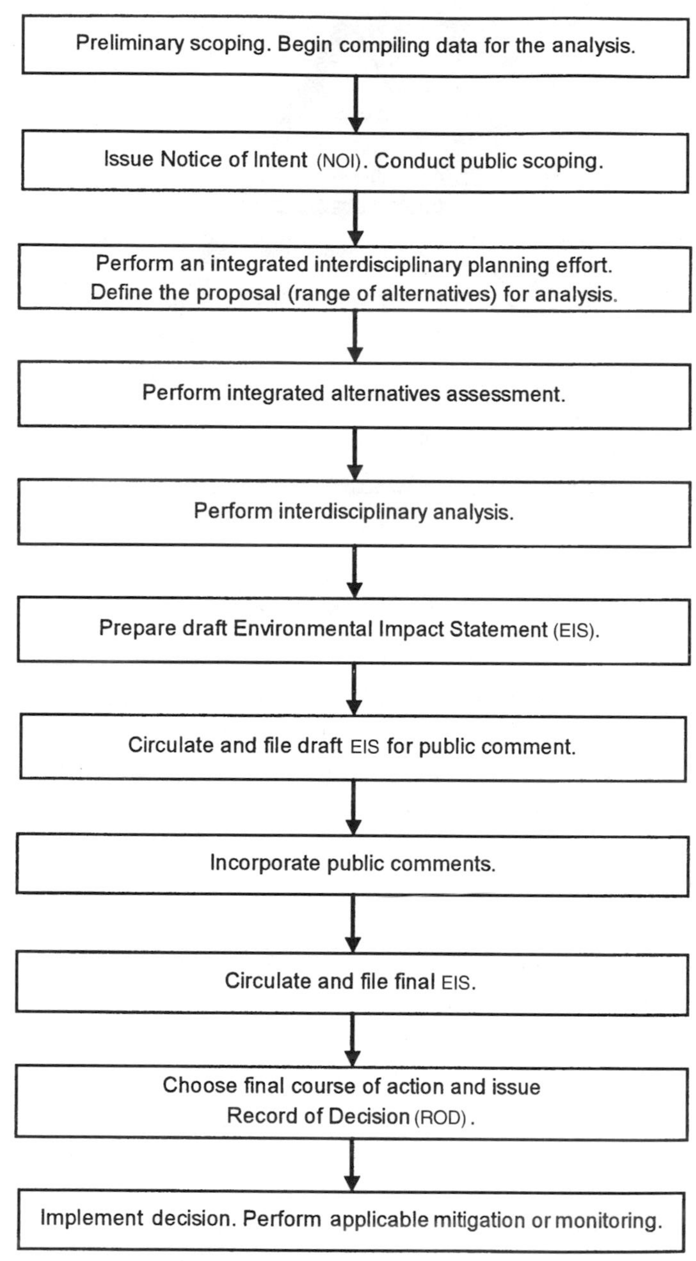

Figure I.4 Principal steps described in Chapters 2 and 3 for preparing an EIS. This process may be performed either in conjunction with a total federal planning (TFP) strategy (outlined in Chapter 5) or as an independent effort.

is now free to pursue the action described in the ROD. Any applicable monitoring and mitigation measures must be implemented.

Closing Thoughts

Before introducing chapter 1, the author wishes to propose three corollaries to Newton's famous laws of motion. These corollaries are cited for more than humor's sake. The reader may note that each has a subtle yet profound implication in terms of safeguarding our environment and the manner in which agencies choose to conduct their EIS planning process:

1. *First Law of Environmental Movement:* A top-level commitment to environmental quality tends to continue in the direction of quality, unless acted upon by other countervailing forces; lack of commitment to environmental excellence tends to promote environmental degradation.
2. *Second Law of Environmental Movement:* The force that an EIS brings to bear in protecting the environment is equal to the mass of forethought expended in the planning process, multiplied by the decisionmaker's commitment to environmental protection.
3. *Third Law of Environmental Movement:* For every project proponent attempting to sidestep the EIS process there is an equal and opposite adversary waiting to litigate.

Disclaimer

The reader is advised to consult the actual regulatory provision for the details and precise wording of the cited regulation. Agency-specific NEPA implementation procedures and orders should also be checked for provisions that may impose requirements beyond those presented in the CEQ's NEPA regulations.

The author wishes to apologize for any inaccuracies that may appear in this book. While the book provides the reader with direction for efficiently complying with the EIS requirements, I stress the importance of seeking the counsel of NEPA and regulatory specialists, and legal counsel, particularly in areas involving complex or controversial issues.

Notes

1. Author unknown.
2. 40 CFR 1500.5, 1501.2, and 1506.4.
3. 40 CFR 1501.2, 1501.7, and 1502.14(a).
4. C.H. Eccleston, introduction to *The NEPA Planning Process: A Comprehensive Guide with Emphasis on Efficiency* (New York: John Wiley & Sons, 1999).

5. Ibid.
6. Department of Energy, *Management of Spent Nuclear Fuel from the k-Basins at the Hanford Site, Richland Washington,* Final, DOE/EIS-0245, (61 FR 3922), 1996.
7. Eccleston, *The NEPA Planning Process,* chapter 4.
8. Society of American Value Engineers (SAVE) International, *Value Methodology Standard,* May 1997.
9. B.L. Lenzer, "Reengineering for Value: Using Total Quality Management and Value Methodology, Tools, and Techniques to Reengineer Effective Organizations," SAVE International Proceedings, May/June 1999.
10. L.D. Miles, *Techniques of Value Analysis and Engineering,* 2nd ed. (New York: McGraw-Hill, 1972).
11. Office of Management and Budget Circular No. A-131.
12. C.H. Eccleston, "Applying Value Engineering and Modern Assessment Tools in Managing NEPA: Improving Effectiveness of the NEPA Scoping and Planning Process," *Environmental Regulations and Permitting* (Winter 1998): 53–63.
13. 40 CFR 1500.1(b), 1500.2(b) and (d), and 1501.4(b).
14. 40 CFR 1502.2(g) and 1502.5.
15. 40 CFR 1502.14.
16. 40 CFR 1501.2 and 1507.2.
17. 40 CFR 1500.2(c), 1500.5, 1501.2, 1501.2(a), 1501.7(a)(6) and (b)(4), 1502.25(a), 1505.2, and 1506.4.
18. 40 CFR 1500.5(a), 1500.5(f), 1501.1(a), and 1501.2.
19. 40 CFR 1502.5.
20. 40 CFR 1502.1.
21. 40 CFR 1502.14(a).
22. 40 CFR 1500.2(e), 1502.1, 1502.14(a) and (c).
23. International Council on Systems Engineering, *Systems Engineering Management Guide.*
24. B.S. Blanchard, *Fundamentals of Systems Engineering* (1991).
25. See http://ndsuext.nodak.edu/iaia/principles/
26. Eccleston, *The NEPA Planning Process,* chapter 5.

1

Initiating the EIS Process: An Integrated and Systematic Approach

Some problems are so complex that you have to be highly intelligent and well informed just to be undecided about them.

—Laurence J. Peter

The traditional scope of project analysis and planning is narrower than that required by the National Environmental Policy Act (NEPA); unless required by regulation, environmental values and factors are seldom afforded the same regard in project planning as more traditional decisionmaking factors. Given the option, project managers tend to focus on a narrowly defined proposed action, investigating only a restricted set of alternatives. In contrast, NEPA's broader mandate requires an open and rigorous investigation of a much larger range of reasonable alternatives. These diametrically opposed agendas can lead to disconnects and impaired federal decisionmaking, not to mention increased costs and delays. This and the next chapter are meant to bridge the gap between these opposing agendas.

One of the few constants in life is that even the best-laid plans are subject to change. *The Art of War,* Sun Tzu, the great Chinese general, wrote:

> Therefore, just as water retains no constant shape, so in warfare there are no constant conditions. The five elements—water, fire, wood, metal, and earth—are not always equally predominant; the four seasons make way for each other in turn. There are short days and long; the moon has its periods of waning and waxing. He who can modify his tactics in relation to his opponent, and thereby succeed in winning, may be called a heaven-born captain.[1]

This insight is equally applicable to planners. A meticulous scoping process, in conjunction with an interdisciplinary planning process that thoroughly investigates all potential alternatives, is essential to providing an agency with the flexibility necessary for responding to changing circumstances.

As part of a well-orchestrated planning process, an agency must perform certain tasks prior to initiating its public scoping process. The goal of this chapter is to arm the reader with a step-by-step methodology for initiating the Environmental Impact Statement (EIS) process, up to but not including the public scoping phase. Sometimes referred to as *preliminary* or *pre-scoping,* these initial tasks can be critical to the success of public scoping and preparation of the EIS that follows. Once this preliminary phase is introduced, the stage is set for describing (in chapter 2) a systematic approach to conducting the formal EIS planning process. The reader should note that much of the discussion presented in this chapter is not limited to the pre-scoping stage and is equally applicable to the rest of the planning process.

The reader should note use of the word *process* in referring to preparation of an EIS. As the term implies, preparation of an EIS is not an isolated event. A process is "A series of actions, changes, or functions that achieve an end or result."[2] While many activities described in this chapter are presented as one-time events, in practice, EIS preparation is an evolving and iterative process; consequently, many associated activities are repeatedly revisited.

As a general-purpose approach, the pre-scoping elements described in this chapter are designed for use in large-scale or complex planning effort. Professional discretion and judgment must be exercised, as not all elements described in this chapter, are intended to be or should be applied to every EIS effort. For smaller or less complex planning efforts, the reader will probably implement only a subset of the elements described in this chapter, which contribute most effectively to their particular planning needs.

1.1 Initiating the EIS

In assessing the state of the environment, former vice president Dan Quayle observed, "It isn't pollution that's harming the environment. It's the impurities in the air and water that are doing it." This section does not address the contentious debate over pollution versus contaminants. Instead, it focuses on the steps that must be taken in initiating an EIS that will lead to informed decisionmaking with respect to actions that may result in environmental harm (e.g., pollution and/or impurities in the air and water). Figure I.1 portrays the principal EIS steps described in chapters 1 and 2. This figure is referred to as each of the pertinent steps are described.

Integrating NEPA Early

An employee responding to a contest calling for real-life "Dilbert incidents" (referring to the hapless engineer in the comic strip) submitted this anecdote: "One day a manager asked me to submit a status report on my project. I asked if tomorrow would be soon enough. To this, my manager retorted, 'If I wanted it tomorrow, I would have waited until tomorrow to ask for it!' " This simple example explains why many projects turn into last-minute crises. The most commonly cited NEPA-related reason for project delays is failure to initiate or properly integrate NEPA into early planning. Often, this occurs because project proponents are not fully cognizant of NEPA's planning requirements or the risks to a schedule associated with failing to initiate NEPA promptly.

Requirement for Preparing an EIS

As required by NEPA, an EIS must be prepared for "every recommendation or report on proposals for legislation and other major federal actions significantly affecting the quality of the human environment."[3] A proposal is defined as existing "at that stage in the development of an action when an agency subject to the Act has a *goal* and is *actively preparing to make a decision on one or more alternative means of accomplishing that goal* and the *effects* can be *meaningfully evaluated*. . . . A proposal may exist in fact as well as by agency declaration that one exists" (Emphasis added. §1508.23).

Preparation of the EIS must begin "as close as possible to the time . . . in which the agency is developing or is presented with a proposal, so that it can be completed in time to be included in any recommendation or report on the proposal" (§1502.5, §1508.23). Moreover, the EIS must be prepared early enough so that it can contribute to decisionmaking and will not be used to rationalize or justify decisions already made (§1502.5).

As described in more detail in section 2.3 of chapter 2, this text discourages the common step of defining a proposed action; this is intended to promote a more objective analysis. Instead, analysts are encouraged to simply define a proposal consisting of a range of reasonable alternatives for later analysis. Such an approach provides planners with a greater opportunity to objectively identify an alternative best meeting all relevant decisionmaking factors.

A word of caution is in order. A new proposal can be quickly absorbed into election-year politics, endangering public support for the proposal. The ramifications of announcing a major or controversial project during an election year should be carefully assessed.

NEPA as an Early-Warning Siren

An EIS provides project advocates with a particularly useful tool for testing the waters of public opposition. This is because NEPA often provides an early, sometimes the first, indication that a proposal is headed for trouble, be it from citizens' groups, regulators, or competing interests. If a proposal cruises through the EIS process with few problems and little controversy, it is unlikely to encounter significant opposition at the later permitting or implementation stage. However, experience indicates that if problems do occur, NEPA is likely to be the first of many obstacles to follow.

Moreover, as an EIS is often the first link in a long, complex chain of federal requirements, it frequently unearths nonpolitical obstacles early in project planning (e.g., technical, coordination, schedule, or budgetary issues).

For this reason, the EIS has been used as a mechanism for floating a proposal to the public and gauging the initial reaction; where early problems are encountered, the decisionmaker may wish to reassess the wisdom of moving forward with the proposal. Should this be the case, the EIS also provides an effective vehicle for identifying less controversial courses of action (alternatives) that would encounter less public resistance.

Described in the following sections are steps that should be considered in integrating a proposal with the EIS planning process.

The Interdisciplinary Team

The interdisciplinary team (IDT) can be viewed as comprising two distinct elements: (1) the interdisciplinary planning (IDP) team and (2) the interdisciplinary analysis (IDA) team (see Figure 1.1). The IDP is responsible for performing the integrated planning effort, which incorporates planning requirements that have often been disconnected from the NEPA process. In contrast, the IDA is that part of the IDT responsible for performing the actual technical and environmental analysis. In reality, many individuals will wear both hats. To simplify the number and use of acronyms, only IDT is used in the text. While *IDP* and *IDA* are not used, the concepts they represent are.

> **Opportunities to Apply the TFP Approach**
>
> If a total federal planning (TFP) approach is adopted (see chapter 5), an interdisciplinary steering team (IST) can be formed to oversee the planning process, ensuring that an open and impartial planning process is conducted that truly explores all reasonable alternatives. The IST oversees and coordinates the entire planning process, from conception to the point where a final decision is made.

If a total federal planning approach is adopted (described in chapter 5), an interdisciplinary steering team (IST) coordinates and oversees the efforts of the IDT (circle located above the IDT icon in Figure 1.1).

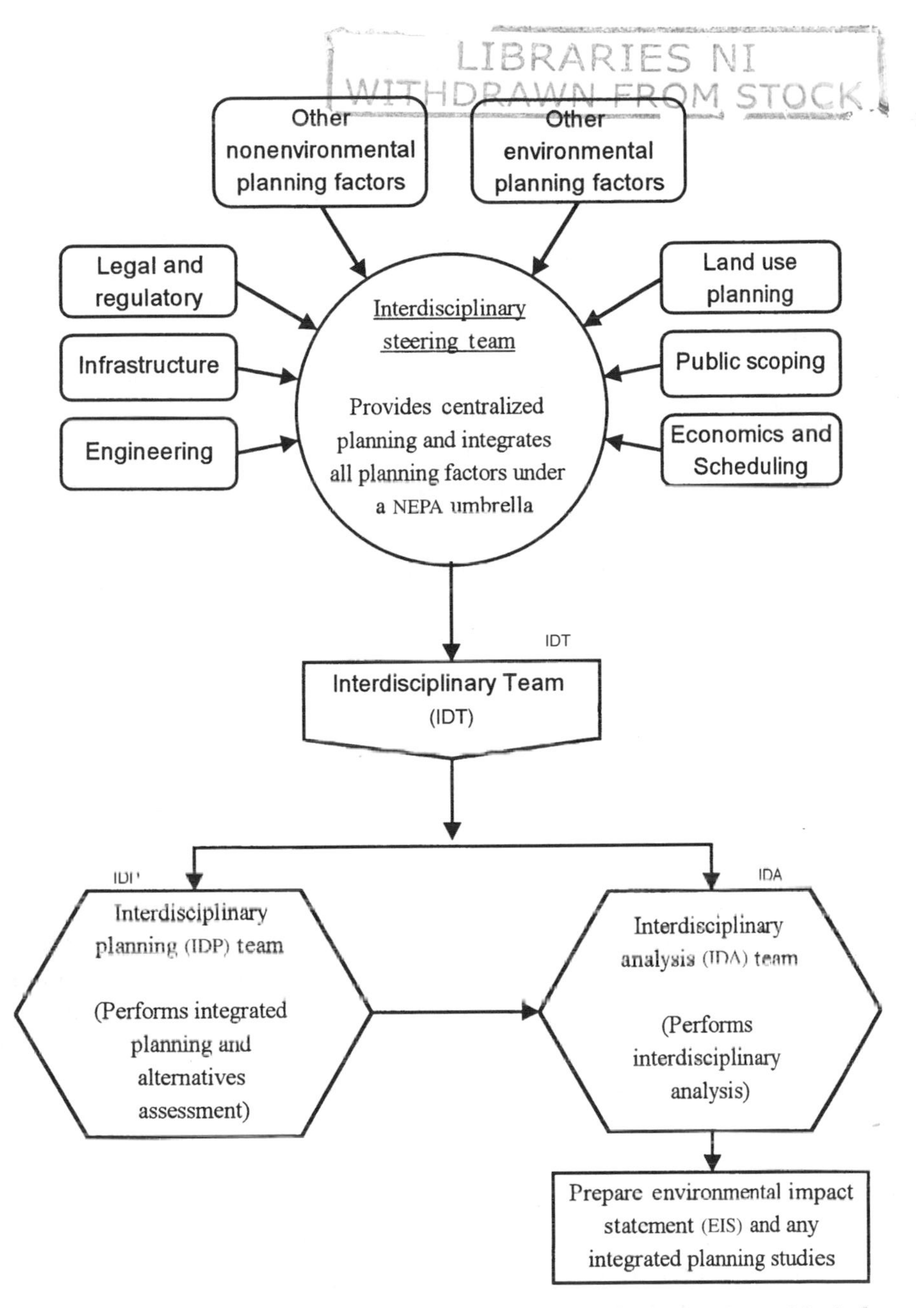

Figure 1.1 Principal steps involved in the environmental impact statement (EIS) planning process. The interdisciplinary team (IDT) is composed of two sub-teams: the interdisciplinary planning team and the interdisciplinary analysis team. If a total federal planning approach (described in chapter 5) is implemented, an interdisciplinary steering team coordinates and oversees the efforts of the IDT and the entire planning effort.

Forming an IDT

Preparation of an EIS normally combines the talents of many disciplines (e.g., planning, science, architectural, engineering and technology, socioeconomics, scheduling, budgeting, and public relations). Planning and analysis must be performed by professionals with credentials and experience appropriate to investigating complex planning and environmental issues. Consistent with NEPA's requirement to perform an interdisciplinary analysis, IDT membership includes professionals representing the principal planning, scientific, and technical disciplines involved.

Revisions and delays are sometimes traced to an agency's internal staff members who have reservations about some aspect of the proposal or the EIS. To preclude such problems, anyone possessing authority to halt or delay the EIS should be included as a direct or indirect participant in its development at the earliest possible time.

Designating an EIS Manager

Once the decision is made to prepare an EIS, an experienced professional should be assigned resources to begin coordinating the effort. The EIS manager is responsible for ensuring that individual team members: (1) understand their roles and responsibilities, (2) complete their assigned tasks on schedule and within budget, and (3) prepare the EIS in compliance with NEPA's regulatory requirements. From a managerial perspective, the EIS manager is responsible for motivating the team, providing day-to-day technical direction, assigning resources and responsibilities, resolving potential problems, coordinating studies and tasks, and ensuring the professional and ethical quality of the analysis.

The manager must be intimately familiar with both the EIS process and environmental issues. Depending on the issues at hand, the manager may need to be competent in scientific disciplines as diverse as biology, geology, soils, air dispersion and hydrologic modeling, and cultural resources. To effectively manage and understand complex environmental interrelationships, most EIS managers have a science or engineering background with experience in diverse technical disciplines.

Kickoff Meeting

Once the core IDT members are identified, a kickoff meeting is held to initiate a pre-scoping effort. At the meeting, team members are fully briefed on the potential scope of the proposal and provided with a specific description of their responsibilities. Schedule and important milestones are fully explained. A copy of the preliminary EIS outline (if available) is provided to each team member, including essential support staff. To the extent that the information is available, the items listed in Table 1.1 should be discussed.

Team members are provided with a description of their work assignments and specific tasks they are expected to perform. An up-to-date list of who is writing each section of the EIS should be maintained and routinely circulated among the entire EIS staff. Progress reports should identifying current activities, costs, and problems should be regularly prepared and circulated.

Coordination between the EIS and Project Staffs

As NEPA is sometimes viewed as a hindrance to their objective, preparation of an EIS may not be embraced enthusiastically by project proponents. Moreover, project proponents frequently attempt to use the EIS as an instrument for implementing their proposed project rather than as a systematic planning process for analyzing and reaching decisions. Not surprisingly, a project office may initiate the EIS late in the game and then scramble to prepare an analysis to meet an impossible deadline; this is but another reason for adopting a TFP strategy, as described later in chapter 5.

Technical coordination meetings should be initiated early and continued throughout each stage of the planning and document preparation process. A commonly observed problem is lack of communication between the EIS team and the project/advocate office. As one EIS task manager stated, "I found it nearly impossible to get the staff to appreciate that the future of their project was dependent on success of the EIS." Mechanisms should be incorporated to ensure sufficient interaction between EIS team members and field or plant personnel. For example, EIS personnel may need to spend time at a plant to learn about equipment and processes. Some projects have gone so far as to ensure

Table 1.1 Items that should be discussed at the EIS kickoff meeting

Description and scope of the proposal
Schedule and budget
Roles and responsibilities
Scoping process
Potentially controversial issues
Principal planning assumptions (having an important bearing on the analysis)
Outline of the EIS (if available)
Preliminary proposal (alternatives and proposed action, if available)
Action items

that a turnaround office is reserved for EIS team members. Likewise, project personnel should be educated about the need and importance of the EIS.

Standardizing Procedures

From the outset, ground rules regarding document standards and style (e.g., standardized section headings, appropriate acronyms and abbreviations, punctuation, methods for referencing material) must be established. Guidelines should also be established for making the EIS readable by the general public. All members should use compatible word-processing software. To save time and reduce errors, agreement should be reached regarding the units of measurement to be used. Because changes in graphics are often one of the largest factors affecting turnaround time, graphical format and style should be decided upon early and all graphics should be included in each stage of the EIS review. A memo specifying document standards should be circulated and initialed by all team members to ensure that they have read and agree to these standards. The technical editor may also want to meet and outline editorial guidelines with the entire EIS staff.

1.2 Pre-Scoping Effort

Benjamin Disraeli once remarked that it is "much easier . . . to be critical than to be correct." In truth, the EIS scoping process must be both critical *and* correct. To this end, an effort should be mounted, prior to public scoping, to identify the preliminary scope of potential actions, impacts and issues, and alternatives that may need to be evaluated. Many of the elements described in this chapter are depicted in Figure 5.5 of chapter 5. The reader should note that, depending on the circumstances, some of the tasks described in this section may actually be performed during the formal public scoping phase described in chapter 2, and vice versa.

Scoping conducted prior to publication of the Notice of Intent (NOI) (referred to as *preliminary* or *pre-scoping*) may supplement but cannot substitute for scoping that follows such publication. EIS meetings or consultations should be held to begin screening the proposal for issues that may need to be addressed. As a streamlining measure, effort should be made to identify issues likely to be nonsignificant. A checklist outlining all important steps that must be conducted during the EIS process should be prepared and maintained.

Often, the perception of impacts and issues may change upon visiting the proposed site(s); therefore, to the extent practical, all team members should personally visit the potential affected area(s). Such visits allow analysts to gain an appreciation for the proposal, its affected environment, and potential environmental issues.

Most project personnel view an EIS as simply covering the "bugs and bunnies" issues, yet it is far more comprehensive than that (e.g., it addresses socioeconomic, cultural, and other issues). For this reason, training sessions should be considered for EIS and project staff and for analysts who will be engaged in the development of the statement.

The types of underlying decisions that may eventually need to be considered must be determined. The response to this simple inquiry is critical to the objective of properly shaping the bounds of the analysis and ensuring that the scope is properly captured. This effort is described in the following section.

Purpose and Need Drive the Scope

A statement of the underlying purpose and need for action is prepared (§1502.13). Defining a clear statement of the underlying need is not a trivial effort—nor should it be treated as such. Yet some proposed projects lack a clearly defined need for action. As explained in more detail in chapter 2, correctly defining the underlying purpose and need can spell the difference between an effective and an ineffective EIS. On the surface, the statement of need may appear intuitively obvious, yet correctly defining the *underlying* need can be a deceptively complicated task. For example, a detailed investigation of certain alternatives may be necessary, while other courses of action may be dismissed based solely on how the statement of purpose and need is defined. Incorrectly defined, the statement of need may increase the level of effort or lead an agency to inadvertently dismiss potentially advantageous alternatives.

While the EIS team members may unanimously agree on the merits of a particular project, consensus regarding the underlying purpose and need is sometimes less than total. Court rulings have pivoted on how such statements were written. The absence of a succinctly expressed need can also adversely affect the public's perception of a proposal.

Identifying Potential Decisions

Responding to another request for real-life "Dilbert" quotes, an employee quoted a manager as saying, "What I need is a list of specific unknown problems we will encounter." The objective of this section is slightly less lofty. The importance of identifying potential problems (or decisions) cannot be understated. As part of its public scoping process (described in chapter 2), the agency must identify the range of actions, alternatives, and impacts to be addressed. Unfortunately, an agency may perform an extensive scoping process and still find, upon completing the EIS, that the analysis does not adequately address decisions that eventually need to be made. Federal agencies have prepared many EISs only to find that they missed the mark because the analysis did not adequately support decisions that eventually needed to be made; even after completing an exhaustive scoping effort, the scope of an analysis sometimes does not support the decisionmaking that follows. Consequently, the NEPA process may be condemned as ineffective because the EIS did not adequately support decisionmaking. This observation is at least partly attributable to agencies diving into EIS preparation without giving much thought to the actual decisions that the analysis may eventually need to support.

Determining What Decisions May Need to Be Made

To prevent the disconnects just described, analysts should carefully consider the scope of future decisionmaking before beginning the analysis. Specifically, an effort should be mounted to address this question: "What types of decisions might need to be considered by the decisionmaker?" The response to this "simple" question is essential in shaping the scope and bounds of the analysis.[4]

Once a decisionmaker accurately assesses the specific decisions that might need to be considered or made on completing the EIS, the agency has a much better handle on the scope of actions and alternatives that must be analyzed. The scope of the EIS should be carefully tailored to reflect the response to this vital question. Such questions may at first appear to be intuitively obvious, even rhetorical, yet on closer examination they are often deceptively complicated. Failure to gain consensus on the scope of decisionmaking may indicate an ambiguous need or that aspects of the proposal are not ripe for decision (§1502.20).

It is important to emphasize that such an exercise must be designed to identify decisions that might need to be considered or made rather than to determine the outcome of such decisionmaking. Carefully adhering to this step may provide an effective and defensible tool that can assist planners in defining the scope of the EIS. The risk of having to revise or supplement the EIS because it does not adequately support decisionmaking is also reduced.

Decision-Based Scoping

The author has developed an approach, referred to as *decision-based scoping* (DBS), which is markedly different from the way scoping efforts are normally conducted.[5] Most scoping efforts begin by identifying the *actions* that comprise the proposed action and alternatives. Under DBS, emphasis is placed on first identifying potential *decisions* that eventually may need to be considered. Once potential decisions are identified, possible actions and alternatives naturally follow.

One approach to facilitating a *DBS* effort involves conducting a facilitated interdisciplinary workshop using selected value engineering (VE) techniques to assist practitioners in identifying potential decisions. For example, a facilitator challenges the group to identify and prioritize such decisions.

To facilitate the *DBS* task, an innovative tool, referred to as a *decision-identification tree* (DIT), is introduced in this section to assist practitioners in identifying potential decisionpoints. First, a word of caution is in order. The DIT is not intended for use in determining the actual outcome of a given decision. Instead, it provides a tool for identifying potential points that may eventually need to be considered and that the EIS must be designed to support.

DECISION-IDENTIFICATION TREE. Figure 1.2 provides an example of a DIT. The box to the left of the vertical dashed line denotes potential factors or decisionpoints that are outside the scope and control of the EIS decisionmaking process. Such factors or uncertainties are important to capture, as these might influence potential decisions, which are mapped to the right of the vertical dashed line.

The horizontal axis of the DIT is referred to as the *will-Axis* because it indicates potential decisionpoints designated in terms of the following question: "Will or might decisionmakers be faced with having to consider and make decisions with respect to the following course of action?" Specific decisions that might need to be considered are denoted by diamonds along the will-axis. Highest-level (i.e., most significant, fundamental, or important) decisions are drawn on the left side of the will-axis and proceed toward less important decisions as one moves to the right.

A decision to pursue a given course of action often triggers subsequent lower-level considerations (i.e., dependent decisionpoints). Accordingly, the DIT also builds downward along the vertical axis. Where a choice to pursue a given decision precipitates the need to consider lower-level considerations, the triggered decisionpoints are mapped downward, from most important to least important, along the vertical axis. The vertical axis is referred to as the *what-axis* because it denotes the question, "What types of decisions might decisionmakers be faced with having to consider if a decision were made to pursue the course of action denoted on the will-axis?"

The following example illustrates the value of preparing a DIT.

EXAMPLE OF CONSTRUCTING A DIT. Suppose an agency has a need to modernize and possibly expand a dangerous liquid waste (DLW) management operation at one of its installations. For the purposes of this example, the term *DLW* is used to signify a generic rather than a legally defined waste type.

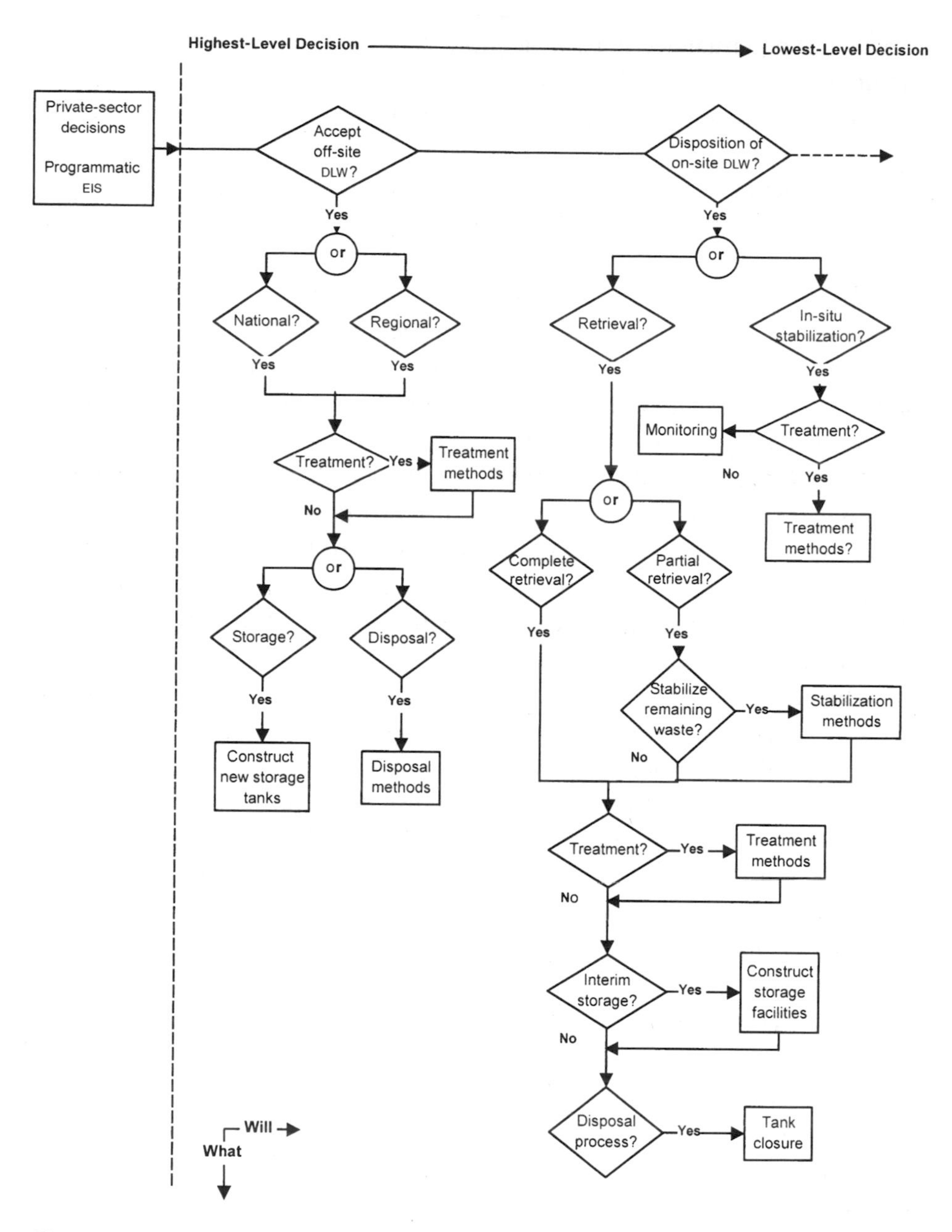

Figure 1.2 Example of a decision-identification tree for a waste management program managing dangerous liquid waste (DLW)

Further assume that a decision regarding future acceptance of offsite DLW is identified as the most fundamental decision that the agency might be faced with considering. Accordingly, the diamond-shaped icon labeled "Accept off-site DLW?" is entered as the first element along the will-axis (Figure 1.2).

Assuming a decision is made to accept off-site DLW, the highest-level (i.e., most significant or important) decision that could be triggered involves determining the extent to which this waste would be accepted (e.g., accepting waste, on either a restricted regional basis or from across the nation). The *or* gate indicates that one but not both of these decisions would need to be made. It is important to note that the DIT does not necessarily follow the standard logic convention used in many other types of logic diagrams; the *yes* response indicated at the bottom of the two diamonds simply denotes the logic path that would be taken if a decision (i.e., yes) were made to accept waste on either a national or regional basis (but not both). A *no* path is unnecessary, as such a decision simply indicates that further decisionmaking along that path would stop; hence, a *no* path, in this case, is unnecessary.

The next-lower-level decision involves the need for potential treatment. Various treatment technologies would need to be evaluated (see box labeled "Treatment methods" at the right) if a decision is eventually made to treat the off-site waste. Two potential decisionmaking paths, labeled "Storage" and "Disposal," branch downward from the diamond labeled "Treatment." A decision to pursue either a storage or a disposal option would require consideration of either "Construction of new storage tanks" or potential "Disposal methods."

Once decisionpoints are completely mapped along the what-axis, the process is repeated starting at the top of the will-axis. In this example, a decision regarding disposal of on-site DLW waste is considered the next-highest-level decision that would need to be contemplated; the procedure for mapping the lower-level decisions along the what-axis is performed in a fashion similar to that just described.

COMPLETING THE DECISION-BASED SCOPING EFFORT. Once the DIT is completed, it is reviewed by the decisionmaker, who should thoroughly understand the latitude of decisionmaking ability that will be available based on the scope depicted therein. Now the effort turns to the more standard task of identifying the scope of facilities, operations, activities, and specific alternatives that need to be evaluated.

The DIT provides an effective tool for identifying the range of decisions—and, consequently, potential actions—for later analysis. The utility of this tool becomes increasingly apparent as the complexity of the EIS planning process expands. By reducing risks associated with uncertainty, this approach can result in long-term cost savings.

Integrating the EIS with Other Requirements

"Agencies shall cooperate with State and local agencies to the fullest extent possible" (1506.2[b]). Such cooperation shall to the fullest extent possible include: (1) joint planning processes, (2) joint environmental research and studies, (3) joint public hearings (except where otherwise provided by statute), and (4) joint environmental assessments. The EIS process is to be integrated with all other related environmental regulatory requirements, permits, agreements, and policies, so that the processes runs concurrently rather than consecutively.[6] The reader should note that this discussion is applicable to the entire EIS planning process and not simply the preliminary scoping phase discussed in this chapter.

To facilitate such coordination, a detailed schedule of environmental compliance requirements for each element of the project or program should be prepared. As described in the following sections, an effort is mounted to identify cooperating agencies as well as state and local requirements.

Identifying Cooperating Agencies

The agency having overall responsibility for preparing an EIS is known as the *lead agency* (§1501.5, §1508.16). Agencies assisting the lead agency are referred to as *cooperating agencies.* A review should be performed to identify any other agencies that may be involved or have an interest in the scope of the EIS. Potential cooperating agencies should be contacted at the earliest possible time. Cooperating agencies should identify a contact responsible for coordinating the EIS effort with the lead agency.

Once a lead agency is designated, responsibilities for funding, data collection, analysis, and other tasks are assigned. It is recommended that the lead and cooperating agencies sign a memorandum of understanding designating such responsibilities.

Integrating State and Local Requirements

Many state and local governments have their own environmental or equivalent EIS planning requirements (e.g., State Environmental Policy Act [SEPA]) that must be satisfied in addition to NEPA. Federal agencies are directed to cooperate with state and local agencies in reducing duplication between NEPA, and state and local requirements. Where state or local environmental requirements do not conflict with NEPA, federal officials are directed to prepare an EIS that fulfills all requirements.[7] The EIS manager should explore the possibility of consolidating federal, state, and local planning and environmental requirements. A SEPA document can also be incorporated by reference into the EIS.

Identifying Interim Actions

To minimize project delays and advance the goal of comprehensive, well-ordered planning, an early effort should be mounted to review actions that may need to be undertaken prior to completion of the EIS. Actions are first investigated to confirm that they actually fall within the scope of the EIS; such actions are then reviewed in terms of the interim action criteria presented in §1506.1. Emphasis is placed on identifying potential problems so that appropriate contingency measures can be developed. For example, a project's schedule and funding source may need to be modified if the action does not comply with interim action criteria.

A Tool for Managing Interim Actions

An interim action justification memorandum can be prepared for individual actions that have been reviewed and found to qualify for interim action status. This memorandum briefly describes the proposal and provides evidence demonstrating that it meets interim action eligibility criteria presented in §1506.1. This step has many uses, not the least of which is that it can provide clear evidence that actions were adequately reviewed prior to implementation. The memorandum may be appended to any NEPA document that is later prepared for the relevant interim action.

Identifying Nonfederal Actions

Certain actions undertaken by a federal agency may federalize (for the purposes of NEPA) a nonfederal action that would otherwise not be subject to NEPA's requirements; the scope of an EIS may need to include such nonfederal actions.[8] Nonfederal actions may also need to be evaluated from the standpoint of the cumulative impact analysis. Accordingly, effort should be made to ensure that the scope includes applicable private actions.

Determining When a Nonfederal Action Must Be Analyzed

Many questions have been raised regarding the applicability of NEPA to state and private projects when such actions are conducted solely or largely in support of federal agency needs. More specifically, at what point do state or private actions undertaken to support the needs of a federal agency become federalized, triggering the requirements of NEPA? The author and Barbara Williamson have developed a general-purpose tool for addressing this question.[9] Specifically, it can be used in determining when a nonfederal action is federalized to the point where it either (1) triggers preparation of NEPA documentation (EA or EIS), or (2) needs to be included as part of the scope of a NEPA document that has been triggered by a related federal action.

Five criteria that provide the basis for constructing a decisionmaking tool are briefly described in the following sections. The reader is directed to our paper for additional details regarding the basis and appropriate use of this tool.

FEDERAL CONTRACT OR FINANCIAL ASSISTANCE. Projects supported by federal payment for services rendered may require preparation of an EIS. In cases where federal funding subjected a state or private project to the requirements of NEPA, the funding was generally active, as opposed to a passive deferral of payment. Normally, a substantial percentage or amount of federal funding is necessary to trigger a NEPA review.

FEDERAL CONTROL. Federal control over a state or private project may be sufficient to federalize an otherwise nonfederal action. A federal action, such as federal approval of a lease, license, permit, certificate, or other entitlement that enables a private or state action to take place, may be subject to an EIS. In such circumstances, overt federal agency action in furtherance of the nonfederal project is necessary before an action is federalized.

FEDERAL ENABLEMENT. Federal enablement entails federal agency decisions that must take place prior to actions being taken by nonfederal entities. In cases involving federal enablement, the federal agency decision is required by statute and is a legal precondition authorizing another party to proceed with an action affecting the environment. When an agency has discretion in its enabling decision to consider environmental consequences and that decision forms the legal predicate for another's environmental impact, preparation of an EIS or other NEPA review is warranted because the agency has substantially contributed to the environmental impact.

CONTINUING AGENCY INVOLVEMENT. Some courts also review whether there is continuing agency involvement in a challenged project such that termination or modification of this involvement would terminate or significantly impact the project. Because NEPA requires only federal agencies (not states or private parties) to consider environmental impacts, nonfederal actions must sufficiently involve federal actions to be subject to an EIS.

CAUSATION. The legal principle referred to as *causation* is an additional factor that is sometimes used by the courts in determining whether a nonfederal action has become federalized for the purposes of NEPA. A nonfederal action may be federalized if the nonfederal action would not take place were it not for specific actions undertaken by a federal agency—that is, but for the federal action, the nonfederal action would not occur. These but-for actions by themselves, however, do not necessarily trigger NEPA. Rather, the federal action must also be substantially interrelated with the otherwise nonfederal action.

A General-Purpose Tool

Based on the five preceding criteria, Table 1.2 provides a systematic tool for determining when a nonfederal project is federalized for the purposes of preparing NEPA documentation. This process is also represented graphically by the flowchart in Figure 1.3.

Applying the Tool in Making Determinations

Application of the tool shown in Figure 1.3 begins by reviewing the relationship between the nonfederal action in question and the federal agency's involvement. One begins at the top of this figure by answering the first question: "Is there a federal action significantly interrelated to the nonfederal action such that but for the federal action, the nonfederal action would not occur?" If the response is *no,* the user continues down through the remaining tests. A response of *no* to all five tests supports a decision that the nonfederal action is not subject to NEPA's requirements. A response of *yes* to any single test is sufficient to support a decision that the nonfederal action is subject to the requirements of NEPA.

Professional judgment must be exercised in cases where the answer to any test is not clearly obvious. While the flowchart is intended to be used as a general-purpose tool, it may not cover every conceivable condition. Our paper describes important caveats on the appropriate use of this tool.

Preparing the NOI

Once the agency has a good understanding of the preliminary scope of the proposal, an effort should be mounted to prepare a draft NOI. A list of stakeholders and other entities to whom the notice will be mailed or in other

Table 1.2 Criteria for determining when an activity conducted by a nonfederal entity may become federalized for the purposes of NEPA

Is the federal action substantially interrelated with a nonfederal action to such an extent that but for the federal action, the nonfederal action would not take place?
Is there continuing federal involvement in a nonfederal action to such an extent that termination or modification of this involvement would terminate or significantly affect the nonfederal project?
Would the nonfederal action involve a substantial degree of federal control?
Would the nonfederal action involve a substantial degree of financial support by way of a federal contract, grant, loan, or other financial assistance?
Would the nonfederal action be enabled through a federal lease, license, permit, or other entitlement?

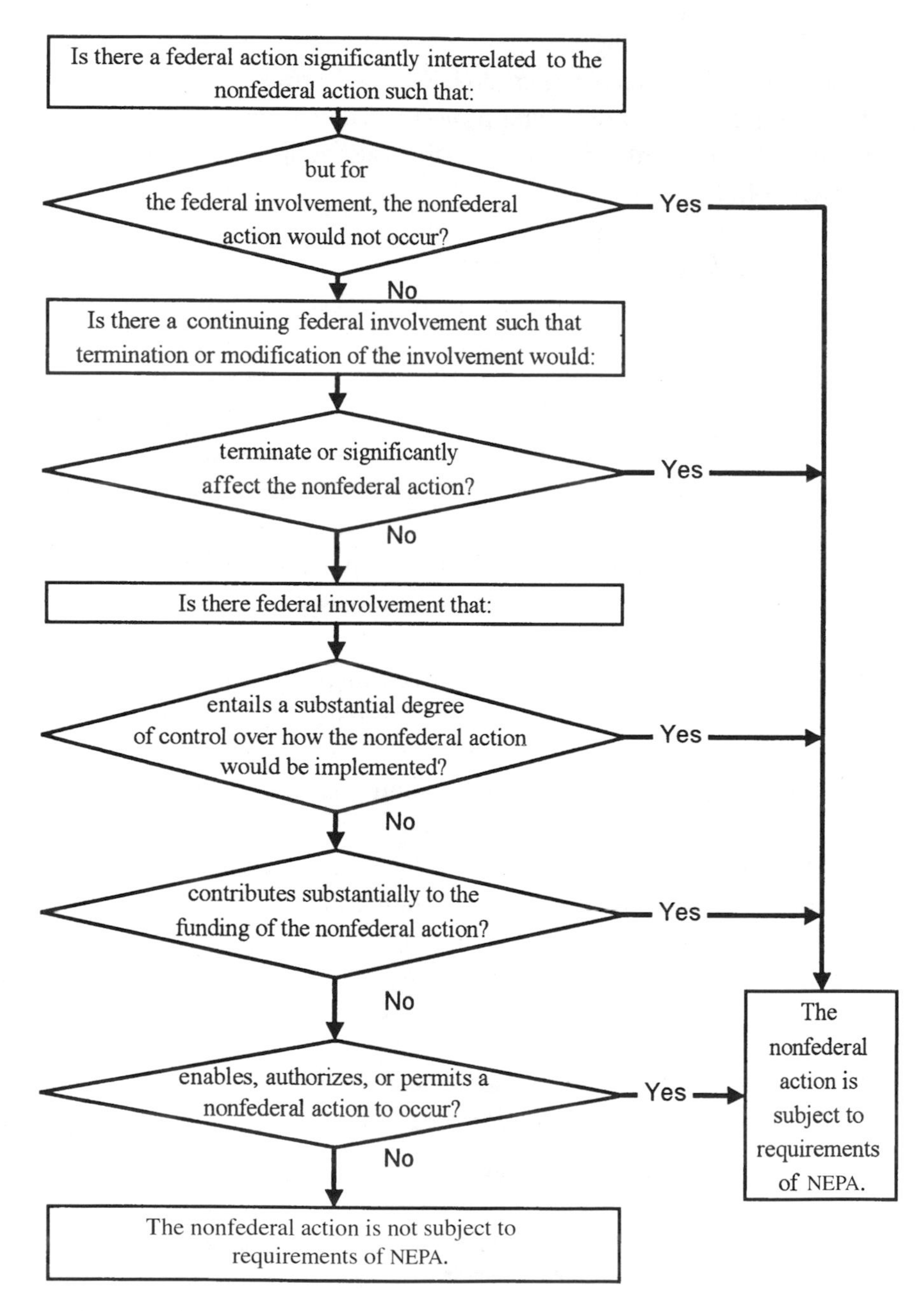

Figure 1.3 Tool for determining when NEPA applies to nonfederal actions

ways made publicly available is also prepared. The NOI undergoes a thorough internal review prior to being published in the *Federal Register.* Section 3.1 presents requirements that this document must meet.

1.3 Tools for Managing the EIS Process

Modern management techniques and principles can spell the difference between an EIS that supports a high-quality planning process and one that does not. This section describes modern management tools and techniques that can be applied to effectively manage the EIS effort. Professional judgment must be exercised in determining those tools and steps that are most practically applied based on the circumstances.

Preparing a Management Action Plan

A management action plan (MAP), also called a project management plan, should be prepared and updated as necessary. The MAP describes roles, responsibilities, and work assignments. Appropriate milestones, such as public meetings and resolution of public comments, are identified. This plan describes how deliverables such as the NOI, notice of availability (NOA), and the draft and final EIS will be prepared.

A functional roles and responsibilities matrix, such as the example shown in Table 1.3, may be included with the MAP, as it is useful in defining responsibilities and establishing organizational boundaries, especially when more than one entity or organization is involved. Table 1.4 identifies principal topics that may be included in the MAP.

Annotated Outline and Schedule

An annotated outline of the EIS may be prepared and included in the MAP. The outline should be written so that each member of the EIS staff clearly understands what is to be provided. It is normally composed of four elements: (1) outline of the topics and specific issues to be addressed in the EIS, (2) page limits for each section, (3) individual(s) responsible for preparing each section, and (4) brief description of how each section will be prepared and what it will contain. Analysts should refer regularly to the annotated outline. As necessary, the outline should be updated to reflect changing circumstances.

The EIS team members estimate the amount of time and effort required to compete their assignments; using these data, a schedule is prepared showing significant deliverable dates and milestones for measuring progress. The schedule lists specific actions and milestones for each EIS team member. Table 1.5 presents all time limits prescribed in the regulations for complying with various aspects of the EIS process. Revisions to the schedule should be imme-

Table 1.3 Example of a functional roles and responsibilities matrix

Name	Title	Organization	Responsibility
Dr. Treelover	Federal Agency Manager	Federal Agency	Manager, decisionmaker
Dr. Fellover	EIS Manager	Environmental Regulatory Department	Manager, EIS
Mr. Imsogreat	Principal Engineer	ACME Environmental Consulting Services Ltd., Dept. B2	Alternatives
Mr. Beaker	Principal Scientist	ACME Environmental Consulting Services Ltd., Dept. B3	Environmental consequences
Dr. Shark	Air Quality Specialist	ACME Environmental Consulting Services Ltd., Dept. B2	Air quality issues, affected environment
Ms. Cantfly	Senior Biologist	Ace Environmental Consulting Corporation, Dept.A1	Biological issues, affected environment
Mr. Ed	Principal Geologist	Ace Environmental Consulting Corporation	Waste management issues,affected environment
Mr. Roberts	Regulatory Specialist	Environmental Regulatory Department	Regulatory issues
Dr. Askmenot	Hydrologists	Environmental Impact Assessment Division	Hydrological issues, affected environment, modeling
Ms. No	Technical editor	Environmental Support Division	Editing and document production
•	•	•	•
•	•	•	•
•	•	•	•

diately distributed to the team. Principal assumptions, length of review periods, number of reviews, and the number of comments expected to be received should be estimated.

Funding and Budget

Funding for the NEPA process is often an afterthought. Practitioners report that project delays sometimes occur because EIS funding was not factored into early project planning. Agency officials should request funding as soon as the decision is made to prepare an EIS. A time-phased budget representing

Table 1.4 General outline for a management action plan

1. Brief description of the proposal
2. Annotated outline of the EIS
3. Roles and responsibilities
4. Description of how the planning process will be coordinated and conducted
5. Principal cost and schedule justification assumptions
6. Schedule outlining significant milestones
7. Work breakdown schedule (if appropriate)
8. Change request process (if appropriate)
9. Progress reports procedure

resources applied to the baseline schedule may be advantageous. A cumulative time-phased budget provides a baseline against which actual costs can be compared and performance measured.

Some MAPs contain a work breakdown structure that provides a basis for budgeting and controlling cost. The work breakdown structure describes how various efforts will be organized and the hierarchy of work elements. It should also describe all pertinent functions (e.g., project management, scoping, public involvement, and preparation of the draft and final EIS).

Table 1.5 Prescribed time periods for preparing an EIS

- After consultation with the lead agency, the EPA may extend or reduce prescribed regulatory periods. However, if the lead agency does not concur with the extension of time, EPA may not extend it for more than 30 days. (§1506.10[d])
- If an agency circulates a summary of an EIS and receives a timely request for the entire statement and for additional time to comment, the time for that requestor shall be extended by at least 15 days beyond the minimum period. (§1502.19[d])
- If a draft EIS is to be considered at a public hearing, the agency should make the statement available to the public at least 15 days in advance (unless the purpose of the hearing is to provide information for the draft EIS). (§1502.19[c]2)
- No decision on a proposed action shall be made or recorded by a federal agency until the later of the following dates:
  - –Ninety days after the publiction of the notice for a draft EIS, or
  - –Thirty days after publication of the notice for a final EIS
  - If the final EIS is filed within 90 after the draft EIS is filed with the EPA, the minimum 30- and 90-day periods may run concurrently. However, subject to §1506.10[d], agencies shall allow not less than 45 days for comments on draft EISs. (§1506.10[c])
  - a legislative EIS may be transmitted to Congress up to 30 days later than its accompany ing legislative proposal, to allow time for completion of an accurate statement that can serve as the basis for public and Congressional debate. (§1506.8[a])
- An EIS should normally require less than one year to complete. (40 Questions, 335)

A method for implementing change requests may also need to be defined. A change control procedure normally involves tracking cost deviations from the baseline plan, identifying an appropriate course of action, and obtaining approval for changes. Change requests may result from factors as diverse as a change in scope or assumptions, new regulatory requirements, and unforeseen environmental issues.

Managing Schedule and Budget

Considerable coordination is required in maintaining the schedule and controlling the budget. The EIS manager (or designate) monitors the team's charges and progress. Records should be maintained for any beyond-scope effort that must be performed.

The EIS manager should obtain a commitment from each team member to complete his or her portion of the EIS on schedule. If such a commitment cannot be obtained, the manager should seek the services of other professionals.

As is often human nature, veteran EIS managers report that team members frequently postpone important tasks until the last minute. This is especially true when members are juggling responsibilities among more than one project. Invariably, one or more EIS team members will request additional time. As a compensation measure, the EIS manager should consider developing a schedule based on a due date that is one or more weeks ahead of the actual drop-dead deliverable date. The manager may find it advantageous not to publicize the actual date by which the EIS *must* be finalized; this approach builds in a buffer for instances where team members request additional time or to cover other delays.

Developing a Public Involvement Strategy

In another recent solicitation for real-life "Dilbert" quotes, a manager had stated, "We know that communication is a problem, but the company is not going to discuss it with the employees." Perhaps to the dismay of this manager, a public involvement strategy outlining an approach for involving the public in the EIS process should be prepared.

The agency may find it advantageous to identify methods specifically designed to ensure that the public participates in the scoping process. For instance, the agency may hold scoping meetings in public malls or at community gatherings, such as a county fair. To stimulate interest, an exhibit or model of the proposal may be displayed in a mall or town square; agency representatives can solicit comments from citizens who stop to view the exhibit.

Highly controversial or technical proposals are generally difficult for the public to understand and comment on. Moreover, people are sometimes suspicious, or too intimidated, to participate in scoping. Special measures may need to be taken to induce the public to participate. The public involvement strategy should specifically address such problems. The importance of this step cannot be overstated, particularly for controversial proposals. Developing an effective means for dealing with the public is a vital step that can save an agency endless hours of grief, schedule delays, and misdirected effort, not to mention a tarnished reputation.

Dealing with Opposition

A successful strategy should be oriented toward identifying and dealing with public opposition. In particularly controversial cases, the agency may want to consider obtaining the services of an impartial public relations specialist to educate the public regarding both the positive and negative aspects of the proposal.

One method that can be used to defuse hostility is simply to continually stress that the agency is actively (and genuinely) seeking ideas for alternatives and design changes that would eliminate or mitigate potential impacts. The agency should clearly explain that no decision has been made and that NEPA provides the public with the opportunity to influence future decisions.

Identifying Potential Adversaries

Early input from the public may also allow the agency to modify a proposal in a manner that minimizes later controversy. To this end, agency officials may want to meet with community leaders prior to making the proposal public. Discussions with public interest groups should be held as necessary.

An effort should always be made to identify potential adversaries in an effort to bring them on board. Providing public interest groups and potential adversaries with an opportunity to voice their opinions and concerns early in the planning stage may prevent significant delays later. The reader should note that under no circumstances should agency personnel attempt to silence points of view that differ from that of the agency.

Scoping Plan

A good scoping plan can ensure that a comprehensive scoping process is performed. The scoping plan should be tailored to the specific proposal for which the EIS is being prepared. The plan outlines the entire scoping process, including when and where meetings and hearings will be held, and the appropriate format to be followed in managing those meetings. The plan also includes results from the public involvement strategy described above. Emphasis is placed on ensuring that all interested parties are represented and given an opportunity to participate and provide input.

An effort should be made to identify specific information that will be desired from the public. The plan should identify how, when, and by whom input will be requested from agencies, organizations, and individuals. The medium most appropriate for informing the public about the details of the proposal is also identified.

Advertisements

Draft media and distribution lists, news releases, and presentations should be prepared as outlined in the scoping plan. In the past, comments have been solicited from the public via advertisements in local newspapers, radio, television, direct mail, professional journals, telephone recordings, public meetings, and from booths at community events. More recently, the Internet is being used as well. A method describing how comments will be recorded and reviewed should also be defined. Remember that the agency must make "diligent efforts" to involve the public and may be successfully challenged if it doesn't (§1506.6[a]).

The Administrative Record

Mark Twain once mused:

> When I was younger I could remember anything, whether it had happened or not; but my faculties are decaying now, and soon it shall be so I cannot remember any but the things that never happened. It is sad to go to pieces like this, but we all have to do it.

It is imperative that an agency maintain an institutional memory of its EIS process. A thorough and well prepared administrative record (ADREC) is crucial in demonstrating that the agency has properly weighed all pertinent factors in reaching a final decision. Alternatively, an adversary may find it advantageous to scrutinize the agency's ADREC in an attempt to find weaknesses or even lack of evidence that the agency fully discharged its EIS responsibilities.

Although an ADREC is important in demonstrating compliance, the EIS must be capable of standing on its own. In one case, the ADREC contained an important analysis that was not cited in the EIS. When challenged, the agency lost because the EIS analysis was incomplete. Political embarrassment aside, losing a case can substantially delay a project and escalate costs.

Defining the Administrative Record

It is difficult to define exactly what constitutes the administrative record because NEPA does not establish a formal fact-finding process (i.e., a detailed procedure for establishing facts or reaching a final decision). For the purposes of judicial review, an agency's ADREC can be viewed as the body of documents and information considered or relied upon in the process of reaching a final decision. This body of information includes the entire record that existed at the time the decision was made and not simply the portion of the record read by the decisionmaker.[10] At a minimum, the ADREC should include "all relevant studies or data used or published by the agency" in preparing the statement.[11] E-mail may also be part of the ADREC.

Because NEPA is a full-disclosure act, an agency's ADREC is usually considered to be open for public review. Many people are involved in shaping plans, assumptions, and internal decisions before the EIS reaches the decisionmaker; in the event of a legal challenge, this entire process is considered part of the agency's ADREC.

Internal Conflict and the Administrative Record

The ADREC should demonstrate that the agency was taking a hard look at the proposal and was actively seeking alternative views. A record that indicates healthy debate within the agency's planning circles (diverse views, conflicts, evolving ideas) can be useful in demonstrating that the agency diligently struggled to reach a good decision. As a judge is trained and works daily with conflicts, an ADREC that reveals a diversity of viewpoints can work in the agency's favor rather than to its detriment. Lack of internal conflict may indicate that the decision was already made and the EIS was prepared simply to document that decision.

Importance of an Auditable Trail

The following two situations illustrate the importance of maintaining a traceable record of the agency's thought process and decisions. First, an independent specialist was assigned to audit a federal agency's programmatic EIS. On reviewing it, this expert could not understand how certain conclusions had been reached. Questions were also raised regarding the origin of some of the data and assumptions used in the analysis. An investigation was conducted to determine the basis for the conclusions and the origin of the data. Professional staff turnover had been high and records had not been kept up to date. Substantial portions of the data and analysis were completely untraceable. While the conclusions and data may have been valid, they were nevertheless untraceable. Consequently, a significant portion of the analysis had to be redone to ensure that the analysis was accurate and defensible.

The second example involves issuance of a permit under Section 404 of the Clean Water Act by the U.S. Army Corps of Engineers (Corps) for construction of a dam and reservoir. In challenging the Corps, the plaintiffs alleged that the analysis prepared for the permit did not adequately cover impacts of an existing liquid petroleum pipeline that would cross under the proposed reservoir. The Corps had assumed the pipeline would be relocated but did not make relocation a condition of the Section 404 permit. The court found that the ADREC did not support the assumption that the pipeline would be relocated and that the agency failed to consider the impacts of such a pipeline.

Litigation and the Administrative Record

An agency should not depend solely on environmentally sound decisions as a means of demonstrating due diligence with respect to NEPA compliance. If the agency is challenged, a court may require the agency to turn over files and documentation related to the suit.

In many cases, a plaintiff's ultimate strategy is not to win a legal suit but simply to delay the project or to stimulate public opposition. In countering such tactics, it is important that agencies maintain an auditable trail that can quickly demonstrate compliance to a court. Conversely, a potential adversary should zero in on areas where the agency's administrative record appears weak.

With respect to the adequacy of an EIS, the court's role is to determine if the authors have made an adequate and objective effort, judged in light of the rule of reason, to complie with and present all reasonable alternatives and significant information to the decisionmaker.[12] A court's review of an agency's decisionmaking process is governed by Administrative Procedure Act.[13]

While federal agencies routinely deal with complex technical issues, judges typically lack such expertise. Consequently, it is generally imprudent for judges to consider testimony and documentary evidence unless this material was first presented and considered by the agency in question. With few exceptions, courts are expected to confine their review to the ADREC compiled by the agency.[14] Only in limited circumstances may a court expand its review beyond the record or permit discovery.[15]

DISCOVERY. Discovery is the legal process adversarial parties use during litigation to obtain information they do not have access to but need to support their case. For example, a plaintiff may petition a defendant for documents and information pertinent to its case. The defendant may then be required to provide such documents and information to the plaintiff according to a schedule set by the court.

RELEASING THE ADMINISTRATIVE RECORD. Under no circumstances should any applicable material be withheld. An aggressive litigator is unlikely to accept the agency's word that it has compiled a complete record for review. The Freedom of Information Act can also provide a useful tool for gaining access to an agency's internal documents.

A well-documented ADREC can reduce the chances that a court will allow discovery. An agency that appears to be secretive is likely to raise a judge's concerns, and no agency wants to lose credibility with the judge. A plaintiff may be granted discovery if it can be demonstrated that there is reason to believe the ADREC is incomplete and the agency hasn't been forthcoming.

GOING BEYOND THE ADMINISTRATIVE RECORD. The court cannot meet its responsibility to conduct a substantial inquiry if it must accept an agency's word that it considered all relevant matters. In some cases, the court must look beyond the administrative record to ensure that the agency did so. Such reviews, however, should be limited to collecting background information or "for the limited purposes of ascertaining whether the agency considered all the relevant factors or fully explicated its course of conduct or grounds of decision. . . . Consideration of the evidence to determine the correctness or wisdom of the agency's decision is not permitted, even if the court has also examined the administrative record."[16]

Allegations that an EIS does not investigate all reasonable alternatives, overlooks significant impacts, or sweeps "stubborn problems or serious criticism . . . under the rug" may raise questions sufficient to justify introduction of new evidence outside the ADREC, including expert testimony. Such evidence may be introduced where challenges involve the adequacy of the analysis and cases involving an agency's finding that an EIS is not required.[17]

When necessary to establish the reasons for an agency's decision, the court may inquire outside the ADREC, introducing affidavits and testimony. Courts may also do so when it appears that an agency used material outside the record. Moreover, courts may inquire outside the ADREC when there is reason to believe that the agency acted improperly or in bad faith with respect to its knowledge.[18]

A court may not, however, undertake its own examination where it finds that an agency's inquiry was inadequate or is not supported by the ADREC. Instead, the court's duty is to remand the matter to the agency for further consideration.[19]

Establishing an Administrative Record

As mandated by the Federal Records Act, agencies must retain copies of all draft documents that show a substantial change in the agency's thinking. Depending on the complexity of the planning process, an agency may want to consider developing a formal system to capture all important aspects of the

EIS process as part of its ADREC. A formal record system can be a vital link in demonstrating that an open, public, rigorous, interdisciplinary, and systematic approach has been performed.[20] Professional judgment must be exercised in determining which steps described in the following subsections are most practically applied based on the complexity and particular circumstances of the EIS process.

A point-of-contact should be assigned responsibility for compiling the ADREC, which should be reviewed periodically to ensure that it demonstrates implementation of a thorough and systematic process. It is recommended that all documents prepared for or used in the EIS planning process be filed in a centralized storage area. For example, a specific filing cabinet(s) may be designated for exclusive use in storing the ADREC for a given EIS. Some agencies also make backup copies of critical documents that are stored in an alternate location.

The EIS manager should keep a running chronology of events and progress. In support of this task, procedures should be established to ensure that all EIS personnel maintain records of important events, discussions, direction, decisions, and conclusions reached. Important oral communications should also be recorded, as should the results of relevant meetings. Public and interagency discussions should be formally documented. Discretion, however, should be exercised in documenting meetings and conversations, as they may be made publicly available either by a court or by way of the Freedom of Information Act.

The ADREC system should be capable of accurately reconstructing the investigation of the actions, alternatives, issues, and impacts that were considered. The investigation of both significant and nonsignificant impacts should be documented. A short description of who, when, and what was considered, as well as important conclusions that were reached, should be recorded. The ADREC also should retain relevant correspondence, such as official memos, meeting minutes, internal memorandums, public comments, and analysis reports.

ARTS AND COMTRACK DATA BASE. If the ADREC is particularly voluminous, the EIS manager may want to consider implementing an administrative record tracking system (ARTS) computer data base as a tool for tracking and managing important sources of information (e.g., technical reports and studies, personal correspondence, telephone calls). As specific issues are investigated, the results are entered into the ARTS data base, which maintains a brief summary of each review. The system can be used to demonstrate not only that pertinent impacts and issues were considered but when, where, and by whom.

An electronic data base approach has an important advantage in that it provides the capability to automatically generate reports. Keyword searches can be performed. By sorting on various fields, reports can be quickly generated to assist the team in scheduling, tracking, and ascertaining progress to date. In the advent of a challenge, the data base can also be quickly queried to generate reports useful in demonstrating due diligence.

Similarly, where a project elicits a voluminous number of scoping or draft EIS comments, the EIS manager may want to consider constructing a comment tracking (COMTRACK) database to capture, track, and manage comments and responses. This is especially true for large or controversial EISs, where comments may run into the thousands. The Glen Canyon Dam EIS, for example, received over 17,000 scoping comments from citizens in nearly all the 50 states.[21]

CROSS-REFERENCING DOCUMENTS. Using a computer data base or by other means, the EIS team may find it advantageous to develop a master file that cross-references issues, impacts, and alternatives that have been considered against the physical location of the corresponding source documents. Such a file is particularly useful where there is potential for significant turnover in the technical staff; it can easily be maintained within an ARTS data base. Table 1.6 illustrates how such a system can be logically constructed so as to effectively manage a large amount of data.

If the analysis is particularly complex, the IDT may want to consider

Table 1.6 Example of environmental Impacts and Issues cross-referenced against their corresponding document location

Alternatives Considered	File No.	Cabinet No.
No action alternative	File #1	2
Waste transportation alternative	File #18	1
Incineration alternative 1	File #13	3
Incineration alternative 2	Memoranda	3
•	•	•
•	•	•

Impact Considered	File No.	Cabinet No.
SO_2 impacts on air quality	File #2B	6
SO_2 impacts on wildlife	Letter #65	4
Noise impacts on wildlife	File #21-1	4
Fugitive dust impacts wildlife	File #16	4
•	•	•
•	•	•
•	•	•

assigning unique reference numbers to individual impacts and issues whose status is maintained in a data base; this approach can facilitate recording specific actions, impacts, and issues, by whom and when they were considered, and the outcome of the reviews. Environmental review forms may be distributed to the IDT members for recording their individual reviews, which can then be easily entered into the data base. Having such a system in place may obviate the need to regenerate EIS work someday.

Selecting an EIS Contractor

Ideally, an EIS is prepared directly by the responsible agency. In practice, this is often not the case; limitations in personnel, resources, and expertise may compel agencies to contract some or most of their work to EIS contractors.

Managing an EIS Contractor

If the agency obtains the services of an outside contractor, it should designate a point-of-contact with EIS experience to manage and oversee the effort from start to finish. The agency's EIS manager is responsible for making decisions, providing direction, ensuring the accuracy and adequacy of the analysis, and providing data to the contractor. The EIS manager is also responsible for ensuring that the contractor stays within the assigned scope of work. In one case, a NEPA manager was complacent because the support contractor's track record was exemplary. However, the manager failed to provide sufficient oversight for what turned out to be an inexperienced team. As a result, the analysis had to be rewritten, resulting in a delay and substantially increased costs.

Statement of Work

As appropriate, the agency's procurement officer should be contacted for assistance in preparing a statement of work. This statement specifies the exact scope of work, including the specific expertise and resources that the outside contractor is expected to provide. The statement of work clearly defines what is expected to be included in the draft and final EIS. Whenever feasible, an outline of the EIS should be included with the statement of work. A schedule should also be included, specifying contractor milestones as well all important actions that the agency is responsible for accomplishing. If selection of a contractor involves a competitive bid, care should be taken to protect information that may be procurement sensitive.

PERFORMANCE-BASED STATEMENT OF WORK. A performance-based statement of work may be advantageous, as it tells the EIS contractor what is to be produced in lieu of prescribing how to perform the work. While conceptually simple, practitioners report that preparing the document in sufficient detail to ensure that the final result meets the agency's expectations can be a daunting task.

An introductory section of the statement of work should describe the agency's concept of what is expected. For example, an introduction might explain that the agency is directly responsible for preparing the EIS, thus emphasizing that the contractor's role is limited to providing assistance. Information that is to be contained in the deliverable should be described in detail. An example of the work product may be attached. Specifying parameters, such as minimum staffing requirements and how often a particular activity should be performed, should be avoided unless absolutely necessary; such specifications may inhibit innovative management practices that the contractor can use to reduce costs.

Shopping for a Contractor

The agency should *shop* for a contractor experienced in preparing EISs. It is equally important to obtain a contractor possessing knowledge and expertise related to the proposal and, in particular, with the potential environmental issues.

The agency is required to ensure that the analysis is interdisciplinary. To this end, the agency must secure a contractor that possesses the appropriate interdisciplinary expertise necessary for performing the analysis. The EIS manager should verify that the contractor has specialists representing all appropriate environmental disciplines or has access to such expertise. The agency should also verify that the consultant is internally capable of or has access to computer modeling, monitoring instruments, laboratory facilities, or other technical resources that may be needed.

Selecting a Contractor

The agency reviews the contractor's proposal in terms of the experience, qualifications, and availability of the professional staff designated to work on the analysis. The bait-and-switch trick has sometimes been used in the consulting world.[22] This is a ploy in which the services of experienced or recognized professionals are proposed in an effort to gain a contract. On obtaining the contract, the consulting company switches these key individuals with less experienced personnel so that the experienced personnel can be reassigned to other marketing tasks. As a countermeasure, the agency may want to stipulate that key individuals identified in the contract or statement of work may not be replaced without written concurrence from the agency.[23]

The agency and the selected contractor should agree on policies and procedures to be used. All significant guidance and direction to the contractor should be provided in writing.

CONFLICT OF INTEREST. The agency is responsible for verifying that there is no conflict of interest between the potential EIS contractor and the proposal (§1506.5[c]). The agency is responsible for preparing a disclosure statement, to be signed by the contractor, indicating that the contractor has no financial or other interest in the outcome of the project.

Interpreting the conflict-of-interest provision may not always be straightforward. Witness the following case. Plaintiffs challenged a Department of Transportation EIS for a proposed highway interchange. The plaintiffs argued that the agency failed to comply with NEPA by allowing a private contractor with a conflict of interest (expectation of future work based on the agency's unvarying practice of awarding the final design contract to the company preparing the EIS) to assist in preparing the EIS. The plaintiffs also alleged that the contractor failed to execute the required conflict-of-interest disclosure statement until after the EIS had been filed. The court concluded that the contractor had no contractual agreement or guarantee of future work at the time the EIS was prepared. The court also found that the degree of oversight exercised by the agency was sufficient to prevent defects that might arise from such a conflict. The court held that the ADREC demonstrated that the agency exercised proper managerial oversight (preparing many sections of the EIS without contractor assistance), thoroughly reviewed all of the contractor's data and analysis, and used the contractor personnel only for technical expertise.[24]

1.4 Compiling Data

Collecting pertinent information is a vital component of virtually all EIS efforts. As is often the case, the EIS planning schedule is not driven by the time required to prepare the statement but instead by the time needed to compile data and complete field studies as input for the analysis. Agencies are expected to make a diligent effort to obtain data for the analysis. A literature search should be initiated as soon as practical to identify existing NEPA documents, scientific studies, engineering reports, and other technical support data.

It is important to note that data collection is driven by the ultimate scope of the analysis. While data collection should be initiated as early as feasible, it should not get too far ahead of the scoping process. In one instance, an individual reported that problems arose because data collection was ongoing even though the initial scope was ill defined.[25]

Data Accuracy

The agency is responsible for ensuring the accuracy of data used in the EIS. Data should be checked to ensure that it conforms to accepted scientific standards. A procedure should be established for ensuring the validity and accuracy of data. For example, the Sierra Club claimed that the data relied on in a U.S. Forest Service EIS for predicting population levels of various animal species was faulty. The court concluded that the Forest Service failed to respond to the criticisms of two respected experts who objected to the use of 10-year-old data in the analysis; this raised serious questions about the reliability of the environmental impact projections. The court found that the data used by the Forest Service was arbitrary and capricious.[26]

Gap Analysis

The agency must determine specifically what data are currently available and what data need to be generated. This step can be crucial in preventing schedule delays. Consider the following example. An EIS manager was assigned to a high-profile project. Several months were spent developing a plan and budget, preparing a scoping plan, and outlining potential alternatives. A detailed EIS schedule was prepared. However, no effort was made to identify input data requirements for the EIS. On issuing the NOI, the agency selected a contractor to prepare the EIS. The EIS contractor did not realize that input data had not been generated. At the kickoff meeting, the contractor made a request to the agency for technical input data necessary for preparing the analysis. Caught by surprise, the agency scheduled an emergency meeting to determine what data needed to be generated before the analysis could begin. It was determined that it would take a *year* to generate the required engineering studies and input data. Thus, on the first day of the kickoff meeting, the agency's project was already one year behind schedule!

The agency may want to consider performing a gap analysis, which can assist analysts in determining how much technical information currently exists and what new information needs to be generated. Emphasis is placed on identifying information that truly contribute to eventual decisionmaking as opposed to accumulating "extraneous background data" (§1500.2[b]).

Data Collection

Based on the gap analysis, the lead agency determines if the additional effort and expense is justified in collecting or generating the missing data. Where important information is incomplete or unavailable, the EIS team should consult the CEQ regulations for dealing with such circumstances (§1502.22). The regulatory requirements concerning incomplete or unavailable information are described in chapter 3. Because of the expense and complexity involved, agencies may find it desirable to apply a sliding-scale approach, in which the level of effort expended in obtaining information is commensurate with the potential significance of the impacts. Thus, more effort and resources are expended in generating data where impacts are potentially significant; correspondingly, less effort is expended where the potential for impacts is deemed marginally significant.

Data requested from a project, engineering, or support office should be carefully tracked. In one case, plant personnel were not aware of a request for data and did not receive a copy of the request for several months; this miscommunication resulted in a significant schedule delay. All data requests should be tracked to ensure that plant or field personnel receive copies of the requests promptly. As field and plant personnel are sometimes unfamiliar with the expectations associated with supplying information, it may also be advisable to assign a point-of-contact responsible for ensuring that the effort is coordinated and that appropriate information is collected.

Common Types of Data

Tables 1.7, 1.8, and 1.9 list categories of data typically required in a NEPA analysis. A vast amount of environmental data are routinely collected by both public and private parties. Most data used in an EIS come from existing sources (e.g., government agencies and contractors, universities, libraries, scientific studies). Much of this information is never publicly disseminated.

For example, agencies such as the U.S. Fish and Wildlife Service may already have compiled a list of species within a potentially affected area. Table 1.7 depicts commonly required alternatives and engineering data.

Table 1.7 Commonly required alternatives and engineering data

- Description, schematics, diagrams, and artist's renderings of the proposal
- Footprints (e.g., size of disturbed land area)
- Description of connected or related actions
- Infrastructure additions or modifications
- Potential mitigation measures
- Cost and schedule
- Transportation routes
- Construction and operational resource usage for each alternative (e.g., electricity, water, diesel fuel, construction materials)
- Construction and operational personnel requirements

Table 1.8 Commonly required source term data

- Air (chemical, radiological, particulate) emissions (e.g., quantities, rates, concentrations, and release points)
- Surface water (chemical, radiological, solids) releases (e.g., quantities, rates, concentrations, and release points)
- Groundwater (chemical, radiological) releases (e.g., quantities, rates, concentrations, and release points)
- Noise levels
- Hazardous, nonhazardous, and radioactive wastes (e.g., types, mass, volume, concentrations)
- Chemical or radioactive materials (e.g., types, mass, volume, concentrations, activities)
- Potential accidents (e.g., probabilities and source terms)

Tables 1.8 and 1.9 depict commonly required source term information and data on the affected environment. The actual data requirements depend on the specific location, context, and nature of the proposal.

Monitoring

Pre-monitoring, sampling, and environmental measurements must sometimes be conducted to supply baseline information for describing the affected environment and performing the analysis. Monitoring may involve measuring attributes as diverse as existing air and water contaminant levels, existing noise levels, and cataloging indigenous species and habitat.

As the monitoring effort is sometimes the most expensive aspect of the analysis, the EIS manager should ensure sufficient funding. To reduce cost, the agency should also consider the possibility of planning, coordinating, and sharing the cost of the monitoring effort with other agencies.

Table 1.9 Commonly required data on the affected environment

- Maps (topographic, biological, geological, hydrological)
- Groundwater quality, flow rates, and aquifer characteristics
- Surface water quality, flow rates, and potential for flooding
- Geological surface and subsurface structure, well logs, and potential environmental hazards
- Meteorological data
- Ambient air quality data
- Ecology and habitats
- Sensitive and endangered species
- Socioeconomic data (e.g., traffic rates, housing, minorities, infrastructure)

Some biological studies may need to be performed during certain times of the year. Other studies may need to be conducted over a period long enough to account for monthly, seasonal, or even annual variations. The EIS schedule should be planned accordingly.

Field Studies

Field studies may need to be conducted to collect data necessary for the analysis. A representative list of studies is shown in Table 1.10.

Summary

The intent of this chapter is to provide the reader with a comprehensive guide detailing the principal steps and tasks that may need to be considered and, if appropriate, performed as part of an agency's pre-scoping process. Factors as diverse as agency culture, funding and existing data, amount of experience the agency has with the proposal, schedule constraints, and the size and complexity of the planning effort will dictate which steps and tasks are most practical and appropriate.

Table 1.11 summarizes these principal steps and tasks. This table can be used to assist practitioners in preparing a schedule and budget and as a road map for planning and managing the EIS pre-scoping effort. A similar table, outlining the principal steps and tasks involved in preparing the EIS, is included in the summary of chapter 2.

Professional judgment is exercised in determining those items most applicable to the existing circumstances. The reader should also note that items are not necessarily performed in the same order as indicated in the table. A more detailed version of this checklist, designed for use in planning and coordinating an EIS, can be obtained through ecclestonc@aol.com.

Table 1.10 Examples of field studies commonly conducted to provide data for the EIS analysis

- Biological surveys and wetland studies to identify potentially affected species (especially sensitive and endangered)
- Geological surveys to determine soil characteristics and geological hazards
- Surface and groundwater studies to determine hydrologic characteristics
- Meteorological studies and air quality monitoring data for input into air quality models
- Archaeological and cultural resource surveys

Table 1.11 Checklist of steps and tasks that should be considered and, if appropriate, performed as part of an agency's pre-scoping effort

- Determine that an EIS must be prepared. If an EIS is required:
  - –Designate an EIS manager.
  - –Form an interdisciplinary team (IDT).
  - –Identify and contact cooperating agencies.
- Hold a kickoff meeting. Provide IDT members with a description of their work assignments and tasks.
- Begin preparing and circulated progress reports.
- Initiate technical coordination meetings with other entities; these will continue throughout the EIS effort.
- Standardize ground rules and procedures for preparing the EIS. Prepare and circulate a document standards memo to be read and signed by all EIS team members.
- Prepare and maintain a checklist outlining all important steps that must be conducted during the EIS process.
- Identify potential decisions that may ultimately need to be considered by the decisionmaker; if appropriate, construct a decision-identification tree (DIT).
- Prepare a statement of the underlying purpose and need for taking action.
- Conduct an on-site field trip to familiarize analysts with the proposal, affected environment, and potential issues.
- Begin integrating applicable state or local environmental requirements with the federal EIS.
- Identify and review interim actions in terms of criteria specified in §1506.1. Prepare an interim action justification for related actions that may need to proceed before the EIS process is completed.
- Begin identifying nonfederal actions that may need to be included within the scope of the EIS.
- Prepare draft NOI and distribution list.
- Prepare a management action plan (MAP).
  - –Functional roles and responsibilities matrix
  - –Annotated outline and schedule
  - –Budget and allocation of funding
  - –Work breakdown structure
- Develop a public involvement strategy:
  - –Obtain services of a public relations specialist.
  - –Identify potential adversaries.
- Develop scoping plan.
  - –Prepare media distribution list.
  - –Prepare notices and advertisements.
- Develop administrative record (ADREC) system.
- Develop data base systems (if applicable) that may be needed to manage the EIS process:
  - –Administrative record tracking (ARTS) data base
  - –Comment tracking (COMTRACK) data base
- Prepare and distribute environmental review forms to the IDT staff.

Table 1.11 Checklist of steps and tasks that should be considered and, if appropriate, performed as part of an agency's pre-scoping effort *(cont.)*

- Select an EIS contractor (if applicable):
  - –Assign an agency official to oversee the EIS contractor.
  - –Prepare statement of work.
  - –Shop for an EIS contractor.
  - –Prepare a conflict-of-interest statement.
  - –Select an EIS contractor.
- Institute data compilation and monitoring:
  - –Perform literature search.
  - –Establish procedure for ensuring data accuracy.
  - –Perform gap analysis.
  - –Identify incomplete or unavailable data in terms of requirements specified in §1502.22.
  - –Conduct applicable monitoring.
  - –Conduct applicable field studies.
- Prepare notice of intent (NOI).

Notes

1. Sun Tzu, *The Art of War,* ed. James Clavell (New York: Dell Publishing, 1983), 29.
2. *Webster's II New Riverside University Dictionary* (Boston: River Publishing, 1988).
3. *National Environmental Policy Act of 1969,* Section 102(c), as amended (Public Law 91-190, 42 U.S.C. 4321–4347, 1 January 1970, as amended by Public Law 94-52, 3 July 1975, Public Law 94-83, 9 August 1975, and Public Law 97-258, § 4(b), 13 September 1982).
4. Charles Eccleston, forthcoming.
5. Charles Eccleston, forthcoming.
6. 40 CFR 1500.2(c), 1500.4(k), and 1502.25.
7. 40 CFR 1506.2(c).
8. C. H. Eccleston and B. D. Williamson, "Determining When NEPA Applies to Nonfederal Activities," *Federal Facilities Environmental Journal* 7, no. 4, (Winter 1997): 83–94.
9. Ibid.
10. *Haynes v. United States,* 891 F.2d 235 (9th Cir. 1989).
11. *County of Suffolk v. Secretary of the Interior,* 562 F.2d 1368 (2nd Cir. 1977).
12. Ibid.
13. *Administrative Procedure Act,* 5 U.S.C. 706.
14. *Crotin v. Department of Agriculture,* 919 F.2d 439 (7th Cir. 1990).
15. Ibid.
16. *Asarco, Inc. v. EPA,* 616 F.2d 1153 (9th Cir. 1980); *Environmental Defense Fund, Inc. v. Castle,* 657 F.2d 275 (D.C. Cir. 1981).
17. *County of Suffolk v. Secretary of the Interior.*
18. *Animal Defense Council v. Hodel,* 840 F.2d 1432 (9th Cir. 1988).
19. *Asarco, Inc. v. EPA; Environmental Defense Fund, Inc. v. Castle.*

20. 40 CFR 1501.2(a) and 1502.14(a).
21. T. Randle, "Case Study: The Glen Canyon Dam EIS Process," *Federal Facilities Environmental Journal* (Winter 1993–94): 471–477.
22. Presentation on NEPA consulting given at a CEQ-sponsored seminar titled "Implementation of the National Environmental Policy Act on Federal Lands and Facilities," Duke University, 25–29 April 1994.
23. Personal communication, R. L. Colley and R. G. Upchurch, Waste Management Federal Services of Hanford, Inc., procurement department and contract administration, respectively, 1999.
24. *Associations Working for Aurora's Residential Environment v. Colorado Department of Transportation,* 1998 U.S., App. (10th Cir. 1998), as reported in the Department of Energy's *Lessons Learned,* December 1998, 13.
25. U.S. Department of Energy, *NEPA Lessons Learned,* 18, 1 March 1999, 13.
26. *Sierra Club v. Department of Agriculture,* 116 F.3d 1482 (7th Cir. May 28, 1997).

2

Preparing the EIS: An Integrated and Systematic Approach

Predictions are notoriously difficult to make—especially when they concern the future.

—Mark Twain

While the total federal planning (TFP) strategy presented in chapter 5 advances an approach for reengineering the federal planning apparatus into a single integrated framework, this chapter is focused on effectively managing and conducting the EIS planning process. Building on the foundation laid in chapter 1, this chapter presents the reader with a systematic and comprehensive description of the entire step-by-step process for managing and preparing an environmental impact statement (EIS). The intent is not to rehash the National Environmental Policy Act's (NEPA) regulatory requirements concerning what must be done; a detailed review of the EIS regulatory requirements is beyond the scope of this chapter. It is assumed that the reader is acquainted with or has access to a text describing such requirements. For a detailed description of NEPA's regulatory requirements, the reader is referred to the author's companion book, *The NEPA Planning Process: A Comprehensive Guide with Emphasis on Efficiency.*[1] While the companion book describes *what* must be done, this chapter focuses on *how* the regulatory requirements are most effectively implemented.

Often things are not as simple as they appear on the surface. Many actions have both beneficial and adverse effects. For example, some public interest groups are calling for the destruction or overtopping of dams to preserve fish runs. Such an action is widely regarded as an environmentally positive impact. However, the increased water flow might also adversely affect or even destroy wetlands downstream—an adverse effect. Similarly, construction of a sewage treatment plant can be viewed as a measure necessary for preserving water quality, yet this very act may result in effluent discharges that may contaminate a river. It is for such reasons that an EIS is intended to rigorously investigate and flush out all potential impacts so that the agency can make an informed decision.

All to often, NEPA is viewed as a document preparation process. While the resultant document is an integral component, preparation of EISs (even excellent ones) is not why NEPA was enacted. First and foremost, NEPA establishes a comprehensive planning and decisionmaking process. The ultimate goal is not to produce a document but to provide the basis for an informed, planned, and carefully considered decision. Properly implemented, preparation of an EIS is merely a mechanism, and perhaps the final element, used to record the results of a rigorous and comprehensive planning process. To this end, this chapter presents the EIS process from the perspective of an integrated and comprehensive planning process.

With this in mind, this chapter is specifically designed to provide project managers, planners, NEPA practitioners, analysts, and decisionmakers with a comprehensive framework for conducting and managing the entire EIS planning process. Rigor is tempered with common sense. A general-purpose approach for conducting the process is described that can be readily adapted to large or complex planning efforts. Modern tools and techniques for enhancing the effectiveness of the EIS process are also introduced. Chapter 3 presents a detailed description of all significant documentation requirements that the EIS must meet.

Not all steps or elements described in this chapter are necessarily applicable to every EIS effort; as always, professional judgment must be exercised in determining those steps most appropriate to the proposal at hand. For smaller-scale planning efforts, the reader should choose those elements most practically integrated with their specific needs.

Implementing the TFP strategy described in chapter 5 depends on many factors, not the least of which is the complexity of the agency's planning process, organizational culture, and political considerations. Accordingly, the EIS approach described here is designed to be flexible, such that it can either be fully integrated with a TFP strategy or implemented in a standalone mode independent of TFP. Linkages describing how specific aspects of the EIS approach can be integrated with the TFP strategy are provided. Details for integrating a TFP strategy are left to the reader based on their individual needs and circumstances.

2.1 Public Scoping Process

Mark Twain once noted, "When a person cannot deceive himself, the chances are against his being able to deceive other people." Twain's advice should be heeded as the agency enters the public scoping process.

The principal steps making up the typical EIS process are shown in Figure I.4 of the Introduction. Similarly, Figure 5.6 denotes the principal steps performed as part of the formal scoping process under the TFP strategy. As depicted in Figure 5.3, the TFP approach incorporates measures specifically designed to enhance or optimize both the quality of the data and potential alternatives.

> **Opportunities to Apply Total Federal Planning**
>
> If a total federal planning strategy (TFP) is adopted, an interdisciplinary steering team (IST) is formed with responsibility for overseeing the EIS scoping process. A facilitated workshop can be conducted in which value engineering (VE) principles and techniques are used to assist the EIS team in determining the appropriate scope of the planning process.

Purpose of Public Scoping

Scoping is critical to the objective of accurately focusing the analysis. Agencies must conduct a public scoping process to identify the range of (1) actions, (2) alternatives, and (3) impacts to be addressed in the EIS, which is also commonly referred to as *the statement* (§1501.7 and §1508.25). As the phrase *EIS process* implies, scoping is not a single or isolated event; rather, it is an iterative process that continues, formally or informally, throughout the EIS process. As new information is learned, the scope of the analysis is revised accordingly.

Agencies have a great degree of freedom in how they choose to exercise this responsibility. Potential methodologies include public meetings, facilitated sessions, workshops, public opinion surveys, citizen advisory committees, and a combination thereof. The least desirable methodology often involves public meetings or hearings, with each side postured as a potential adversary.

Scoping is frequently seen as an effort to identify as many topics or issues for analysis as possible. More properly viewed, scoping provides a mechanism for screening and narrowing a potentially universal scope of investigation so that analysts can concentrate on issues that are truly significant. Thus, scoping is separating the wheat from the chaff.

A sliding-scale approach is recommended for focusing efforts on issues of most concern; under a sliding-scale approach, the level of effort devoted to a particular topic or issue is commensurate with its importance. Areas considered of high interest or importance are afforded correspondingly greater attention. The entire public involvement and scoping process should be carefully documented as part of the agency's administrative record (ADREC) (§1501.7 and §1506).

Scoping Objectives

The importance of scoping can be as advantageous to the project engineer as to the EIS manager. For example, one individual recently reported that the NEPA process helped to define and drive the final scope of the project.[2] Table 2.1 lists specific objectives that scoping is intended to accomplish.

Issuing the Notice of Intent

After completing the optional preliminary scoping phase described in chapter 1, the agency is ready to initiate the formal or public scoping phase. Public scoping is formally initiated with publication of the notice of intent (NOI) in the *Federal Register,* announcing the agency's intention to prepare an EIS (§1508.22). It is recommended that public scoping not begin until the proposal has coalesced to the point that it can be coherently explained to affected parties and the public at large; as described earlier in this text, the term *proposal* is interpreted to mean the proposed action (if one is defined) and the range of reasonable alternatives.

The agency must make diligent efforts to involve the public, providing appropriate notice of NEPA-related hearings, public meetings, and the availability of the EIS. Notices must be mailed to parties that have requested notification (§1506.6[b]). The Internet is increasingly used as a means for notifying the public.

Stakeholders include persons or parties who may be affected by the outcome of the decision, can assist in executing a decision, or can impede implementation of a decision. Emphasis is placed on ensuring that all stakeholders are given proper notice. Once the NOI is publicly issued, the agency may begin the formal EIS process.

Table 2.1 Principal objectives of the public scoping process[3]

Identify public concerns and areas where agency expertise may be warranted.
Identify alternatives that will be examined.
Identify significant issues that need to be analyzed.
Eliminate unimportant issues.
Identify problems and solutions early in the process.
Ensure that problems and positive aspects of the proposal are identified early and are properly studied.
Identify potential mitigation measures.
Identify state and local agency requirements, such as permits and land use restrictions.

Involving the Public and Dealing with Hostility

Abraham Lincoln once stated, "You can fool some of the people all the time, and all of the people some of the time, but you cannot fool all of the people all of the time." With this thought in mind, we turn our attention to the topic of involving the public and dealing with hostility. Experience indicates that there is rarely unanimity in views regarding the scope or desirability of a major proposal. Outside entities that participate in the scoping process generally do so because they are opposed to the proposal; the remaining participants who support the proposal often do so because they stand to benefit from it. Not surprisingly, numerous public scoping efforts have dissolved into sessions of frustration, dissension, or outright confrontation.

Establishing Rapport with the Public

Many agencies report that the public is often more inclined to accept a proposal if citizens are given a fair opportunity to voice their opinions and influence the final decision. When the public is afforded little or no opportunity to provide input or to shape the final decision, they often feel resentment and are inclined to oppose or even obstruct the eventual project. If for no other reason, scoping should be viewed as a cooperative venture with the public rather than an adversarial conflict. The agency must strive to address public concerns. In one case, members of the public were particularly concerned that a proposal would lead to elimination of local jobs and, consequently, increased unemployment. The public participation process was instrumental in quelling fears by providing a forum in which to explain that this would not occur.

To avoid stepping blindly into a potential minefield, agency personnel should strive to understand public interests, sentiments, and concerns, particularly where environmentally sensitive issues are involved. It may be advantageous to assign a public relations specialist to help coordinate the scoping effort. A public involvement specialist can be instrumental in coordinating tasks such as preparing the NOI and advertisements, soliciting public comments, and reserving facilities. Early activities may include preparing posters, exhibits, and public advertisements. Fact sheets should be prepared to inform the public on the purpose of the EIS and the potential scope of the proposal.

Scoping Information Package

Prior to beginning the formal scoping process, a public information packet should be prepared. The packet should explain the EIS process, emphasizing that the purpose of scoping is to elicit information and that a final decision will not be made until the EIS is completed. The information packet may include example items listed in Table 2.2.

Public Scoping Meetings

In a real-life "Dilbert" quote, an employee reported a manager as stating, "E-mail is not to be used to pass information or data. It should only be used for company business." Perhaps to the trepidation of this manager, scoping meetings can be seen as providing an excellent vehicle for exchanging information and seeking input from the public; e-mail can also be used as a rapid and cost-effective vehicle for communicating with the public.

Under the Regulations, the lead agency is not required to hold public hearings unless the proposal is highly controversial or it is specifically requested to do so by another agency (§1506.6[c]). While scoping meetings may not be required, it is often considered good practice to hold at least one. Moreover, the agency's NEPA implementation procedures may require meetings or hearings. Professional courtesy dictates that a public announcement be made at least 15 days prior to any scoping meeting or hearing. Table 2.3 provides a checklist of actions that should be performed in preparing a public scoping meeting.

Under no circumstances should one ever assume a casual attitude toward public scoping events. People who attend public hearings and scoping meetings usually have a vested interest in the outcome—local homeowners, employees concerned with their jobs, environmentalists. Attendance tends to increase with the level of controversy. Members of the public who attend scoping meetings are often well educated and technically sophisticated. This tends to be especially true of activists representing environmental interests.

Table 2.2 Items typically included in a scoping information packet

An invitation requesting interested parties to submit comments and recommendations with supporting data and evidence
How and where comments may be submitted, including a name and phone number of a point-of-contact
Brief discussion and purpose of the EIS process, with emphasis on scoping, public participation objectives, opportunities to provide comments and input, and the process for submitting comments
Description of the proposal, including maps, figures, historical and background material, and other supporting material that would assist the public in providing comments on the proposal
Description of the potential environmental impacts and issues; this discussion should identify the environmental resources potentially at risk
A draft schedule and outline for the EIS, if available

Table 2.3 Checklist for planning public scoping meetings

Identify opportunities and means by which interested participants may record their comments, both in writing and orally.
Identify agency and contractor representatives who will attend the scoping meetings.
Identify city(ies) and specific location(s) in which scoping meetings will be held.
Identify a moderator (if appropriate).
Establish a format (in consultation with the moderator) for conducting the meeting.
Identify and reserve a building or facility in which scoping meeting(s) will be held.
Establish a toll-free hotline (if appropriate); a recorded message containing pertinent scoping information may also be useful.
Establish an e-mail address for contacting the agency's representative; include it on all advertisements, invitations, news releases, fact sheets, etc.
Establish a Web site containing pertinent information.
Prepare exhibits (e.g., posters, slide shows, pictures).
Prepare public information fact sheets and question-and-answer sheets.
Identify local and regional newspapers and other media in which advertisement will be placed.
Design and place advertisements with newspapers, radio announcements, and other appropriate media
Prepare and issue media and news releases.
Establish a mailing list of individuals and organizations to which invitations and notifications will be sent.
Prepare and distribute invitations and notifications.
Prepare a briefing for the audience.
Conduct orientation meeting and train staff (e.g., agenda, format, responsibilities)
Inform local law enforcement of the dates and location of the scoping meeting. Obtain services for on-site security (if warranted).
Schedule a court reporter.
Prepare comment forms.
Identify and assemble equipment (e.g., U.S. and federal agency flags, computer and printer, copy and fax machines, view graphs, tape recorder, paper and supplies, registration forms, posters, and exhibits).
Compile handout material (e.g., fact sheets, NOI, list of contacts, comment forms, name tags).
Clear all information for public distribution.

As indicated in Table 2.3, an orientation meeting for agency officials and EIS staff members who will attend the scoping meeting may be advantageous. Where a high level of controversy is expected, the EIS manager may want to hold a dry run, including a question-and-answer session and mock media interviews.

Managing Public Meetings

To fully appreciate public sentiment, the lead agency's decisionmaker should attend scoping meetings. As explained in the next section, the agency may find it advantageous to obtain the services of an independent moderator for conducting public scoping meetings.

INDEPENDENT MODERATOR. Many facilitators are adept at conducting public scoping meetings and defusing hostility. Obtaining the services of a neutral facilitator to conduct scoping meetings can mitigate public hostility, particularly where the agency lacks credibility. The facilitator should be thoroughly briefed on all aspects of the proposal.

The independent facilitator manages the meeting by explaining its purpose and procedures and by establishing ground rules. The public is often confused about the purpose of scoping meetings. Individuals may mistakenly believe that the decision to pursue the proposal will be made at the scoping meeting. The facilitator clearly explains that the purpose of the scoping meeting is to elicit input and comments and not to make a decision concerning the proposal. Participants should be told how much time they will have to speak. Some agencies report that when scoping meetings are videotaped, participants take a more serious attitude and the meetings are more productive than otherwise.

FOCUSING PARTICIPATION. The public scoping meeting should focus on identifying the principal actions, alternatives and mitigation measures, and environmental issues. Emphasis should also be placed on identifying environmental issues that are *not* of concern.

Experience shows that informal meetings conducted in small groups are sometimes the most effective means of identifying environmental issues and eliciting useful comments. For large or complex proposals, the agency may want to consider use of selected value engineering (VE) techniques to assist participants in identifying or screening actions, alternatives, and impacts[4] (see introduction and chapters 1 and 5).

The facilitator may want to consider dividing attendees at large meetings into small discussion groups. Each group may be tasked with a list of problems or issues to consider. One group, for example, might be delegated responsibility for identifying potential alternatives while a second group considers potential mitigation measures. A third group could consider important impacts and issues that need to be analyzed in the EIS. Brainstorming techniques can be applied in identifying potential alternatives and impacts that need to be investigated. The entire group are reassembled so that results may be shared. Special techniques can be applied to eliminate, combine, sort, and rank the potential alternatives, and issues for study.

Finalizing the Scope

On completing the formal scoping phase, the EIS team is responsible for reviewing all comments. The agency must consider all public scoping comments, both written and oral. The EIS manager must ensure that sufficient resources are available to review, evaluate, and address them. If the volume of scoping comments is unusually large, the EIS manager should consider tracking and managing scoping comments and their resolution using a comment tracking (COMTRACK) data base, as described in chapter 1.

The scope of actions, alternatives, and impacts for later investigation must be clearly delineated. Placing appropriate and defensible constraints on the final scope is a step of critical importance in managing schedule and budget. Irrelevant or nonsignificant issues are either deemphasized or entirely dismissed.

Preparing an EIS Implementation Plan

The Council on Environmental Quality (CEQ) encourages publication of a post-scoping document that notifies the public of the results of the EIS scoping process.[5] As viewed by the CEQ, such a document "may be as brief as a list of impacts and alternatives selected for analysis; it may consist of the 'scope of work' produced by the lead and cooperating agencies . . . ; or it may be a special document that describes all the issues and explains why they were selected." Consistent with CEQ's guidance, some agencies prepare an EIS implementation plan on completing the EIS scoping process. If the proposal is later challenged, the implementation plan can assist agency officials in demonstrating that due diligence was exercised in determining the appropriate scope of analysis.

If an implementation plan is prepared, it should be issued as soon as possible on completion of the formal scoping process and should be publicly distributed to organizations and interested members of the public. The implementation plan should assist the agency in meeting the following three objectives:

1. Publicly record the results of the scoping process.
2. Provide a compendium of all scoping comments received and the agency's response to these comments.
3. Provide a plan, schedule, and outline for preparing the EIS; the plan should assign responsibilities to the lead and cooperating agencies.

Contents of the Implementation Plan

The implementation plan discusses the scope of the analysis, outlining the proposal, mitigation measures, and environmental issues and impacts to be investigated. A proposed outline, page limits, assignment of responsibilities, and a schedule for the EIS should be included. A general outline is suggested in Table 2.4. The actual outline and contents should, of course, be tailored to meet the circumstances.

Creeping Scope Syndrome

Once the scope is defined, the EIS manager is responsible for ensuring that the analysis of actions and alternatives remains within defined bounds. As new information or circumstances arise, the EIS manager is frequently confronted with requests to expand the scope of analysis. One of the most common errors that EIS managers report making is allowing the scope to increase when this is not absolutely necessary. What might at first appear to be only a slight increase in scope may result in substantially increased effort. This creeping scope syndrome has been responsible for many cost overruns and missed deadlines.

2.2 Integrated Interdisciplinary Planning Effort

Thomas Huxley once observed, "In scientific work, those who refuse to go beyond fact rarely get as far as facts." With respect to EIS planning, Huxley's observation is especially insightful. Properly executed, preparation of an EIS is merely a mechanism, perhaps the final step, for documenting a thorough and rigorous planning process. Anything short of this goal is cause for concern. As a detailed description of NEPA's requirements for performing an integrated plan-

Table 2.4 Suggested outline for an EIS implementation plan

- Description of the proposal (including alternatives and mitigation measures)
- Potentially significant impacts and issues to be investigated
- Nonsignificant issues that can be either deemphasized or entirely eliminated
- Actions and alternatives that will not be described in detail or can be entirely eliminated
- Detailed outline of the EIS
- Page limits for each section of the EIS
- A schedule and plan integrating the EIS with other agency planning and decisionmaking time limits
- Outline of the EIS
- Responsibilities of the lead and cooperating agencies and the EIS contractor (if one is used)
- Compendium of all scoping comments received and the agency's response to the comments
- Identification of documents related to the proposal, with emphasis on other NEPA documents that can be used in tiering or incorporated by reference into the EIS

ning effort are beyond the scope of this chapter, the reader is referred to the companion book, *The NEPA Planning Process,* which treats this subject in detail.[6]

> Opportunities to Apply Total Federal Planning
>
> If a total federal planning strategy (TFP) is adopted, an interdisciplinary Steering Team (IST) is assigned responsibility for overseeing and coordinating the entire planning effort (see chapter 5). It is left to the reader to determine the most effective method for integrating each facet of the TFP approach with the internal steps outlined in this section.

Integrated Planning

Coordinating, integrating, and combining an EIS with other analyses and planning processes is stressed throughout the Regulations. The goal is to integrate what might otherwise be a diverse body of independent environmental statutes and disjointed planning requirements into a single, holistic, and interrelated planning procedure with an eye to environmental protection. The reader is referred to chapter 1, which contains a more thorough discussion on integrating the EIS with other requirements.

Integrated planning facilitates decisionmaking and promotes efficiency by reducing the risk of later disconnects and accompanying project delays. For example, the Army Corps of Engineers recently reported that integrating its Section 404 wetlands process with the Department of Energy's NEPA process provided the public with a more comprehensive analysis than would have occurred under Section 404 by itself. Moreover, this integration helped resolve wetland mitigation concerns, upon which a mitigated FONSI was later based.

Interdisciplinary versus Multidisciplinary

Agencies are required to prepare an EIS "using an interdisciplinary approach that ensures integrated use of the natural and social sciences and the environmental design arts." As explained in chapter 1, the interdisciplinary team (IDT) is responsible for ensuring that a coordinated and comprehensive planning effort is performed. The disciplines of the IDT staff are to be "appropriate to the scope and issues identified in the scoping process" (§1502.6).

A distinction needs to be drawn between the terms *interdisciplinary* and *multidisciplinary.* The latter denotes a process in which specialists representing pertinent disciplines perform their assigned task with little or no interaction. Such practice leads to problems and disconnects, and fails to meet NEPA's legal requirement for conducting an interdisciplinary planning process. In contrast, *interdisciplinary* denotes a process in which specialists interface and work together in assessing issues.

NEPA's interdisciplinary requirement places a burden on the EIS manager to ensure that the analysis is conducted by professional specialists in relevant disciplines who work together to investigate common environmental issues. While the members don't necessarily have to work in the same office or even the same building, they must work together in performing an integrated planning analysis. Some agencies have designated a centralized "war room" for preparing the EIS. Documents, reference materials, files, and technical data are maintained in this room. Maps, diagrams, progress schedules, and other reference data may be mounted on walls. Computers containing information pertinent to the analysis are available for use. The IDT either works or consults regularly in this room.

An IDT may require the talents of specialists representing a diverse array of technical disciplines. Such disciplines typically include urban and environmental planners, land use planners, infrastructure specialists, regulatory specialists, toxicologists, health physicists, biologists, geologists, hydrologists, archaeologists, and sociologists, to name just a few. The interdisciplinary planning process also requires the talents of other professionals, including designers, engineers, economists, and lawyers. Support staff includes technical editors and graphic artists.

Identifying Laws, Permits, and Licenses

A fifth-grader once remarked that "there are 26 vitamins in all, but some of the letters have yet to be discovered." In this sense, permits, laws, and other planning requirements are a little bit like vitamins—they are apt to be overlooked unless a concerted effort is made to identify them. As part of the scoping process, agencies are required to identify environmental review and consultation requirements. Regulatory specialists should be consulted in an effort to identify related permits and reviews that must be scheduled concurrently with the EIS. As requirements are identified, the lead agency should contact outside agencies to coordinate activities and address any duplication of efforts. Table 2.5 lists four items specifically called out in the Regulations as requirements to be integrated with the EIS effort (§1502.25[a]).

A permit, license, or other required approval or authorization (e.g., land or habitat disruption, air emission, water discharge) is often an indication of the type and severity of the potential impacts. Compliance with permitting requirements may be addressed from the perspective of potential measures for mitigating the impacts. While compliance with other environmental laws or regulations may minimize potentially adverse impacts, one should not arbitrarily conclude that such compliance will mitigate impacts to the point of nonsignificance or that the impacts will be acceptable.

Using Expert Systems to Identify Planning Requirements

Identifying and integrating a multitude of complex and evolving environmental and nonenvironmental laws, regulations, codes, procedures, and internal agency requirements can be an overwhelming task, even for seasoned experts. Moreover, the applicable set of requirements nearly always varies with the size, type, and location of an activity. At least one commercially available expert system software has been developed, based on the Simpson-Wilberg technique, that promises to substantially simplify this task. Such systems provide managers and practitioners with a tool for rapidly and consistently identifying applicable project, environmental, safety, and health requirements.

Other requirements that commonly need to be integrated with the EIS process are listed in Table 2.6. Some of the most important requirements are described in the following sections.

Fish and Wildlife Coordination Act

The Fish and Wildlife Coordination Act requires: (1) consultation with the U.S. Fish and Wildlife Service and appropriate state agencies to assess potential impacts on wildlife resources, and (2) modification of plans by "justifiable means and measures" that are deemed necessary to protect fish and wildlife. The Fish and Wildlife Service should be contacted to coordinate applicable requirements. Alternatives and mitigation measures should be investigated for avoiding or reducing potential impacts.

National Historic Preservation Act

The National Historic Preservation Act (NHPA) was enacted to preserve the nation's cultural resources.[7] The Council on Historic Preservation has promulgated regulations for implementing the NHPA.[8] Section 106 of the NHPA regulations directs federal agencies to "take into account" the effects of its

Table 2.5 Environmental statutes and requirements that are to be integrated with the EIS effort

Fish and Wildlife Coordination Act (16 U.S.C. 661 et seq.)
National Historic Preservation Act of 1966 (16 U.S.C. 470 et seq.)
Endangered Species Act of 1973 (16 U.S.C. 1531 et seq.)
Other environmental review laws and executive orders

Table 2.6 A partial list of additional environmental statutes and requirements that may need to be integrated with the EIS effort

Protection of Wetlands (Executive Order 11990)
Floodplain Management (Executive Order 11988)
Wild and Scenic Rivers Act of 1968 (16 U.S.C. 1271–1287)
Costal Zone Management Act of 1972 (16 U.S.C. 1451 et seq.)
Farmland Protection Policy Act of 1981 (7 U.S.C. 4201 et seq.)
American Indian Religious Freedom Act of 1978 (42 U.S.C. 1996)
Pollution Prevention Act of 1990 (P.L. 101–508, 6601 et seq.)
Environmental Justice (Executive Order 12898)
Biodiversity and ecosystem management

actions on sites eligible for or listed in the National Register of Historic Places.

Agencies should conduct a review to identify known historic properties and any previous cultural resource surveys that have been performed. The State Historic Preservation Officer (SHPO) should be consulted to determine the existence of resources either in or eligible for inclusion in the National Register of Historic Places. A determination should be made, in consultation with the SHPO, regarding the need to conduct additional survey work.

As appropriate, alternatives and mitigation measures should be investigated for avoiding or reducing potential impacts. Analysis of cultural resources is performed by a cultural resource specialist in coordination with potentially affected American Indian tribes. If historic properties cannot be avoided, mitigation is required. For example, in one recent case, a practitioner reported that properly integrating NEPA with the NHPA facilitated the appropriate relocation of proposed project boundaries, thus avoiding historic properties.

RECENT CHANGES IN SECTION 106. The advisory Council on Historic Preservation has recently revised its regulations (36 CFR Part 800) implementing Section 106 of the NHPA.[9] The revised regulation implements the 1992 NHPA amendments and streamlines the previous regulations. A new section, 36 CFR 800.8, allows agencies to comply with Section 106 requirements within the NEPA process.

Under 36 CFR 800.8, an agency may use an environmental assessment (EA) or EIS to comply with Section 106 in lieu of the procedures set forth in 36 CFR 800.3 and 36 CFR 800.6, provided the agency notifies the public and the Council on Historic Preservation, and meets certain established standards.

State and local governments, Native American tribes, and the public are often directly involved in federal activities affecting historic properties. The revised regulations provide for a Tribal Historic Preservation Officer (THPO) to substitute for the SHPO when a tribal official assumes the responsibilities of the SHPO with respect to tribal lands.

Archaeological and Historic Preservation Act

The Archaeological and Historic Preservation Act (AHPA) of 1974, as amended, requires preservation of historic and archaeological data that may be in jeopardy as a result of federal actions. The Archaeological Resource Protection Act (ARPA) requires a permit from the U.S. Department of the Interior for excavation or removal of archaeological resources from public or Native American lands. Criminal penalties are established for illegal excavation or removal of archaeological items.

American Indian Religious Freedom Act

The American Indian Religious Freedom Act et seq. establishes a national policy protecting the right of Native Americans to exercise traditional religious and ceremonial rites.[10] Consultation with Native American tribes is required where actions may infringe on religious rites or ceremonial sites. The Native American Graves Protection and Repatriation Act protects Native American graves, human remains, and funerary objects, and ensures reparation of these items.[11]

Endangered Species Act

The Endangered Species Act (ESA) applies to both federal and private actions. Under Section 9 of ESA, it is illegal to "take" any endangered species. The term *take* includes killing, harming, harassing, or capturing a member of a threatened or endangered species. This requirement also safeguards habitat. Section 7 requires federal agencies to:

- Obtain a permit prior to monitoring, capturing, killing, or performing scientific studies on threatened or endangered species.
- Prepare (if warranted) a biological assessment on such proposals.
- Perform (if applicable) a formal consultation with the U.S. Fish and Wildlife Service on potential impacts to species and habitat.

The regional office of the U.S. Fish and Wildlife Service should be contacted to determine if any federally protected species or habitat could be affected by the proposal. If there is a potential for affecting endangered marine or anadromous species, the regional office of the National Marine Fisheries Service should be contacted. In preparing the EIS, alternatives and mitigation measures should be investigated for avoiding or reducing potential impacts.

EXECUTIVE ORDER 13112. Executive Order 13112 applies to federal agency actions that may affect the status of invasive species.[12] It specifically applies to species not native to a particular ecosystem "whose introduction does or is likely to cause economic or environmental harm or harm to human health." Agencies are prohibited from authorizing or funding actions that may contribute to the introduction or spread of invasive species. Under this order, federal agencies are directed to:

1. Prevent the introduction of invasive species.
2. Detect, respond quickly to, and control invasive species populations.
3. Monitor invasive species populations.
4. Provide for restoration of native species and habitat conditions where invasions have occurred.
5. Research and develop technologies to control the introduction of invasive species.
6. Promote public education.

Floodplains and Wetlands

Executive Orders 11988 (Floodplain Management) and 11990 (Protection of Wetlands) require federal agencies to consider impacts of their actions on these ecologies through existing review procedures such as NEPA. Preparation of a floodplain or wetlands assessment is often required for actions that may affect such areas. The findings of this assessment should be coordinated with and incorporated into the EIS.

Flood insurance rate maps and flood hazard boundary maps, and consultation with applicable government agency personnel, can be invaluable in determining if an activity may be located within a floodplain. Wetlands can be identified by consulting:

- Wetlands specialists and federal agency personnel
- U.S. Fish and Wildlife Service national wetlands inventory
- U.S. Department of Agriculture, Natural Resources Conservation Service local soil identification maps
- U.S. Geological Survey topographical maps
- State and local wetland inventories, land use plans, maps, and inventories

2.3 Alternatives Synthesis and Assessment Phase

The investigation of alternative courses of action lies at the heart of NEPA planning (§1502.14). As a detailed discussion of NEPA's requirements for performing an alternatives analysis is beyond the scope of this chapter, the reader is referred to the author's companion book for a detailed discussion of the topic.[13]

Investigation of alternatives is performed under the auspices of the IDT. Seasoned practitioners report that the task of accurately identifying and describing the set of reasonable alternatives is often the most technically demanding aspect of preparing an EIS. Once the alternatives are correctly identified and described, the rest of the analysis is often relatively straightforward.

Section 102(2)(a) of NEPA requires federal agencies to "utilize a systematic, interdisciplinary approach which will insure the integrated use of the natural and social sciences and the environmental design arts in planning and in decisionmaking which may have an impact on man's environment." In this context, the term *systematic* describes a process that follows a logical and ordered step-by-step methodology in which each stage of the process builds on previous stages. In furtherance of this requirement and to provide a more comprehensive planning framework, a phased approach incorporating selected systems engineering (SE) principles for defining and synthesizing a set of alternatives for later analysis is introduced in this section.

Identification and assessment of alternatives is a dynamic process that tends to evolve over time. While the steps described in the following procedure are generally presented as single or one-time events, in practice, alternatives synthesis tends to be an iterative process that is revisited many times throughout the EIS process. As the plan matures, the agency may need to consider additional or fewer alternatives as the options and their respective impacts become better understood. For example, the investigation of alternatives may need to be broadened to include new options identified after circulation of the draft EIS (§1503.4[a][1] and [2]).

As always, professional judgment must be exercised in determining the applicability of the following approach to the agency's specific circumstances. Not all steps are applicable to every EIS.

Regardless of the actual approach taken (e.g., SE or others), it is essential to note that NEPA is a planning process, not a process for preparing a document. Agencies are mandated under NEPA to employ a systematic approach. A systematic procedure in one form or another is to be utilized in synthesizing a reasonable set of alternatives for analysis.

Applying a Modified Systems Engineering Approach

For detailed or complex proposals, a modified SE approach can be applied in the planning, identification, and synthesis of EIS alternatives. The word *modified* indicates that the approach incorporates only those SE steps deemed of most value in synthesizing a set of EIS alternatives. Thus, this modified SE approach incorporates some but not all aspects of a standard SE procedure. An overview of the standard SE process was briefly presented in the introduction to this text. This approach consists of seven discrete steps.

> **Opportunities to Apply Total Federal Planning in Synthesizing Alternatives**
>
> If a total federal planning strategy (TFP) is adopted, Chapter 5 describes how the interdisciplinary steering team (IST) coordinates various steps in performing an alternatives analysis, although not necessarily in the order shown in Figure 5.7.
>
> Under the TFP approach, the IST oversees, defines, and provides the specifications and data requirements for the technical specialists who are responsible for developing descriptions of alternatives and generating technical data; the data package must be prepared in detail sufficient to support the integrated analysis that will be conducted by the interdisciplinary team (IDT). On completion, this data package is transmitted to the IST, according to a specified schedule, for review and validation. The reader is referred to the section titled "Phase 3: Data Development and Alternatives Design Stage" (Figure 5.7), which describes this step in more detail.

The modified SE approach is depicted by the boxes shown below the dark horizontal line in Figure 2.1. The line separates input and output functions from the actual steps that make up the Alternatives Synthesis and Assessment Phase. As depicted by the box lying on the left side of the figure (above the thick horizontal line), the Integrated Interdisciplinary Planning Effort (see section 2.2) provides input used in defining the alternatives. This input includes the purpose and need, requirements and constraints, and comments from stakeholders and the public scoping process. Similarly, alternatives defined during this phase (*a*lternatives *s*ynthesis and *a*ssessment *p*hase) provide input for performing the Interdisciplinary Analysis Phase (see section 2.4), depicted by the box lying on the right side of the figure (above the dark horizontal line).

While these steps are shown as occurring sequentially, some may actually be performed in concert with one another; in practice, some of the steps may actually be ongoing efforts that continue throughout the course of the NEPA

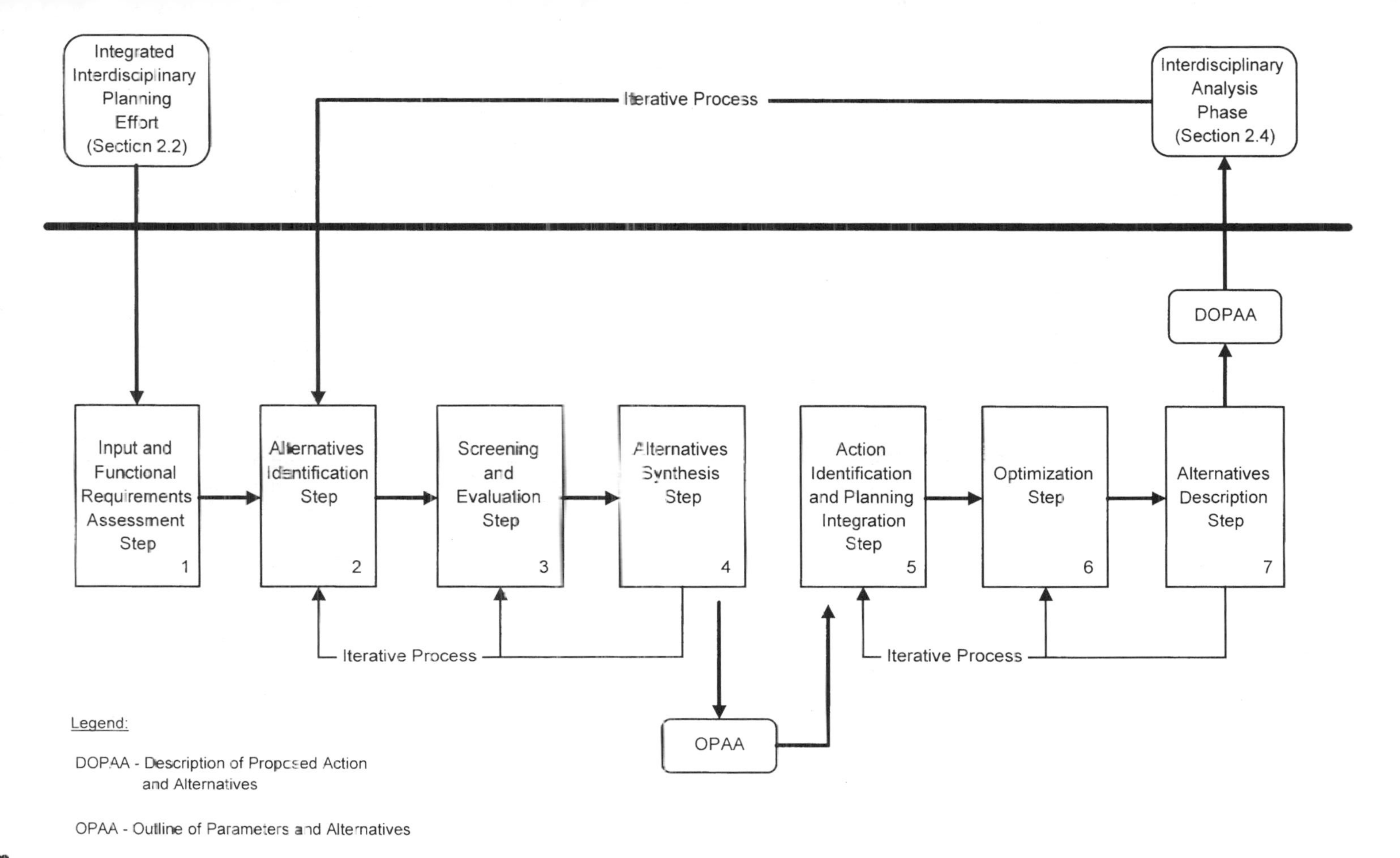

Figure 2.1 Alternatives synthesis and assessment phase. A modified systems engineering approach is applied in planning, screening, defining, and describing EIS alternatives

planning process. Designed with flexibility in mind, this procedure may be modified and tailored to meet specific missions or circumstances. Details of how this SE approach is best integrated into a specific EIS planning process are left to the reader, based on their individual needs and circumstances; professional judgement must be exercised in adapting this approach to a specific project. The reader should note that the SE approach may not be appropriate for every proposal (even complex ones). For example, depending on the type of project, an SE approach to developing alternatives may not be appropriate for a specific forestry or land use proposal. The reader may wish to obtain the services of a NEPA specialist and SE engineer who can provide guidance in integrating the selected SE steps with the EIS process.

Each of the steps shown in Figure 2.1 is described in the following sections. Figure 2.2 provides a more detailed flowchart.

Input and Functional Requirements Assessment Step

A function can be viewed as a task or action that must be accomplished. Functions normally consist of a verb and a noun, such as "supply water," "store waste," and "drill well." A procedure known as *functional analysis* is particularly useful in establishing a baseline of essential functions that must be performed to meet the agency's purpose and need.

Using input from the integrated interdisciplinary planning effort (section 2.2), an effort is made to further define and collect pertinent input data (e.g., functional requirements) that may have a bearing on the identification, screening, and assessment of potential alternatives (see box labeled #1 in Figure 2.1).

Requirements from all sources and stakeholders are obtained; some of these may be performance requirements that will determine if an alternative satisfies its ultimate mission objectives. Others may be constraints, such as a requirement that a facility needs to be relocated within one year or cannot be built within a Native American burial ground. Analysts assess requirements from all sources in identifying criteria for each requirement (otherwise they may forever be pressed to do better). Thus, with agreement between stakeholders and the performing organization, a baseline of the planning requirements is established. This provides a comparison benchmark useful in determining trade-offs between competing alternatives and completion criteria.

The statement of purpose and need, and other planning factors, helps the IDT in identifying the functional and performance requirements that will be defined. From these requirements, functional specifications can be derived.

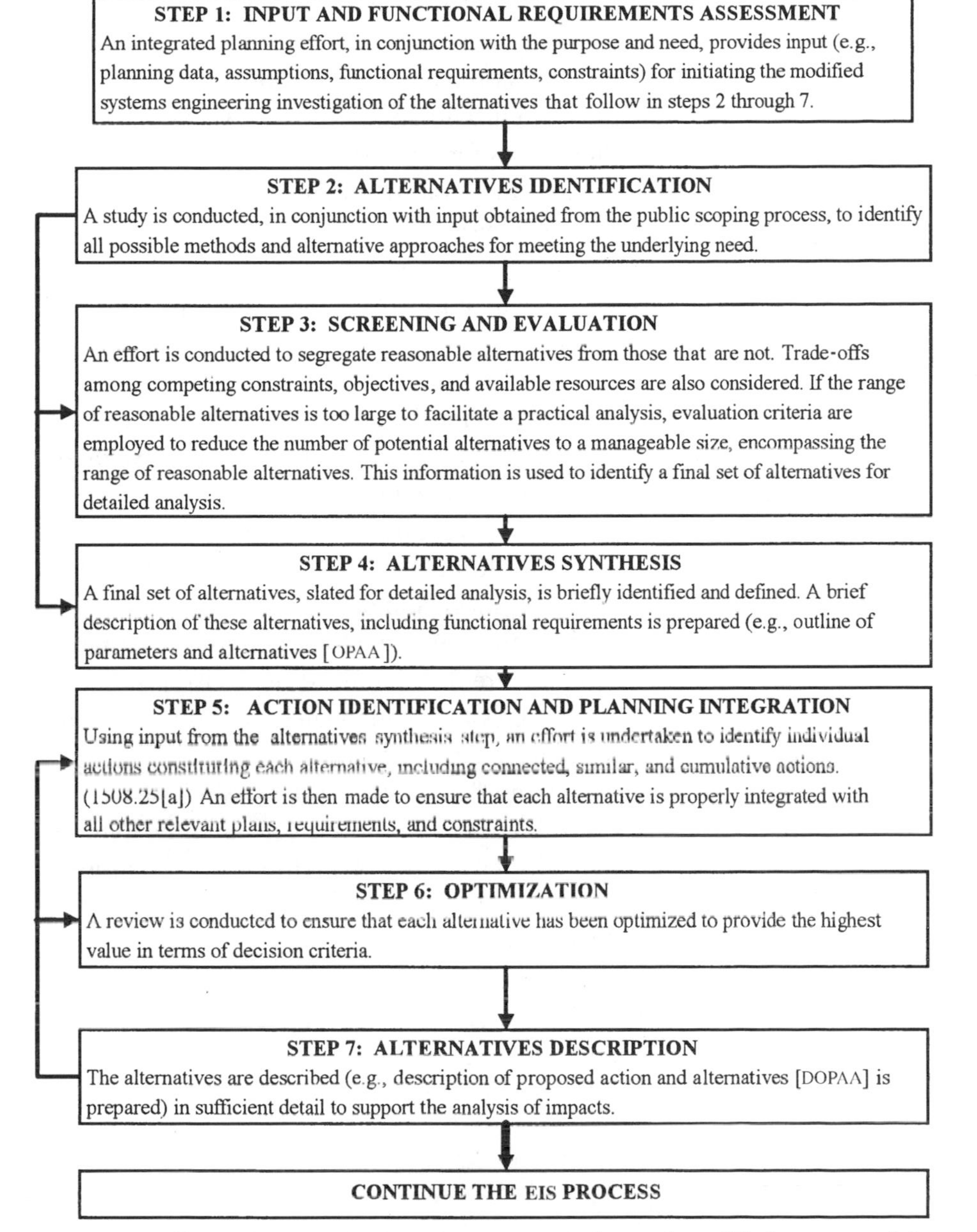

Figure 2.2 Alternatives synthesis and assessment phase. A modified SE approach is applied in planning, screening, and describing the EIS alternatives

The input data and functional requirements may include related plans and policies, technical specifications, objectives and constraints, measures of effectiveness, regulatory agreements and requirements, and, of course, public input and sentiment. Relevant decisionmaking factors are also identified. As described below, the statement of purpose and need provides an important basis for defining a set of alternatives for detailed study. The statement also provides a basis for defining the functional requirements and performance levels against which alternatives can be evaluated.

From a decisionmaking perspective, this step helps defines what must be considered and accomplished and why it must be done. This step can be subdivided into four principal tasks:

1. Identify all inputs and requirements (e.g., planning data, assumptions, functional requirements, constraints) from all sources, including stakeholders and the public scoping process.
2. Organize these requirements so that performance levels are related to functional requirements. Performance levels include such terms as contaminate levels, volume, mass, and distances.
3. Identify constraints and limitations to which potential alternatives must conform. Examples are laws and regulations, procedures, treaties, agreements, congressional mandates, schedule commitments, and annual funding allocations.
4. Allocate the functional requirements with their desired or required performance levels to the entity responsible for devising solutions that meet the requirements, given the constraints and limitations.

Alternatives Identification Step

Once relevant input, functional requirements, and constraints have been defined, in conjunction with comments received from public scoping, the IDT begins a detailed investigation of potential alternatives for meeting the underlying need (see box #2 in Figure 2.1). By definition, NEPA recognizes three types of alternatives:

1. No action alternative
2. Other reasonable courses of action (including the proposed action)
3. Mitigation measures (not in the proposed action) (§1508.25[b])

An interdisciplinary effort is used in identifying all possible methods for and approaches to meeting the agency's underlying need. This effort should exhaustively pursue all potential avenues such that no stone is left unturned.

Emphasis is placed on identifying best-value solutions within the defined constraints.

> **Opportunities to Apply Total Federal Planning**
>
> If a total federal planning strategy (TFP) is adopted, the reader is referred to the section titled "Phase 2: Formal Scoping Stage" (Figure 5.6) in chapter 5 for a discussion of the process and techniques that can be used to identify potential alternatives.

A Proposal Versus the Proposed Action

Project engineers are often under a great deal of pressure to ensure that their project is implemented on schedule and within budget. Experience indicates that once a proposed action is identified, it often takes on a life of its own. The author refers to this phenomenon as *project inertia.* Consequently, analysis of alternatives may not be given treatment equal to that of the proposed action.

To promote a more objective analysis, this text discourages the common step of defining a proposed action. Instead, a proposal consisting of a set of reasonable alternatives (including no action) is identified and defined for later analysis. Forces tending to influence an analysis in a particular direction are therefore minimized because efforts are not concentrated in favor of a particular course of action. Ultimately, this approach provides greater opportunity to actually identify an alternative best meeting all relevant planning requirements and decisionmaking factors. Of course, the agency must identify a preferred alternative, but this step should be done only once the alternatives are thoroughly and impartially evaluated.

Unless otherwise indicated, the term *proposal* is interpreted in this text to mean the set of reasonable alternatives and the proposed action (if one is defined). While the term *proposed action* is used in both the Act and the Regulations, more than one lawyer experienced in NEPA case law indicates it is unlikely that an agency would be successfully challenged simply on the basis that it did not define a proposed action, provided this track was taken to facilitate a more objective analysis of the reasonable alternatives. This guidance is applicable to most circumstances. However, it is important to note that circumstances may arise where defining a proposed action could be advantageous. As always, professional judgment in conjunction with factors such as agency culture and political considerations should be considered in determining if a proposed action should be defined. A model is provided in the following sections to assist practitioners in defining a range of reasonable alternatives.

Schmidt's Model for Defining the Scope of Alternatives

As espoused by Schmidt, the statement of purpose and need can be critical in successfully identifying the range of reasonable alternatives for later analysis.[14] If the statement is too broad, the number of alternatives may be virtually unlimited. If the statement is too narrow, it may negate consideration of better alternatives. Lee provides a number of examples and expands on Schmidt's model.[15]

Properly defined, the statement of purpose and need provides a powerful focusing tool for reducing the number of reasonable alternatives to a manageable set. A methodology for defining a reasonable range of alternatives for detailed analysis is explained in the following sections.

STATEMENT OF PURPOSE AND NEED. The Regulations state that an EIS shall "briefly specify the *underlying purpose and need* to which the agency is responding in proposing the alternatives, including the proposed action" (§1502.13, emphasis added). One might well ask why both terms—*purpose* and *need*—were used. As viewed by Schmidt, the terms describe two related yet different concepts. *Need,* as defined in *Webster*'s, is "a lack of something useful, required, or desired."[16] Simply put, a need can be viewed as something lacking or desired.

As an example, suppose an agency proposes to construct a water well and treatment facility to provide drinking water to a nearby facility. Rather than state that there is a need to "improve the quality of the water supply system," the statement might more properly indicate a need to "obtain a usable quantity of drinking water." In terms of reasonable alternatives, the two statements are actually quite different. The range of reasonable alternatives for improving "the quality of the water supply system" may be much broader than the more specific need of obtaining a "usable quantity of drinking water." In general, there are many ways of improving water quality but fewer alternatives for obtaining a usable quantity. The latter definition also provides a more specific performance parameter against which alternatives can be evaluated.

In contrast to the definition of need, *Webster's* defines *purpose* as a "goal" or "an end or aim to be kept in view in any plan, measure, exertion, or operation."[17] More simply, a purpose is a goal or an objective(s) to be met; it may involve specific engineering, economic, environmental, or other objectives. Using the last example, an agency might have a need to obtain a usable supply of potable water. Its purpose or objectives might be to: (1) obtain an economic supply of water, (2) which is potable, (3) providing 350,000 liters per day, (4) within the next two years, and (5) requiring a minimal amount of maintenance. Such specifications are better viewed as objectives (purposes) than as part of the underlying need (i.e., a lack of something that is required)

PROPOSED ACTION VERSUS ALTERNATIVES. Recall that the Regulations require an agency to "briefly specify the underlying purpose and need to which the agency is responding in proposing the alternatives, *including* the proposed action" (§1502.13, emphasis added). Note use of the word *including* in referring to the proposed action. By definition, this provision does not direct agencies to define an underlying purpose and need for the proposed action. Rather, this provision explicitly directs agencies to specify the purpose and need for proposing the alternatives—the proposed action being viewed simply as one of the alternatives.

While this text does not advocate identification of a proposed action per se, this is nevertheless a concept in common use. But what does the phrase *proposed action* really mean? As implied in the Regulations, a proposed action is simply the agency's favored approach for satisfying something lacking or desired (i.e., underlying need). What, then, are alternatives? Under Schmidt's model, alternatives are not alternatives to the proposed action. The alternatives are simply contending means for achieving the underlying need. Alternatives to the proposed action may be quite different from those intended to meet the agency's underlying need.

THE DIFFERENCE BETWEEN NEED AND UNDERLYING NEED. A distinct difference exists between a superfluous or ambiguous need and the agency's fundamental or underlying need. Precisely, what is the difference? Let us suppose that an agency defines a need "to remove a highly explosive tank of waste." This is a need, but is it really the underlying need? It is instructive to ask why the agency needs to remove the tank. The answer to this question might be "to protect human health and environmental quality from the consequences of an explosion." Thus, in this case, the agency's underlying need for action is actually to protect human health and environmental quality from the consequences of an explosion.

Note the difference between the two statements of need. Now, alternatives for protecting human health and environmental quality may be quite different than alternative methods for removing a highly explosive tank of waste. In fact, an alternative that does not involve removal, such as in-situ stabilization (e.g., grouting or vitrifying the waste in place), might actually be safer and provide better environmental protection than alternatives involving removal of an explosive tank of waste.

Alternatively, the agency's underlying need might be "to facilitate reuse of the land for residential purposes on and in the vicinity of the waste tank."

In this case, the range of alternatives might appropriately focus on removal of the tank and its waste contents.

NEED DEFINES THE RANGE OF ALTERNATIVES. Alternatives that meet the underlying need are more reasonable that are those that simply meet a need. Alternatives that do not meet the underlying need may be viewed as unreasonable and may be dismissed from detailed analysis. Thus, Schmidt's model provides an effective and defensible tool for screening out what might otherwise be a diverse or even unbounded range of alternatives.

A broadly written statement of underlying need can be advantageous, as it tends to compel analysts to consider more diverse approaches for satisfying the underlying need; the disadvantage is that it can increase the complexity of the analysis by raising the number of alternatives to be investigated. Conversely, if the number of reasonable alternatives slated for analysis is unduly large, the range can be reduced by writing a more narrowly focused statement of need. In reality, preparation of the statement of underlying need is often an iterative exercise.

The reader is cautioned, however, that it is inappropriate to define the statement so narrowly that only an agency's proposed action can meet the statement of purpose and need. While the statement of underlying need provides a valuable tool for defining the range of analyzed alternatives, it must be prepared and used judiciously.

EXAMPLE. Suppose that an agency determines that an underlying need exists for flood control. The EIS alternatives involve methods of controlling floods, with a proposed action involving construction of a concrete dam.

In this case, alternatives to the underlying need for controlling floods may be quite different than those for constructing dams. For example, alternatives to the proposed action (i.e., constructing a dam) are likely to focus on alternative ways of constructing a dam (e.g., concrete, earthen, location). In contrast, alternatives for flood control are more likely to focus on diverse ways of controlling flooding (e.g., dams, levies, enhancing wetlands, increasing vegetated areas, diverting water into a nearby reservoir).

Now consider how a change in the definition of underlying need can affect this range of alternatives. Suppose the agency had identified the underlying need as "limiting the damages and impacts from flooding." This definition might now include options such as land use controls that limit development in a floodplain.

In general, alternatives can involve options as diverse as siting, modes of transportation, technologies, and leasing a service or facility from the private sector. The federal government typically has much more authority and control than do private applicants; accordingly, a federal agency's broader mandate may necessitate consideration of a wider range of alternatives than would need to be considered by an applicant alone. Moreover, the federal government has a responsibility to act in the public's interest and not simply that of the applicants.

NO-ACTION ALTERNATIVE. Let us apply this model to the no-action alternative using the example involving flood control. The no-action alternative is sometimes viewed as not taking the proposed action. Under this interpretation, if the proposed action is to build a dam, the no-action alternative is simply not to build the dam.

Using Schmidt's model, the no-action alternative is more properly viewed as not meeting the underlying need. Thus, if the underlying need is to control floods, the no-action alternative essentially involves taking no action to control floods. Under this interpretation, the analysis of no action tends to focus more sharply on the implications of what would happen if the underlying need is not met. In this example, the impacts of taking no action would be flooding! Now, what are the impacts of flooding? What are the impacts of constructing a dam to control flooding—both adverse and beneficial? While both interpretations may eventually lead to the same conclusions, the second interpretation is often useful in helping the IDT to more effectively focus attention on describing the impacts of taking no action.

PURPOSE PROVIDES A BASIS FOR DECISIONMAKING. Recall that purposes are objectives that the agency wishes to achieve. Numerous alternatives may need to be analyzed, but not all of these are likely to meet the agency's objectives. Consequently, the purpose tends to come into play during the decisionmaking process after the EIS is completed. The purpose provides a basis for screening the analyzed alternatives in an effort to select the one best meeting the agency's goals or objectives.

The reader should note that splitting purpose and need into their two semantic roles is consistent with a dual-stage NEPA process (i.e., preparation of the EIS followed by a record of decision [ROD]). The reader is cautioned that such a process is not followed in every circumstance by every agency. For example, not every circumstance involves a first stage (investigating alternatives that meet the need) followed by a second stage (determining which alternative is best). In such cases, the concepts of purpose and need may need to be modified.

CASE LAW AND UNDERLYING NEED. To date, in about a dozen court cases a strong match was made between the agency's statement of underlying need and the corresponding range of analyzed alternatives.[18] Federal agencies have won every one of these cases; no agency has yet lost a case where the agency built a strong link between its statement of underlying need and the corresponding range of alternatives that were evaluated. It follows that agencies are less likely to win in court when such a match is not present. The aforementioned methodology appears to be the most defensible approach devised for defining the range of reasonable alternatives.

It must be emphasized that the statement of purpose and need is not to be written so as to slant the range of alternatives in favor of an agency's desired agenda. Witness a case in which the Federal Highway Administration prepared an EIS for a proposed toll road. The EIS was found inadequate by the court because it failed to justify the current and future need for the toll road and did not provide sufficient information to discriminate among the alternatives. Analysis of future transportation needs in all of the alternatives was based on socioeconomic forecasts that assumed construction of a highway similar to the proposed toll road; therefore, only the toll road could satisfy the forecasted needs.[19]

A change in purpose and need may require development of a new set of alternatives. In a related case, an EIS prepared by the Federal Highway Administration was found to be defective because the statement of purpose and need was narrowed without a corresponding rescoping of alternatives between the draft and final EIS stages. This change added a constraint on road capacity that only one alternative could meet. The court concluded that the statement of purpose and need can be changed as long as a range of alternatives also remains open. However, an agency abuses its discretion when a range of alternatives is developed consistent with the underlying need which is subsequently changed so as to eliminate all but one of the original alternatives.[20]

Similarly, the U.S. Forest Service canceled a contract to sell timber. The Forest Service then reopened the contract for rebid. The earlier EIS had only considered alternatives for meeting the original contractual obligations. Plaintiffs claimed that a new and broader set of alternatives needed to be considered, as the contract could now encompass a wider scope of possibilities. The court concluded that the change in purpose and need required consideration of a new set of alternatives.[21]

ALTERNATIVES AND APPLICANTS. A number of cases involving purpose and need have centered on actions where an applicant has requested approval from a federal agency. An applicant's purpose and need may be different from that of the permitting agency. The federal government normally possesses more authority and control over how an action can be implemented than does an applicant. Consequently, an agency's mandate to consider a reasonable range of alternatives may require consideration of a wider range of alternatives than would be necessary for an applicant alone. Also, the federal government has a responsibility to act in the public's interest and not simply that of the applicants. In one case, a court concluded that an agency needed to consider a broader range of alternatives, not simply alternatives that were deemed reasonable from the applicant's perspective.

Screening and Evaluation Step

A large number of alternatives may confuse or even overwhelm the audience (including the decisionmaker), to say nothing of increasing the cost and complexity of the analysis. Eliminating extraneous or essentially duplicate alternatives is perhaps the single most effective method for constraining schedule and cost.

Alternatives can be dismissed for being technically or economically infeasible, speculative in nature, or premised on unproven methods. Prudence must be exercised in eliminating alternatives. In one case, a court did not agree with an agency's rejection of an alternative as being economically unfeasible; the ADREC failed to demonstrate why the alternative was considered economically unfeasible. Such a demonstration can be difficult (e.g., preparation of economic analysis, profit-margin studies).

> Opportunities to Apply Total Federal Planning
>
> If a total federal planning strategy (TFP) is adopted, the reader is referred to the section titled "Phase 2: Formal Scoping Stage" (Figure 5.6), which discusses the use of value engineering principles and techniques for screening and selecting a final set of alternatives for detailed analysis.

A review is conducted to screen and evaluate potential alternatives (see box labeled #3 in Figure 2.1). The alternatives should be assessed against the key functions and performance levels established earlier. This phase may consider various trade-offs among competing constraints, objectives, and available resources. If the range of reasonable alternatives is too large to facilitate a practical analysis, predefined evaluation criteria can be employed to reduce the number of alternatives to a smaller set that still encompasses the entire range of reasonable alternatives.

The Rule of Reason

The rule of reason has been invoked in providing direction for defining an appropriate scope of alternatives. Specifically, only reasonable alternatives are required to be analyzed. Here, the phrase *reasonable alternatives* is interpreted to mean those alternatives that are practical or feasible from a technical and economic standpoint. Consistent with the rule of reason, common sense is exercised in discriminating reasonable alternatives from those that are not.[22]

Case law provides useful guidance for determining what constitutes common sense. Specifically, alternatives may be dismissed as being too exotic, technically infeasible, or premised on unproven methods. However, alternatives cannot be dismissed from analysis simply because they lie outside the legal jurisdiction of an agency or conflict with a local or federal law (§1502.14[c]).[23] In one case, a court did not agree with an agency's rejection of an alternative as being economically unfeasible because the ADREC did not clearly demonstrate why the alternative was economically unfeasible; such a demonstration may require an in-depth analysis (e.g., requiring economic analysis, profit-margin studies, or modeling).

Defining Screening Criteria

Before the alternatives are screened, criteria should be developed to aid in objectively eliminating redundant, unnecessary, and unreasonable alternatives from detailed study. Table 2.7 depicts general categories of evaluation factors that might be applied in screening and selecting a reasonable range alternatives for detailed analysis. Project-specific evaluation criteria would be defined for the applicable categories indicated in the table.

Defining a Range of Alternatives for Analysis

The agency's objective is to reduce what might otherwise be a vast number of alternatives to a reasonable set covering a full range of possibilities. Alternatives that are actually analyzed in detail, in addition to simply being reasonable, are often referred to as the *analyzed alternatives*. The number of analyzed alternatives is often smaller than the set of all possible reasonable alternatives but still represents the full range of potential courses of action.

High-level decisionmakers generally have greater authority to make broad-based decisions than decisionmakers at lower levels. This observation is reflected in the following requirement, which implies that the range of alternatives should vary with the level of discretion the decisionmaker may exercise in reaching a final decision:

> The range of alternatives discussed in environmental impact statements shall encompass those to be considered by the ultimate agency decisionmaker. (§1502.2 [e])

Table 2.7 Categories of evaluation criteria that may be useful in screening alternatives

Satisfies the agency's statement of underlying need.
Meets specified functional requirements or constraints.
Shows promise for avoiding or reducing environmental impacts.
Is economically practical or feasible.
Is technically practical or feasible.
Is reasonable and practical from the standpoint of common sense.
Meets reasonably defined schedule requirements.
Complies with applicable laws and regulatory requirements. (*Note:* An alternative cannot be dismissed simply because it is not within the agency's jurisdiction.)

Schmidt's model is useful in narrowing the final set of alternatives for detailed analysis. It must be reemphasized that reasonable alternatives may not be excluded from analysis merely because they do not conform with the agency's desires. Perhaps, most importantly, an EIS must seriously consider alternatives that "would avoid or minimize adverse impacts or enhance the quality of the human environment" (§1502.1). The rationale used in eliminating alternatives from further study should always be documented as part of the ADREC.

The reader should note that a limited number of circumstances may not lend themselves to a range of alternatives. For example, the act of granting a permit might encompass only an up-down (i.e., yes-no) choice; the agency can either grant or deny the permit. Here, the alternatives simply consist of taking no action (not granting the permit) or granting the permit. The range is two—hardly what one normally thinks of as a "range."

RULE OF SEVEN. It is vital that an analysis appropriately represent a range of reasonable alternatives but not be unwieldy. The psychologist-mathematician George Miller first observed that there is a limit to the number of items that the human mind can accurately manage.[24] This is true, regardless of the discipline or types of items involved: mathematical variables, computer program routines, operational tasks, or planning options.

Miller found that the ability to comprehend and manage information decreases rapidly as the number of objects or concepts approaches 7 + 2. Above this number, the ability to juggle, comprehend, and manage items decreases exponentially. This concept has been referred to as the *magical number seven* or the *rule of seven.*

Consistent with Miller's findings, the author recommends that the IDT attempt to limit the number of analyzed alternatives to no more than 7 ± 2. Normally, it is not difficult to choose alternatives judiciously such that they cover the entire spectrum of reasonable courses of action while conforming to the rule of seven. For example, it may be possible to combine two or more alternatives into one. Of course, the goal of excellent planning (to say nothing of legal sufficiency) should never be compromised simply to satisfy the rule of seven. Yet, in most cases, seven alternatives is more than sufficient to provide a decisionmaker with a reasonable range of options while meeting NEPA's legal requirements.

Alternatives Synthesis Step

Once a final set of alternatives is identified, the IDT is in a position to begin briefly outlining them (see box #4 in Figure 2.1) in advance of more detailed analysis.

As with the statement of purpose and need, correctly defining a proposal (range of analyzed alternatives) can be less than straightforward. Ambiguous or broadly defined actions should be avoided, as they often invite problems. For example, the public's reaction may be to demand that a large array of alternatives be investigated, making an already difficult process more so. The scope of potential environmental issues also tends to decrease where the potential actions and alternatives are well defined (less ambiguous). Conversely, practitioners should not describe alternatives with such specificity (as in the case of the plutonium stabilization EIS described in the introduction that the agency is boxed into a corner with little flexibility to maneuver as circumstances change; a balance between these two extremes must be struck.

Preparing Outline of Parameters and Alternatives

An outline of parameters and alternatives (OPAA) briefly defines the spectrum of reasonable alternatives in preparation for a more detailed description to follow (see box labeled "OPAA" in Figure 2.1). Decisionmakers and planners (and the public, if the OPAA is made publicly available) are also afforded an opportunity to review the alternatives identified for detailed analysis to ensure that the best alternatives have been selected and that the analysis truly covers a full range of possibilities.

DESCRIPTION OF THE OUTLINE OF PARAMETERS AND ALTERNATIVES. The OPAA can save an agency endless hours of rework by properly identifying the alternatives at the outset. An OPAA is brief document, typically running between one to a dozen pages or so. It briefly identifies the principal actions constituting each alternative as well as the parameters and specifications useful in discriminating among these alternatives. As appropriate, the OPAA briefly defines and describes the parameters indicated in Table 2.8.

For example, an OPAA for a proposal to construct and operate a cogeneration plant to produce both electricity and steam might identify the set of alternatives portrayed in Table 2.9. The reader should note that while the set of alternatives exceeds seven, it nevertheless is smaller than the maximum allowed (nine) under the rule of seven.

A brief description, a few paragraphs to a page or so in length, accompanies each of these alternatives (see second item in Table 2.8). This description provides information (e.g., technical parameters, mitigation measures) essential for initiating an effort to describe the alternatives in detail.

Once the OPAA is prepared, it should be reviewed by all pertinent parties for consensus regarding the final scope of alternatives slated for later analysis. If the OPAA is prepared before completion of public scoping, it may need to be revised to reflect public comments.

Action Identification and Planning Integration Step

The agency's final course of action invariably consists of a set of individual or component actions. For example, an alternative involving construction of a small gas-fired power plant might embody an array of component actions that includes but is not limited to site clearing and excavation, construction of access roads, a gas pipeline and power line corridor, parking lots, petroleum storage tanks, support buildings, drilling a water supply well, construction of the power plant, mitigation measures, and the individual actions necessary to operate the plant. To adequately evaluate potential impacts, these individual component actions must be identified and described.

To further the objective of comprehensive planning, an effort is mounted to ensure that all other planning factors and requirements are identified and

Table 2.8 Outline of a generic OPAA

Site location(s): Identify potential sites at which the proposal could be implemented.

Alternatives: Brief description of each alternative, including principal construction and operational activities that would take place.

Parameters: Brief description of the succinct attributes, footprints (e.g., land disturbance and physical size), and project parameters that characterize each alternative. Identification of such parameters is useful in defining and discriminating among alternatives. A review that verifies and compares these parameters to the functional requirements and performance levels defined earlier should also be made.

Mitigation measures: A list of potential mitigation measures.

Table 2.9 Outline of an OPAA for proposed power plant

No-action alternative
Energy conservation alternative
60-megawatt coal-fired fluidized bed cogeneration plant
150-megawatt coal-fired fluidized bed cogeneration plant
60-megawatt gas-fired boiler and steam turbine cogeneration plant
150-megawatt gas-fired boiler and steam turbine cogeneration plant
60-megawatt gas-fired combined cycle cogeneration plant
150-megawatt gas-fired combined cycle cogeneration plant

accounted for. Factors such as scientific, technological, regulatory, economic, and schedule considerations must be properly integrated with each alternative. All related projects, plans, and infrastructure requirements must be appropriately considered.

Types of Actions Defined under NEPA

An action is not an environmental impact. An environmental impact is the result of an action's disturbance on some environmental resource; thus, actions precipitate environmental impacts. NEPA recognizes three types of action: (1) connected, (2) cumulative, and (3) similar. These actions are defined as follows:

(1) *Connected actions,* which means that they are closely related and therefore should be *discussed in the same impact statement.* Actions are connected if they:
   - (i) Automatically trigger other actions which may require environmental impact statements.
   - (ii) Cannot or will not proceed unless other actions are taken previously or simultaneously.
   - (iii) Are interdependent parts of a larger action and depend on the larger action for their justification.

(2) *Cumulative actions,* which when viewed with other proposed actions have cumulatively significant impacts and should therefore be *discussed in the same impact statement.*

(3) *Similar actions,* which when viewed with other reasonably foreseeable or proposed agency actions, have similarities that provide a basis for evaluating their environmental consequences together, such as common timing or geography. An agency may *wish to analyze these actions in the same impact statement.* It should do so when the best way to

assess adequately the combined impacts of similar actions or reasonable alternatives to such actions is to treat them in a *single impact statement.* (§1508.25[a], emphasis added)

The reader is referred to *The NEPA Planning Process* for a more detailed treatment of this topic.[25]

Identifying Actions

Using input from the OPAA, an effort is mounted to identify the detailed activities and operations that constitute each of the alternatives (see box #5 in Figure 2.1). This effort includes identifying connected, similar, or cumulative actions associated with each alternative (1508.25[a]). It is vital that all actions are correctly identified because these actions drive the environmental analysis. Emphasis is placed on identifying actions that are sometimes overlooked (utilities, wells and septic systems, transportation, support buildings, and access roads). The effort may also involve preliminary identification of the indirect and cumulative impacts that might otherwise be overlooked. An effort is mounted to ensure that the proposal is appropriately integrated with all other programs, plans, actions, operations, requirements, and constraints.

Related and future actions can often be identified by contacting other federal, state, and local agencies to identify future plans. Future actions planned by private entities may be more difficult to assess; state and local agencies are often a reliable source in identifying nonfederal actions.

Optimization Step

Prior to preparing a detailed description of each alternative, analysts need to identify potential methods that could be employed to optimize the alternatives so as to provide the highest value in terms of future decisionmaking criteria (see box #6 in Figure 2.1). Tools such as utility curves may provide a particularly useful methodology for assessing key parameters in terms of optimal value.

> **Opportunities to Apply Total Federal Planning**
>
> If a total federal planning strategy (TFP) is adopted, the reader is referred to the section titled "Phase 4: Optimization and Data Validation Stage" (Figure 5.8), which describes a phase devoted to optimizing the outcome of the planning process and ensuring the validity of the data and analysis.

Emphasis is placed on identifying and incorporating design features that would minimize environmental impacts. This step is important for more than environmental reasons. Reducing environmental impacts at this stage may eliminate the need for costly permitting and diminish controversy at a later date.

This effort should not be limited to optimizing environmental factors. The goal of excellent planning, as promoted in this text, dictates that, wherever feasible, an effort be made to optimize all decisionmaking factors (e.g., environmental, economic, technical, societal) within the funding, time, and operational constraints. This is a concept basic to total quality management.

Investigating Mitigation Measures

The EIS must describe reasonable policies, procedures, standards, and practices that could be employed to mitigate potential environmental impacts. As defined in the Regulations, mitigation measures can include:

(a) Avoiding the impact altogether by not taking a certain action or parts of an action.

(b) Minimizing impacts by limiting the degree or magnitude of the action and its implementation.

(c) Rectifying the impact by repairing, rehabilitating, or restoring the affected environment.

(d) Reducing or eliminating the impact over time by preservation and maintenance operations during the life of the action.

(e) Compensating for the impact by replacing or providing substitute resources or environments. (§1508.20)

The scope of this investigation must address the range of potential impacts. Once an action is shown to involve significant impacts, mitigation measures must be considered and analyzed for all impacts, even those not deemed to be significant (§1502.14[f], 1502.16[h], 1508.20).[26] Mitigation measures should be both reasonable and cost-effective. Liaisons representing potentially affected parties, including minority and low-income classes, should be consulted in developing mitigation measures.

The IDT members should maintain a running list of potential mitigation measures as they are identified. These individual measures can then be combined to produce a master list. Collectively, the IDT should review the master list before determining which ones to include in the alternatives description. The step of identifying and investigating mitigation measures will need to be revisited once the analysis is complete and all the impacts evaluated.

Alternatives Description Step

The stage is now set for developing a detailed description of the alternatives. The Alternatives Description step is depicted by box #7 in Figure 2.1. In preparing the detailed description of the alternatives, an interdisciplinary approach should be used in describing the who, what, where, and when, and how of particular actions. As depicted by the feedback loop leading to boxes #5 and #6 (see Figure 2.1), this procedure is iterative.

Description of Proposed Action and Alternatives

The actions must be described in detail sufficient to allow analysts to identify and evaluate potential impacts. The Sufficiency-Test tool presented in section 3.2 of chapter 3 can assist in determining the level of detail that is sufficient to ensure that potential actions have been adequately described.

Some branches of the Department of Defense prepare a document, called the *description of proposed action and alternatives* (DOPAA), which describes the proposal in detail, providing technical input data sufficient for preparing the analysis and assessing potential impacts.[27] Some agencies refer to this document as a *technical support document* or *technical information document.* A DOPAA, in one form or another, is frequently prepared to provide the IDT with technical and engineering data for use in preparing the EIS.

While the OPAA briefly identifies and outlines the range of alternatives for analysis, the DOPAA provides the detailed description of these alternatives. The reader should note that the analysis of the actual environmental effects is not performed in either document (see section 2.4).

The DOPAA describes all actions directly or indirectly related to each alternative and provides input data necessary for performing the analysis (e.g., emissions, effluents, natural resource consumption, footprints); other proposals that are less construction oriented (e.g., watershed management plan) require a different set of data. To the extent possible, each alternative should be described and evaluated over its entire life cycle. Support facilities and activities should be discussed; routine inspections, maintenance, repairs, and upgrades should be considered. To the extent they are reasonably foreseeable, continuing actions need to be projected into the future. Connected actions are discussed in sufficient detail to allow analysts (in later stages of EIS preparation) to identify potentially significant impacts that may result.

As shown in Figure 2.1, the box labeled "DOPAA" provides the technical input for the *I*nterdisciplinary *A*nalysis *P*hase described in section 2.4. The process used for analyzing these alternatives is also described in that section.

2.4 The Interdisciplinary Analysis Phase

In explaining why he supported exporting toxic wastes to Third World countries, a managing economist for the World Bank said, "I've always thought that underpopulated countries in Africa are vastly underpolluted." It was to prevent errors and fallacies such as this, perhaps more than any other reason, that Congress enacted NEPA. This example may also explain why so many nations (including those in the Third World) have followed NEPA's precedent. An interdisciplinary analysis is critical to the goal of providing decisionmakers with accurate information that provides the basis for informed decisionmaking.

Environmental impact analysis is an intrinsically demanding task, often requiring investigation of a complex chain of events. Analysts are continually challenged with the daunting task of anticipating what environmental consequences would result *if* certain actions occur *and* these actions result in the forecasted environmental disturbances, *presuming* environmental receptors respond as predicted, *supposing* specified mitigation measures are correctly implemented, and *assuming* specified restrictions and regulatory requirements are complied with.

> Opportunities to Apply Total Federal Planning
>
> If a TFP approach is followed, Figure 5.7 describes the principal steps performed and the coordination of various planning, analytical, and design functions within an organization.

It is not uncommon to encounter disagreements or conflicting views among specialists in the course of performing the analysis. All such views must be openly examined and discussed. If consensus is not reached, outstanding disagreements or conflicting views need to be disclosed in the EIS. Practitioners are expected to rigorously investigate potential environmental impacts and to document the results in the ADREC, including when and by whom each specific impact was reviewed.

A detailed description of requirements that must be followed in performing an integrated analysis is beyond the scope of this chapter. The reader is referred to *The NEPA Planning Process* for a detailed description of these requirements. To foster a more systematic and interdisciplinary framework, a general-purpose methodology for performing the environmental analysis is presented below.

Preparing an Analysis Plan

Experience indicates that EIS analyses tend to follow the renowned 80/20 rule—that is, approximately 80 percent of the effort is expended on evaluating 20 percent of the environmental issues. For this reason, it is imperative that the EIS team properly target issues of greatest importance before initiating the analysis. The earlier scoping process is critical in providing input necessary for setting such priorities. Failure to properly prepare for and coordinate the environmental investigation has resulted in many analyses having to be revisited or repeated.

To help ensure that a thorough and systematic analysis is performed, and to assist the IDT in efficiently allocating resources, many agencies prepare an analysis plan (ANPLAN) prior to initiating the analysis. The value of such a plan increases with the size or complexity of the analysis. At a minimum, the ANPLAN details what, how, and by whom the analysis will be performed. It describes the scope of investigation, alternatives, and potentially significant environmental impacts that will be evaluated. Data bases, computer codes, and analytical methodologies that will be used should also be documented. A mechanism for ensuring quality control and verifying data may also be included.

Analytical Methodologies

As described in chapter 1, the IDT is responsible for the accuracy of the data and assumptions used in the analysis. Often, more than one methodology exists for evaluating a problem or impact. Consensus must be reached regarding the specific analytical methodologies and computer models employed. Where possible, analysts should use methods and computer models endorsed by the U.S. Environmental Protection Agency (EPA), other governmental agencies, or professional societies.

Factors such as time, cost, and accuracy must be balanced in choosing a particular methodology. A cheaper but less accurate methodology may be sufficient if the results allow the decisionmaker to reach an informed decision. The rationale used in selecting a specific analytic methodology must be clearly documented in the EIS (typically in an appendix) and the ADREC.

Documenting Assumptions

The plan should provide a mechanism for ensuring that all assumptions used in performing the analysis are properly documented. Where an analysis is particularly complex, some agencies distribute record of assumption (ROA) forms either physically or electronically to each analyst. These are filled out as the analysts work. The completed forms are then submitted either physically or electronically to the EIS manager or designee. The individual forms are correlated to form a compendium of all assumptions used in the analysis. All important assumptions should either be documented in the appropriate sections of the EIS or included in an EIS appendix. Table 2.10 depicts a simplified example of such a form.

Table 2.10 Example of a record of assumption (ROA) form

Analytical Topic	Assumption	Rationale
A.1.1	The population baseline for the air impact analysis will be based on year 2000 federal census data supplemented by local and county survey data.	Use of best available data (40 CFR 1502.22).
A.1.2	Air emission impacts from facility releases will be calculated using average annual meteorological conditions.	Accepted agency procedure.
A.1.3	Fugitive dust emissions from earth-moving activities will be estimated using Environmental Protection Agency methodologies.	Accepted methodology for projects involving these types of activities.
A.2.1	Groundwater transport is assumed to be governed by steady-state conditions.	Assumption has been deemed reasonable based on hydrological tests performed in 1999.
A.2.2	The analysis assumes no contaminate leaching through the soil column into the groundwater aquifer.	Geological tests performed in 1999 concluded that the high content of clay minerals would significantly limit contaminate leaching into the aquifer.

Performing the Environmental Analysis: A Systematic Approach

In response to a question about the causes of dust, one fifth-grader suggested that "one of the main causes of dust is janitors." Such an assessment may suffice at the fifth-grade level; if scrutinized by a judge in a court of law, however, such an analysis is likely to be viewed with a more exacting standard. As described in Section 102(2)(a) of NEPA, federal agencies must "utilize a systematic, interdisciplinary approach . . . in planning and in decisionmaking which may have an impact on man's environment." To provide a more systematic and interdisciplinary framework, a general-purpose methodology, consisting of six principal steps, for investigating complex environmental impacts and their relationships is presented in Figure 2.3. The six-step approach is also outlined in more detail in Table 2.11.

The approach is designed to provide the reader with a systematic procedure that is flexible enough to adapt to a wide range of circumstances. Professional discretion may need to be exercised in determining those steps that are most practically applied based on the complexity of the analysis and the

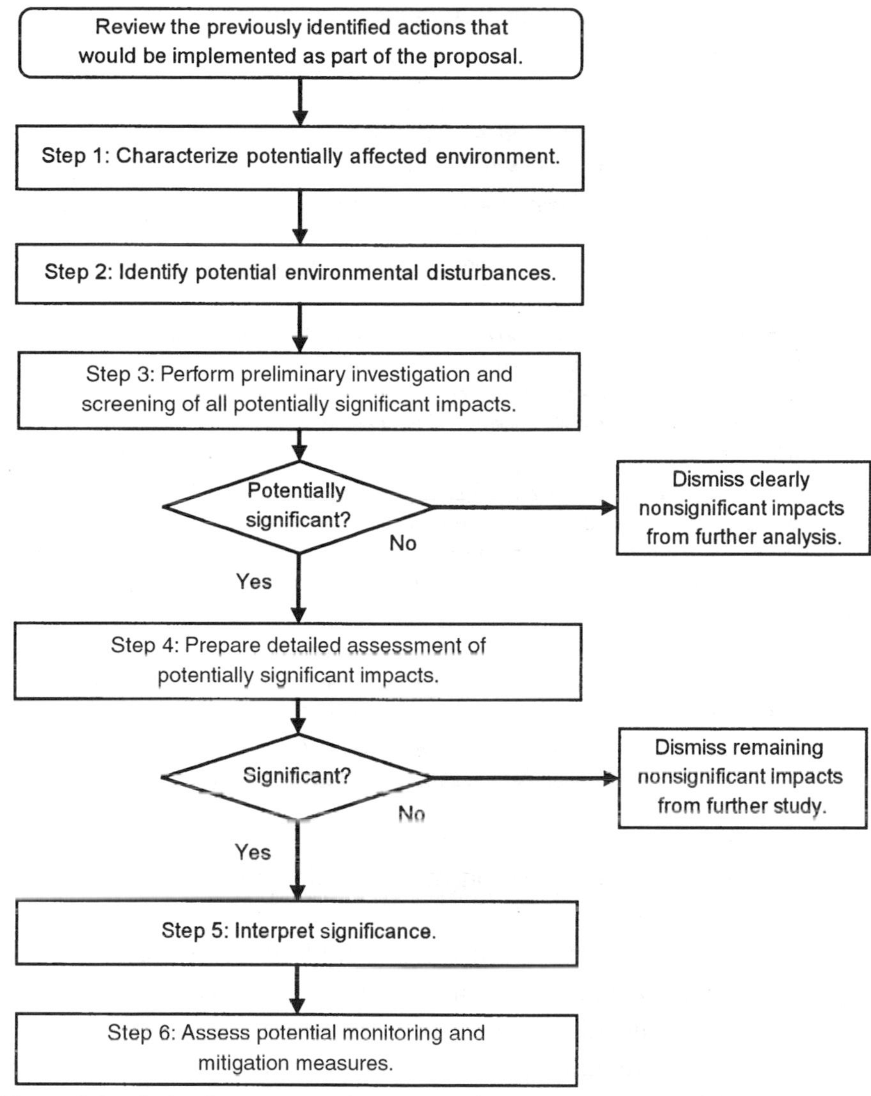

Figure 2.3 A six-phase approach to performing the environmental impact analysis

particular circumstances. Thus, the procedure may need be tailored to the existing circumstances.

Step 1: Characterize Potentially Affected Environment

The environmental baseline (affected environment) refers to the existing physical, biological, cultural, and socioeconomic resources (e.g., air and water quality, hydrology, sensitive species, historic structures, socioeconomics). As input data are a function of the impacts and resources that may be affected,

Table 2.11 Six-step approach to performing the environmental impact analysis

Step 1—*Characterize potentially affected environment:* This step involves collecting data on potentially affected resources. Spatial and temporal boundaries are delineated. Potentially affected resources are identified and described.

Step 2—*Identify potential environmental disturbances:* Environmental disturbances (emission, effluents, noise, liquid discharges, ground disturbances) resulting from potential actions that may be taken and that could affect environmental resources are identified. Cause-and-effect relationships are defined.

Step 3—*Perform preliminary investigation and screening of all potentially significant impacts:* A preliminary investigation is conducted to determine how environmental disturbances would change or affect environmental resources (receptors). The environmental impacts are screened to determine their potential for significance. Impacts deemed clearly nonsignificant are eliminated from further study.

Step 4—*Prepare detailed assessment of potentially significant impacts:* Once significant impacts and issues are identified and clearly nonsignificant impacts dismissed, a more detailed impact analysis is performed to determine how the environmental disturbances would change or affect environmental resources. Significant impacts are described in detail. Remaining nonsignificant impacts are dismissed from further study.

Step 5—*Interpret significance:* On completing the impact analysis, analysts interpret and convey to the reader the significance of potential impacts. These results may be expressed in terms such as human deaths, increased cancers, and loss of species.

Step 6—*Assess potential monitoring and mitigation measures:* Potential mitigation measures are evaluated to determine their effectiveness. If appropriate, a monitoring plan may be prepared.

environmental baseline information must be tailored to support individual analyses.

If more than one agency is involved, the lead agency should assign responsibility for preparing discussions of particular resources to the cooperating agencies possessing jurisdiction or special expertise. Existing data from the U.S. Census Bureau and other agencies can often yield excellent information.

Many practitioners report that the affected environment section of an EIS is sometimes unreasonably detailed, resulting in excessive cost. The EIS manager provides direction and monitors the effort to describe the affected environment, ensuring that this section is prepared cost-effectively. The Sufficiency-Test tool presented in section 3.2 can assist analysts in determining the level of detail that is sufficient to ensure that environmental resources are adequately described.

An agency's internal experts may have a better understanding of the affected environment than an outside EIS contractor. It may be advantageous for the agency to prepare this section of the statement, even if it has retained the services of an independent contractor to prepare the EIS.

Determining Spatial Boundaries

The potentially affected environment is sometimes referred to as the *region of influence* (ROI). The affected environment section delineates the boundaries of the ROI for each resource and why they were chosen. Spatial boundaries may be defined as the distance to which an environmental disturbance can be felt (i.e., ROI). It may be sensible to delineate this boundary as the maximum reasonable distance to which significant effects can be expected. The boundaries and supporting assumptions are clearly identified in the EIS and the ADREC.

Prior to the analysis, the IDT establishes the geographic bounds of the study area (i.e., ROI), recognizing that the size varies as a function of the impact under investigation. These boundaries may change as new information becomes available during the analysis. As is sometimes the case, the affected environment may have no definitive boundaries. Depending on the nature of the activity and resources involved, the spatial dimension may be restricted, while in other cases it may be global in extent (e.g., ozone depletion or global warming issues). For example, a particular activity may be limited to a small location near a river, yet potential impacts downstream of the activity may be felt far beyond the localized area of the activity. Thus, the spatial or geographic boundaries of the analysis varies with the environmental impact and resources under investigation.

Determining Temporal Boundaries

In determining how far impacts should be projected into the future (i.e., temporal boundary), practitioners must consider both the time frame of the specific actions that may be undertaken and the extent to which impacts may continue once an action has ceased. In some cases, the time frame over which the proposal would occur is the most appropriate duration for use in the analysis; in other cases, if future actions are reasonably foreseeable, one may need to expand the time frame. The boundaries and supporting assumptions are clearly identified in the EIS and the ADREC. These boundaries may change as new information becomes available during the analysis. Maps are particularly useful in delineating boundaries.

Step 2: Identify Potential Environmental Disturbances

Prior to the analysis (see section 2.3), all component actions constituting each alternative should be identified and described in detail sufficient to allow analysts to determine the resulting environmental impacts.

Each potential action may produce environmental disturbances (e.g., emissions, effluents, sound waves, ground disturbances, destruction of vegetation). The disturbances must be identified and then described in detail sufficient to allow analysts to investigate how they could affect environmental resources. Pathways and cause-and-effect relationships should be identified even though the precise quantitative response may be unknown. Tools such as network diagrams may be particularly useful in performing this task. Once cause-and-effect relationships are understood, analysts can begin assessing how the environmental resources or systems would respond to the actual disturbances.

Step 3: Perform Preliminary Investigation and Screening of All Potentially Significant Impacts

NEPA recognizes three distinct types of impacts: (1) direct, (2) indirect, (3) cumulative (§1508.25[c]). Once the potential environmental disturbances and cause-and-effect relationships are identified, a preliminary environmental impact analysis is initiated on a resource-by-resource basis. The terms *environmental disturbance* and *environmental impact* should not be confused. An environmental disturbance can be viewed as some type of stressor produced by an action that may change or affect an environmental resource. For example, operation of a power plant might release a certain amount of sulfur dioxide emissions into the atmosphere (i.e., environmental disturbance). Such a release, in itself, is not an environmental impact. The environmental impacts from such a release denote how environmental resources would actually be changed or affected. Thus, the environmental impact of the sulfur dioxide emissions might include effects such as the degree to which air quality would be degraded, the increased incidence in human respiratory illnesses, and effects on other species. Thus, an environmental impact reflects how a change in an environmental resource, caused by an environmental disturbance, would affect humans and environmental quality.

The focus of this step is on determining how environmental disturbances would affect or change the baseline environmental resources. Analysts should be relentless in their search for significant impacts that may have gone unnoticed during the earlier scoping process. Guidance for performing this step is provided in the following sections.

Applying the Rule of Reason

The rule of reason is useful in determining the degree to which impacts are investigated. Impacts deemed to be remote and speculative normally should not be evaluated on the grounds that it is unreasonable. Conversely, reasonably foreseeable significant impacts must be evaluated. Reasonably foreseeable impacts are interpreted broadly to include impacts of actions that are not formally proposed.[28]

Impacts are generally considered reasonably foreseeable if there is a logical connection between the action and its resulting effect. In other words, an impact normally needs to be investigated if there is a reasonably discernible cause-and-effect relationship between the action and its subsequent impact.[29]

"Reasonably Foreseeable" Versus "Remote or Speculative"

Delineating between an impact that is considered "reasonably foreseeable" from one that is "remote or speculative" is not always easy. In one case, an agency admitted that a significant cumulative impact would occur approximately 40 years in the future. There was no uncertainty with respect to this prediction. Arguing that this impact was not reasonably foreseeable, the agency did not evaluate this cumulative impact in the EIS. When challenged, the court concluded that if there is little doubt that an impact will occur, it must be analyzed regardless of when it would occur.

With respect to NEPA, there is no generally agreed-upon definition of the terms *reasonably foreseeable* and *remote or speculative*. This text employs these definitions:

Reasonably foreseeable: Describes an action or impact that one can reasonably predict or anticipate.

Remote and speculative: Describes an action or impact, occurring at some extended distance or time in the future that depends on assumptions or events that are contingent, conjectural, or problematic.

Table 2.12 lists common characteristics that may assist practitioners in discriminating among actions and impacts that are nearly certain, reasonably foreseeable, and remote or speculative. This guidance is based on case law, regulatory inferences, and experience gained by practitioners. These characteristics should be considered together, rather than independently. In complex or controversial circumstances, regulatory specialists or legal counsel should be consulted in making a final determination. *The NEPA Planning Process* contains additional information on this subject.[30]

Performing the Environmental Impact Investigation

Analysts begin by conceptually determining how the environmental disturbances would change potentially affected resources. This task may involve creation of a conceptual model of the relationship among the potential

Table 2.12 Common characteristics that can assist practitioners in discriminating among actions and impacts deemed nearly certain, reasonably foreseeable, and remote or speculative

	Nearly Certain	Reasonably Foreseeable	Remote and Speculative
Interpretation of These Terms	Clear or present causation; one would have to be blind not to see it.	Doesn't require a Nobel laureate to identify the potential action or impact.	Grasping at straws.
Stage of Planning	An official proposal or plan has been prepared.	Preparation of a proposal or plan is under considera-tion or development.	No effort has been made to formulate any proposal or plan.
Degree of Control Over the Action or impact	The agency has a considerable degree of control over the potential action or impact (e.g., alternatives and mitigation measures).	The agency can probably exercise some degree of control over the action or impact.	The agency has little or no control over the action or impact; the temporal and spatial domain allows for intervening actors.
Confidence or Degree of Concern	High degree of confidence in projections; often involves definite concerns with respect to the action or impact.	Some level of confidence in the projections; may involve some concerns with respect to the action or impact.	Little or no confidence in the projections; little concern over the actual action or impact; disagree-ments tend to center around personal values.
Debate	The action or impact is a virtual certainty. However, there are often technical debates over the specifics or details.	Some disagreements or conflicts exist among experts.	Substantial conflict and virtually no agreement exist among experts.
Outside Entities	Often involves clear concerns among outside citizens groups or other entities.	Often involves some concerns among outside entities.	Involves little or virtually no concern or interest among outside entities.
Benefit or Impairment	Some entity(ies), clearly identifiable, will probably either benefit or be impaired by the action or impact.	Some entity may benefit or be impaired by the action or impact; however, it may be hard to identify.	An entity cannot be clearly identified as either benefiting or being impaired.
Documentation	Substantial evidence or documentation to support conclusions exists.	Some evidence or documentation to substantiate conclusions exists.	Little or no credible data exist; a theoretical worst-case analysis is question-able or conjectural.
Evidence	The federal agency has direct knowledge; specific past examples or clear precedents provide a basis for making projections.	General studies of similar examples indicate a reasonable likelihood that the projections will occur.	Vague or implausible examples or precedents predominate.

actions, their environmental disturbances, and the resulting impacts. Once these relationships are established, the stage is set for quantitatively investigating the environmental impacts and their significance. As they may be more difficult to identify and evaluate, an effort should be made to ensure that nonlinear relationships are not overlooked. Direction for describing the impacts is provided in step 4.

As the no-action alternative provides an environmental baseline, analysts often begin by determining its impacts. Thus, analysts have a basis for comparing the effects of pursuing particular actions against the impacts of taking no action. Once the impacts of taking no action are determined, an effort is mounted to determine how different alternatives would affect the study area. Aerial photographs and remote sensing techniques may provide excellent and cost-effective tools for determining how the study area would be affected.

As part of the planning process, the IDT reviews existing regulatory requirements and restrictions, land use maps, planning reports, engineering, and other studies. Analysts should be provided with maps of the potentially affected areas.

> **Opportunities to Apply Total Federal Planning**
>
> If a total federal planning strategy (TFP) is adopted, the interdisciplinary steering team (IST) provides the IDT with the alternatives data package developed by the project office or proponent. To promote an integrated planning process, relevant planning studies, such as risk and cost-benefit analyses, are coordinated and integrated with the environmental analysis. The reader is referred to the section titled "Step 5: Analysis Stage" (Figure 5.9), which describes this step in more detail.

Analysts must evaluate the degree to which potential actions would comply with existing laws and regulations. Applicable programmatic, project-specific, regional, community, and other land use plans should be reviewed. Consultations should also be conducted with outside planning, environmental, or land use agencies.

Screening Impacts

Environmental impacts must be assessed to determine if they are significant or nonsignificant according to specific factors presented in the Regulations (§1508.27). To this end, environmental impacts are screened, segregating effects deemed clearly nonsignificant from those that are not. The IDT concentrates effort on impacts deemed potentially significant.

Step 4: Prepare Detailed Assessment of Potentially Significant Impacts

Once the preliminary analysis (step 3) is completed and the impacts screened to eliminate effects deemed clearly nonsignificant, a detailed investigation is performed to evaluate the remaining potentially significant impacts (see step 4 in Figure 2.3). The assessment of environmental impacts attempts to gauge how a change to an environmental resource would affect humans and environment quality. The severity or degree of the potentially significant impacts dictates the appropriate level of analysis to perform. On completing this detailed analysis, the impacts are again screened to eliminate additional nonsignificant impacts. For more information on interpreting the significance of an impact, the reader is referred to *The NEPA Planning Process.*[31]

Describing Impacts

The goal is to prepare an analysis that clearly conveys to the reader what the disturbance to a given resource actually means. The Sufficiency-Test tool presented in section 3.2 can assist analysts in determining the amount of detail that is sufficient to ensure that environmental impacts are adequately described.

Instead of simply stating that a resource "would be impacted," the analysis must indicate *how* environmental resources would be affected. For instance, it is much more instructive to describe how and to what degree a habitat would be perturbed than to simply state (as is sometimes the case) that the habitat would be impacted.

Impacts should be quantified using a technically appropriate unit of measurement. Potentially affected resources or populations should be clearly delineated and described. Where applicable, the period over which the impact would occur should be indicated, as should its likelihood or probability. At a minimum, analysts should strive to convey the following information to the reader:

- *Benefit or adversity:* The analysis should be presented such that there is no misunderstanding as to whether an impact is beneficial or adverse. It is not uncommon to find that an environmental impact may have characteristics of both attributes.
- *Magnitude:* Where practical, a quantitative measure of the impact should be presented. If quantifible measures are not available, impacts may be described using qualitative descriptors such as *immeasurable, minor, large, substantial.*
- *Duration and timing:* The analysis should indicate the duration over which the impact would persist. Where possible, the analysis should indicate when the impact would begin and end.

It is usually not necessary to categorize each impact as short- or long-term, direct or indirect, or beneficial or adverse. As a rule, an impact should be specifically labeled only to give the reader a better understanding of the effect. To avoid the appearance of bias, analysts sometime downplay discussions of beneficial effects. Such practice is inappropriate, as an analysis of beneficial impacts can be as important and informative to decisionmaking as that of adverse impacts.

Because of common misconceptions and the complexity involved in their analysis, special attention is given in the following sections to the subject of indirect and cumulative effects.

Indirect Impacts

Indirect impacts are defined as those effects that are:

> . . . caused by the action and are *later in time* or *farther removed in distance,* but are still *reasonably foreseeable.* Indirect effects may include *growth inducing effects* and other effects related to induced changes in the pattern of land use, population density or growth rate, and related effects on air and water and other natural systems, including ecosystems. (§1508.8[b], emphasis added)

An absolute dichotomy for distinguishing between direct and indirect impacts does not exist. Consequently, the exact distinction between these two concepts is often gray. For this reason, Table 2.13 is provided to assist the reader in comparing the principal characteristics useful in differentiating between direct and indirect impacts. As indicated, an indirect effect is removed in the time or spacial domain, but not necessarily both.

A growth-inducing effect most commonly falls within the definition of an indirect impact. A growth-inducing effect causes other actions that in turn

Table 2.13 Comparison between direct and indirect impacts

	Direct Impact	Indirect Impact
Time domain	Now	At some time in the future
Space domain	Here	Removed in distance
Growth-Induced Effect	Possible	Frequently

spawn further actions that can affect either the natural or built environment; thus, a chain reaction can result, causing future development to snowball. Even though an agency may not have prepared an official planning document that points to additional growth or development, other documentation may provide a sufficient basis for such projections (i.e., internal agency memos, local real estate development plans, local press releases, letters from proponents of the proposal, county resolutions).

Confusion and differences often arise in attempting to determine if an indirect impact is reasonably foreseeable. Table 2.14 lists common characteristics that practitioners may find helpful in determining if an analysis of an indirect impact is warranted (i.e., if it is reasonably foreseeable, or remote or speculative). These characteristics should be considered together rather than independently.

Table 2.14 Common characteristics useful in determining if an analysis of an indirect impact is warranted

	Reasonably Foreseeable	Remote or Speculative
Interpretation of these Terms	Doesn't require a Nobel laureate to identify the potential for an impact.	Grasping at straws.
Growth-inducing Impacts	Other actions are likely to be triggered by the federal proposal.	Potential for future growth-inducing impacts is vague.
Stage (Growth-Inducing) of Proposal or Plan	Preparation of a growth-inducing proposal or plan is underway or contemplated.	No effort has been made to formulate a growth-inducing proposal or plan.
Degree of Control Agency Has over the Potential Impacts	The agency has some ability to shape or exercise a degree of control over the potential impacts.	The agency has little or no ability to control or influence the potential impacts.
Spatial Domain	The agency has some degree of confidence in its ability to identify the geographic dimensions and resources that would be affected.	The agency has little or no degree of confidence in its ability to identify the geographic dimensions and resources that would be affected.
Temporal Domain	The agency has some degree of confidence in its ability to identify the time frame over which the impact would occur. would occur.	The agency has little or no degree of confidence in its ability to identify the time frame over which the impact

Cumulative Impact Assessment

A cumulative effect is defined as:

> . . . the impact on the environment which results from the incremental impact of the action when added to other past, present, and reasonably foreseeable future actions regardless of what agency (Federal or non-Federal) or person undertakes such other actions. Cumulative impacts can result from individually minor but collectively significant actions taking place over a period of time. (§1508.7)

Cumulative impacts may result from simple additive disturbances or from complex interactive phenomena, which are generally more complex to investigate (see Figure 2.4). Interactive effects may be either countervailing or synergistic. Countervailing effects occur when the net effect is less than the sum of individual effects. Similarly, synergistic effects occur when the net cumulative effect is greater than the sum of the individual contributions. The concept of a synergistic effect versus that of a simpler additive effect is depicted in Figure 2.5.

The importance of preparing a well-constructed cumulative impact assessment is demonstrated by a recent case. The Sierra Club claimed that a U.S. Forest Service EIS had improperly discussed cumulative impacts of various land uses for a resource management plan. The Sierra Club claimed that the EIS simply described a laundry list of individual effects. The court agreed that the cumulative effects analysis did not meet NEPA's regulatory requirements. Specifically, the court found that the cumulative impact analysis failed to include all effects of the various activities that could occur and did not evaluate impacts of various activities in combination with one another.[32]

While a cumulative impact assessment must be performed, the reader should note that in terms of the present state of the art, cumulative impact

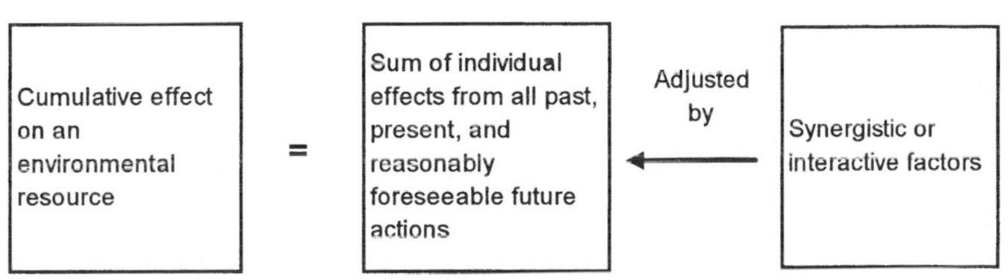

Figure 2.4 A simplified representation of how synergistic factors can affect the analysis of cumulative impacts

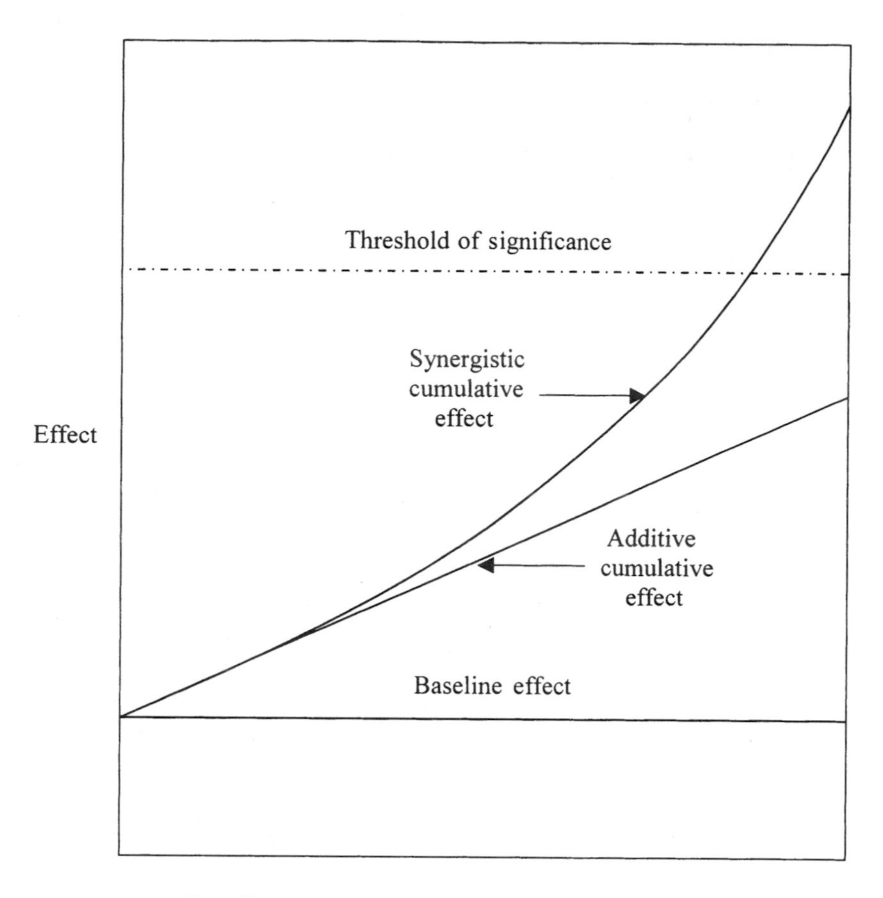

Figure 2.5 Theoretical representation of additive versus synergistic effects

analyses are sometimes too complex to permit a level of analysis equivalent to that performed for direct or indirect impacts. For more information on the addressing cumulative impacts, the reader is referred to *The* NEPA *Planning Process*.[33]

Step 5: Interpret Significance

Once the environmental impact analysis is completed, it is the responsibility of the IDT team to interpret and explain the significance of the environmental impacts. The goal is to convey to the decisionmaker and the public the implications of the environmental impact. Thus, the EIS must clearly convey to the reader the importance of the environmental impacts. This discussion should interpret how, why, and to what degree environmental impacts are significant.

For example, what are the implications of a decrease in air quality standards? What would a change in the incidence in human respiratory illnesses and cancer rates mean to an individual or the population at large? What would the number of fish killed mean in terms of environmental quality and long-term sustainability? What is the importance of a breach in environmental standards? What are the implications of violating land use restrictions?

Step 6: Assess Potential Monitoring and Mitigation Measures

The agency may elect to implement mitigation measures that render a significant impact nonsignificant or otherwise reduces its severity. To this end, the EIS must thoroughly investigate mitigation measures for later consideration by the decisionmaker. The decisionmaker is legally obligated to consider such measures in the course of reaching a final decision. Consistent with a sliding-scale approach, the effort devoted to this investigation should be commensurate with the potential significance of the impacts.

An evaluation of generic measures is normally of limited use. Instead, the investigation should focus on evaluating detailed or site-specific measures. While environmental effects do not have to be mitigated to the point of nonsignificance, the effectiveness of mitigation measures in preventing or reducing impacts must be evaluated.

The CEQ strongly encourages and, in some circumstances, the Regulations dictate that a monitoring program be instituted as the final step in the EIS process (§1505.2[c]). A word of caution is in order. The decisionmaker is responsible for ensuring that sufficient funding is available to support monitoring or mitigation measures that are committed to. As environmental monitoring and mitigation can be an expensive proposition, the author recommend a cost estimate for implementation be included as part of the analysis.

Adaptive Management

Changing conditions are difficult to predict and account for, and can negate environmental protection measures envisioned in the original analysis. A modern approach referred to as *adaptive management* has been promoted by the CEQ as a means for supporting midcourse corrections as new information becomes available. Adaptive management promotes a paradigm consisting of a five-step process: predict, mitigate, implement, monitor, and adapt.[34] Adaptive management can provide agency officials with flexibility for dealing with new or changing circumstances by adjusting actions in light of changing information. Consequently, agencies may be able to begin projects earlier and at lower cost by accepting more uncertainty during the early planning process, while advancing both NEPA's goal and the agency's mission.

Bounding Analyses

A methodology referred to as *bounding* is sometimes used to simplify an environmental impact analysis. A bounding analysis typically uses conservative assumptions and analytical methods to estimate (i.e., bound) the maximum consequences of a potential environmental impact. A bounding methodology described below provides an efficient approach for handling circumstances where:

- An analysis must be completed before the necessary project-specific design data are available.
- A spectrum of related alternatives exist that vary continuously from one alternative to the next.

Background and Problem

The following methodology was originally developed by the author to address a particular problem that required investigating potential steam and power plant alternatives before specific plant designs were available on which to base the analysis; this project involved a federal privatization proposal.

Specifically, a methodology was needed to adequately evaluate a wide range of potential steam and power plant alternatives that might eventually be proposed by a private vendor.

Initially, an approach was considered in which a set of individual plant designs would be evaluated covering a range of reasonable possibilities; the individual plant scenarios would then be analyzed to evaluate potential emissions and environmental impacts. Collectively, these scenarios would bound the effects over a range of potential plant projects that might be proposed.

For example, one plant scenario (alternative) could encompass coal-fired cogeneration plants that use stoker technology and range in size between 50 to 100 megawatts of electric (MWe) generating capacity. A second alternative could encompass plants having similar attributes but ranging in size between 100 to 150 MWe. Ultimately, this approach resulted in identifying a minimum of 18 individual alternatives for eventual analysis. This set of alternatives was deemed necessary to bound the range of possibilities that might be proposed by a private entity.

It quickly became obvious that an analysis and comparison of such a large number of plant alternatives would be very difficult and might verge on being unmanageable. In addition, this approach might lead to an analysis that would be so large and cumbersome that the decisionmakers would be overwhelmed in comparing alternatives and reaching a final decision (see

"Rule of Seven"). Finally, it was unclear what would happen if the project design that was finally selected would fall between two of the analyzed alternatives. Would the project design be adequately covered? Clearly, a more efficient approach was needed.

Continuous Spectrum Analysis

A simpler and more powerful method of evaluating potential plant alternatives involves analyzing impacts over a continuous range of different sizes and types of plants. In contrast to the example above, such an analysis is not treated as an independent evaluation of many discrete plant alternatives. Instead, the analysis is treated as a single analysis of a range of plant sizes and types. In this case, environmental disturbances and impacts are evaluated across a continuous spectrum (CONSPEC) of potential plant sizes and types, ranging from small steam plants to medium-size cogeneration plants. For this reason, the continuous range of plant scenarios was referred to as the CONSPEC *alternative*. For additional details on the use of this methodology, the reader is referred to the author's publication on continuous spectrum analysis.[35]

Example

Figure 2.6 illustrates how the CONSPEC analysis can be used to simplify the comparison and analysis of proposals involving a continuous range of potential alternatives. Both coal- and gas-fired plants are analyzed.

The dashed line represents how the computed annual sulfur dioxide (SO_2) emissions (tons per year) for coal-fired cogeneration plants would vary over a continuous range of plant sizes (ranging from 60 to 240 MWe) that could be constructed under the CONSPEC alternative. The two solid lines that parallel the dashed line delineate the error band.

As indicated in Figure 2.6, projected SO_2 emissions for the CONSPEC alternative involving coal-fired cogeneration plants increases approximately linearly with plant size from approximately 200 tons per year to slightly less than 800 tons per year. In contrast, the projected SO_2 emissions for the CONSPEC alternative involving gas-fired cogeneration plants ranging from 60 and 240 MWe are so low (less than 5 tons per year) that the emission curve is nearly undiscernible, as it virtually lies along the x-axis.

The projected emissions for the alternatives of no action (e.g., continuing to operate an existing plant), renovating an existing steam plant, and constructing a new gas-fired boiler plant are shown plotted on the y-axis. These three alternatives all lie on the zero value of the x-axis, as they would involve

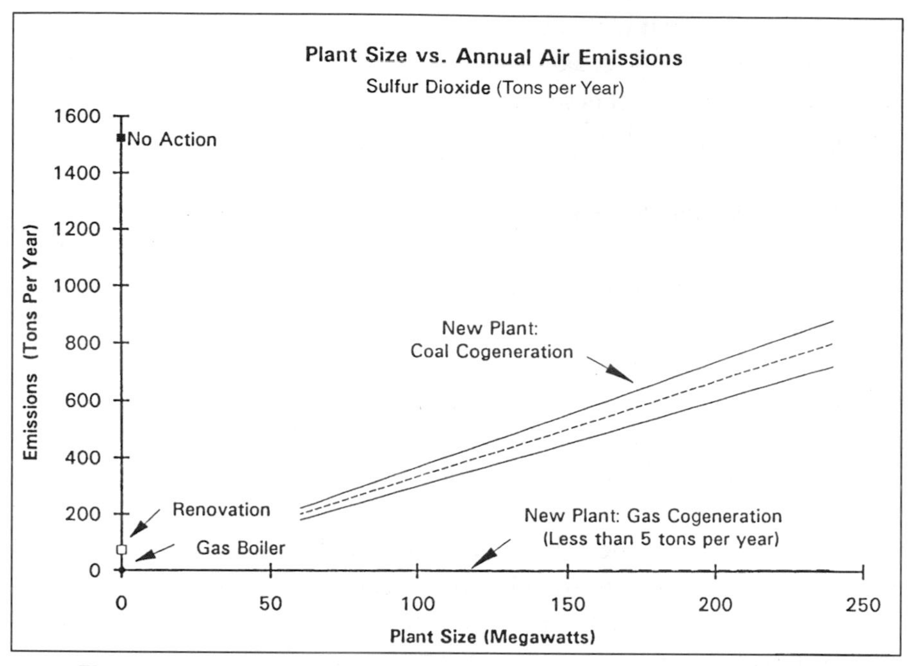

Figure 2.6 CONSPEC analysis of potential coal- and gas-fired alternatives

production of steam but no electric power. The projected emissions produced from continuing to operate an existing steam plant are substantially greater than for any of the other alternatives shown. This is true even though a 240 MWe cogeneration plant would be much larger than an existing steam plant.

The CONSPEC approach is not limited to analysis of air impacts. A similar technique can be applied to many types of environmental impacts (e.g., fuel consumption, water consumption, effluent and waste water discharges, hazardous waste generation, land area disturbances, and changes in the workforce).

Determining the Threshold of Significance

A CONSPEC analysis is also an efficient mechanism for discriminating the point of significance and nonsignificance over a continuous range of plant sizes. For instance, a significant environmental impact might involve an air emission impact that exceeds a federal, state, or local air quality standard, as illustrated by the horizontal line labeled "Significance Level" in Figure 2.7. Any point on or above this line (i.e., significance level) would be considered a significant air impact. As indicated by X_1 in Figure 2.7, the point at which the CONSPEC curve intercepts an air quality emission standard (e.g., SO_2 concentration) would delineate the maximum size of a particular type of plant that could be built without incurring a significant impact.

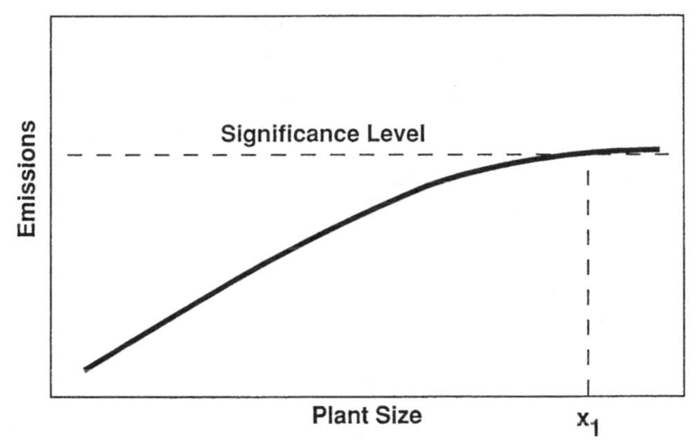

Figure 2.7 Using a CONSPEC curve to determine the threshold of significance

Advantages

This approach is attractive because the discussion and comparison of individual plant alternatives essentially collapse into a consolidated analysis of how impacts vary over a continuous spectrum of potential plant alternatives—the CONSPEC alternative. This can now be compared with the impacts that would result from pursuing either the renovation of an existing plant or the no-action alternative.

Comparisons of plants along the CONSPEC alternative curve can be made by comparing the environmental disturbances and impacts at a particular point (or segment) with those at another; thus, the discussion and analysis focuses on describing the range of impacts that could result from the CONSPEC alternative as opposed to an individual discussion, analysis, and comparison of 18 separate alternatives.

A decisionmaker is no longer faced with the daunting task of comparing, considering, and choosing among 18 plant alternatives. Instead, the decisionmaker is faced with the much simpler case of comparing and choosing between the CONSPEC (i.e., coal- and gas-fired plants), renovation, and the no-action alternatives. With respect to constructing a new plant, the CONSPEC analysis provides the decisionmaker with a method for more effectively comparing how environmental impacts vary over a continuous range of potential plant sizes. This approach avoids many of the problems discussed earlier.

Programmatic EISs

Preparation of a programmatic EIS is normally warranted when decisions made at the national level involve policies or alternatives that would affect an entire federal effort. For example, a programmatic EIS is often required for broad federal actions, such as adoption of new agency programs, policies, or regulations. Similarly, preparation of a site-wide EIS may be warranted for agency sites that support diverse missions and activities. A site-wide EIS can be particularly useful in evaluating individual and cumulative impacts of multiple ongoing and reasonably foreseeable activities.

A programmatic EIS should never be used as a means for avoiding preparation of more site-specific analyses that may be necessary to adequately assess detailed actions. In cases where an agency is challenged for failing to prepare a site-specific analysis, the courts are likely to scrutinize the programmatic EIS closely to determine if it provides an adequate analysis for implementing specific activities.[36] A case in point: The U.S. Forest Service prepared a programmatic EIS and attempted to make land use decisions regarding wilderness versus nonwilderness designations without a site-specific analysis. The court concluded that this was an improper use of the programmatic EIS.[37]

Advantages of a Programmatic EIS

Preparing a programmatic EIS can provide decisionmakers with decisive advantages over piecemeal or segmented approaches. A programmatic analysis can be particularly useful in achieving one or more of the following objectives:

- Providing decisionmakers with a more comprehensive and useful analysis for making high-level or long-term decisions.
- Evaluating impacts of multiple proposed projects.
- Developing a comprehensive land use plan based on an ecosystem approach.
- Providing a platform for tiering project-specific analyses.
- Assessing cumulative impacts of multiple ongoing or proposed site-specific activities.
- Reducing cost and time required to prepare subsequent NEPA documents.

Determining the Appropriate Programmatic Scope

In the past, confusion has arisen with respect to determining the scope, discussion, and analysis most appropriately addressed by a programmatic EIS versus the scope that is best reserved for lower-tier analyses. Neither NEPA nor its regulations provide definitive guidance for resolving this dilemma. Consequently, an inordinate amount of time and resources can be consumed in reaching such a determination. The following section describes a tool for resolving this dilemma. The reader is referred to the author's paper where this tool, including a technical justification for the criteria used, is described in detail.[38]

A Tool for Determining Programmatic Detail and Scope

Eight criteria (or tests) are presented in Table 2.15 for determining the appropriate scope of a programmatic EIS. A flowchart of this process is shown in Figure 2.8.

As indicated in Figure 2.8, each criterion or test is evaluated with respect to the following question: "Would the discussion or analysis . . . ?" The first two tests provide an initial screening mechanism for identifying discussions that are normally best deferred to lower-tier documentation. The remaining six tests provide criteria for determining if discussion of a particular issue, impact, or alternative is appropriate for consideration within a programmatic EIS. A *yes* answer to any of these six tests supports a determination to include the discussion within the scope of the programmatic EIS. A *no* answer to all of these six tests indicates that a particular discussion should be deferred to lower-tier documents.

Table 2.15 Criteria for determining the level of detail and scope appropriate for a programmatic EIS

1. Can the issue be appropriately considered and analyzed at the programmatic level (i.e., is it ripe for decision?) or should it be deferred to a later time when more information is available and events are better understood?
2. Would discussion and analysis of an alternative reduce flexibility (i.e., box the agency into a corner) to adapt or tailor specific lower-tier actions to accommodate a dynamic and often evolving planning process where future events often change or new information becomes available?
3. Would discussion and analysis of the issue substantially improve or change high-level decisions regarding the program or policy, or would it merely provide additional detail? Specifically, would the discussion allow the agency to concentrate on programmatic issues that are truly significant to the action in question?
4. Would the discussion and analysis provide a technical basis (e.g., descriptions, analysis, parameters, constraints, bounds) that would support or facilitate tiering of more detailed documents at a later date?
5. Would the analysis enhance understanding of the potential environmental impacts from a programmatic perspective?
6. Would the discussion and analysis enhance an outside agency's or the public's understanding or ability to provide comments regarding the programmatic implications of the policy or program?
7. Would discussion and analysis identify or develop an alternative approach or course of action that could affect the outcome of the program or policy from a programmatic perspective, or would it merely fine-tune programmatic decisionmaking?
8. Would the discussion and analysis result in programmatic mitigation measures that could be implemented to substantially reduce programmatic environmental impacts?

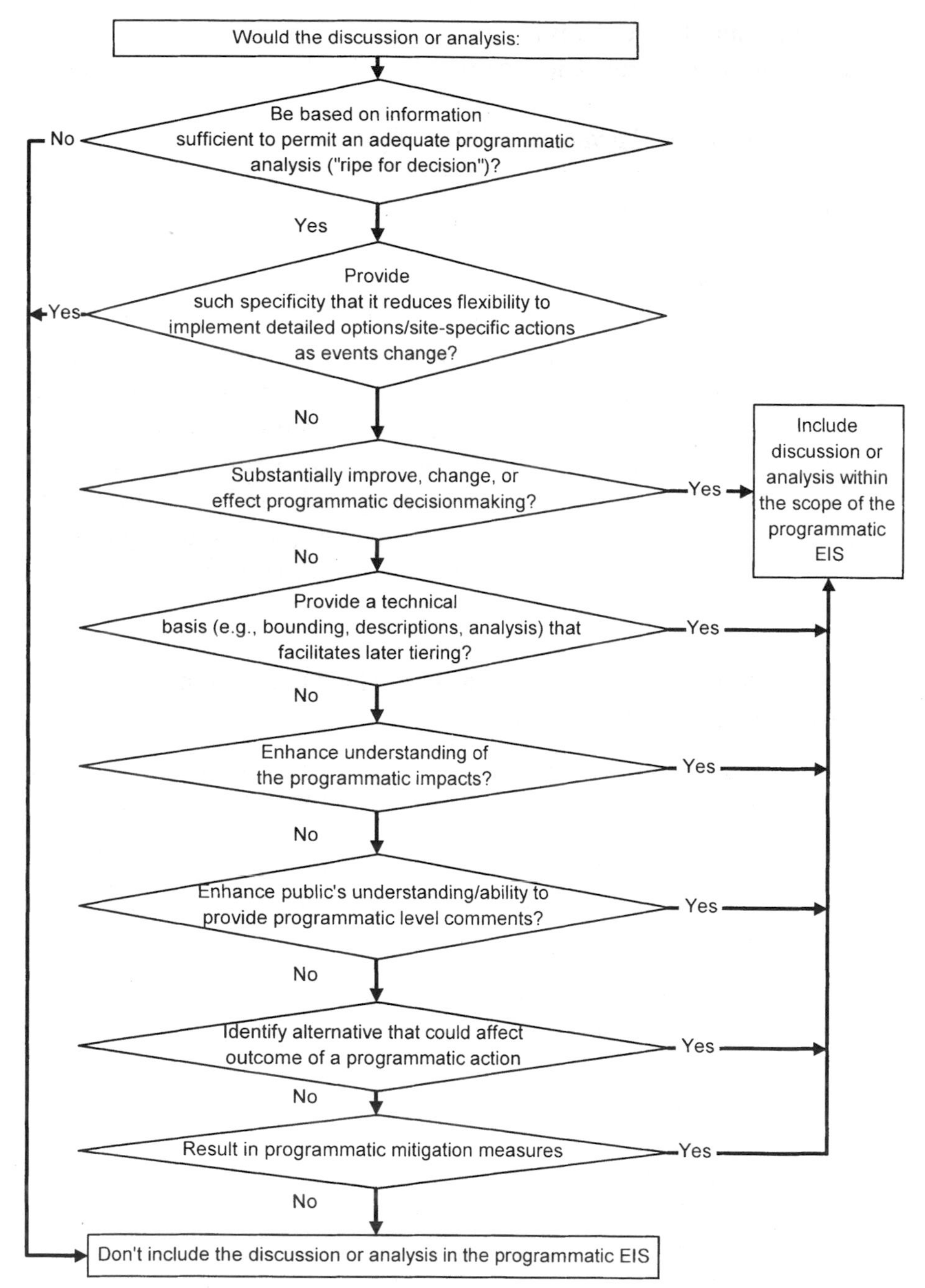

Figure 2.8 Tool for determining the appropriate scope of a programmatic EIS

These criteria can also be used to address a related question regarding the amount of detail that is most appropriate within a programmatic EIS; for example, with respect to the third criterion in Table 2.15, this can be done by substituting for "Would discussion and analysis of the issue . . ." the phrase "Would a more detailed discussion and analysis of the issue . . . ?"

2.5 Preparing the Draft EIS

Marcel Proust once pronounced, "All our final decisions are made in a state of mind that is not going to last." With this thought in mind, the EIS should be designed to provide decisionmakers with a degree of flexibility necessary to accommodate changing circumstances. For example, numerous problems were encountered with the plutonium stabilization EIS (described in the introduction), as it was not designed to readily encompass project changes, even relatively minor ones.

An experienced professional should be assigned to oversee preparation of each major section of the EIS (e.g., affected environment, alternatives, environmental consequences) and given a copy of the EIS outline before beginning work. To minimize revisions, specific direction is provided to analysts regarding preparation of their respective section(s). Page limits should be established and adhered to unless there is a compelling reason to exceed them.

Some EIS managers instruct their staff to do as little formatting as possible (font changes, headings, double columns, etc.). Instead, they are told to use as simple a format as possible. Formatting is left to the editor.

As individual practitioners are most familiar with their own contribution, time and effort may be saved by having them write that portion of the EIS summary section that covers their area of responsibility; to prevent fragmentation, the EIS manager or designee should review the summary for errors, inconsistencies, and to ensure appropriate emphasis is given to each topic.

As a detailed description of the procedural requirements for preparing an EIS is beyond the scope of this book, the reader is referred to chapter 9 of the companion book, *The NEPA Planning Process* for a thorough discussion of the regulatory requirements.[39]

Limiting the Agency's Litigation Risk

To limit the risk of legal challenge, it is recommended that practitioners exercise the following precautions:

- The EIS should present the scientific bases for all conclusions.
- The EIS should explain why reasonable alternatives or analytical methodologies were found to be unreasonable and dismissed.

- Where possible, applicable modeling and scientific procedures developed by EPA other agencies possessing similar jurisdiction, or professional societies should be employed.
- Where there may be a serious threat of legal challenge, the agency should make a special effort to retain scientific experts of the highest caliber to assist in preparing the analysis.[40]

If there is substantial risk of a legal challenge, the EIS manager should consider obtaining the services of independent technical experts who possess established reputations in their field of expertise to prepare the analysis. Emphasis should be placed on choosing experts who are articulate, have experience testifying as expert witness, and project good courtroom demeanor. Independent experts can also establish credibility with the public by way of their professional impartiality.

Incorporation by Reference

Every effort should be made to maximize use of existing material by reference. Table 2.16 provides a list of material commonly incorporated by reference.

Finalizing the Draft EIS

A thorough review of the EIS is critical to the accuracy and defensibility of the statement. Yet, the manner in which the review and comment process is conducted is often one of the principal factors responsible for a protracted EIS schedule. In some arenas, internal review consumes the longest portion of the EIS schedule. Special consideration should be given to identifying inefficiencies and determining methods for expediting the review and comment process.

Table 2.16 Material commonly incorporated by reference

Related NEPA documents
Other planning and analysis documents
Environmental permitting documents (e.g., air, water, and Resource Conservation and Recovery Act [RCRA] permits)
Safety analyses and studies
Biological, soils, geological, hydrological, air quality, meteorological, socioeconomic, and other environmental data and reports
Data bases, maps, drawings, and charts

Identifying the Agency's Preferred Alternative

If, at the draft EIS stage, the lead agency has a preferred alternative, this must be disclosed in the draft statement. In any event, the preferred alternative must be identified in the final EIS unless another law prohibits expression of such a preference (§1502.14[e]). The lead agency's official having "line responsibility for preparing the EIS and assuring its adequacy" is responsible for identifying the agency's preferred alternative.[42]

> **Opportunities to Apply Total Federal Planning**
>
> If a Total Federal Planning Strategy (TFP) is adopted, the reader is referred to the section titled "Step 7: Review and Decisionmaking Stage" (Figure 5.10) for a discussion of a process and analytical techniques that can be used to objectively determine the agency's preferred alternative.

In some cases (particularly complex ones), the preferred alternative is not always apparent; application of a modified value engineering (VE) session using tools such as nominal group techniques may provide the decisionmaker with a rapid and defensible approach for assessing the analyzed alternatives in an effort to identify the one best meeting the agency's purpose and need.

Internal Agency Review

On its completion, the draft EIS is distributed for internal agency review. In reality, numerous working versions of the document are normally circulated for internal review prior to public release. Copies slated for internal review should be stamped "draft predecisional" or some equivalent designation. Likewise, a date and revision number should be included so that comments can be correctly tied to the appropriate working version of the statement. Line numbers should also be included to facilitate comment review and incorporation. Finally, the EIS manager should be mindful that failure to allow for applicable public clearance procedures and, if appropriate, a classification review, has been responsible for more than one missed deadline.

A dozen or more people are commonly involved in the internal review of even a small EIS. It is recommended that personnel not directly involved with the EIS be included among the reviewers; including such reviewers helps assure that the final document is clear and can be understood by the general public.

All agreements reached between the reviewers and EIS preparers should be documented as part of the agency's ADREC. Such documentation can help subsequent reviewers come up to speed with respect to what has been reviewed and changed. With the results of the review in writing, the EIS manager can avoid covering old ground each time the review team is assembled.

Environmental impacts and issues are a function of the specific actions that would take place. However, reviewers sometimes proceed directly to the section related to their expertise, skipping the affected environment, description of the proposal, and other relevant topics. Lacking this background, reviewers are unlikely to fully comprehend the scope of potential actions or assumptions on which the analysis is based; unchecked reviewers may fail to identify environmental impacts that may have been overlooked or inadequately evaluated. Where feasible, reviewers should be instructed to read all relevant sections of the EIS.

REVIEW METHODS. Many methods have been successfully used to expedite internal review. One involves first circulating the statement for internal review. Reviewers should be instructed (whenever possible) to prepare tentative comment resolutions. The reviewers and agency officials having some measure of approval authority over the document are then assembled in a room for comment resolution. The EIS manager takes them through the document page by page, addressing each comment. A copy of the statement can also be maintained in a computer (e.g., lap top) so that changes can be made in real time as they are resolved. This approach has the advantage that all officials possessing approval authority are present as the comments are resolved. Since the members are party to the review and agree as a group to the changes, they should have no reservations about approving the document for public circulation. In some cases, an entire review process has been completed in less than one week.

Circulating the Draft for Public Review

On completion of the internal review, copies of the draft EIS are printed for public distribution. While the Regulations use the term *draft* in referring to this stage of the EIS process, this is actually a misnomer. The statement is not a draft in the traditional sense. The EIS must in fact be complete, meet applicable regulatory requirements, and be capable of standing on its own merit. It is referred to as a draft because it does not yet incorporate comments from the public review cycle. Accordingly, the purpose of the final EIS is to respond to comments received on the draft statement. In reality, the agency often continues to improve and revise the draft, even while it is being circulated for public comment; the reader should note that a major change or revision during this period could trigger the necessity for sending the draft out for a second public review.

Public Entities

A mailing list of individuals, groups, and organizations to whom a copy of the EIS will be distributed is prepared. Prior to filing the draft, copies of the statement are transmitted to interested citizens, groups, and public reading rooms. At a minimum, the EIS must be furnished to the following parties:

- Any federal agency having jurisdiction by law or special expertise with any of the environmental impacts involved, and any federal, state, or local agency authorized to develop and enforce environmental standards
- The applicant (if any)
- Any person, organization, or agency requesting the entire EIS
- In the case of a final EIS, any person, organization, or agency that submitted substantive comments on the draft (§1502.19)

The draft EIS is made available to the public at least 15 days prior to any public hearing or meeting in which the statement is considered. The agency may also find it desirable to hold one or more public meetings or hearings to solicit public comments. Where practical, the EIS and supporting material is made available to the public free of charge or at a fee that covers only the actual costs of reproducing copies (§1506.6[f]).

Filing the Draft EIS

Both the draft and final EIS must be transmitted to the EPA, which has issued detailed direction for filing EISs.[43]

Minimum Comment Period

The minimum comment review period for a draft EIS is 45 days (§1506.10[c]). This is a minimum. A longer period should be seriously considered if it would furthers the purposes of NEPA. For example, the DOE has reported that approximately 40 percent of its draft EISs have exceeded the minimum 45-day comment period. Between 1994 and 1998, the average DOE comment period for project-specific EISs was 52 days. For 13 percent of these EISs, the comment period was extended by an average of 30 days.[44]

The comment period should be established with care. The EIS manager should consider whether the minimum comment period is appropriate given the level of public interest, document or project complexity, and project schedule constraints. If an extension is deemed necessary, the following guidance is offered:

- Strive to announce any extension in the public comment period quickly so that stakeholders may take full advantage of the additional time.
- Use quick and effective methods (e.g., phone, mail, e-mail) to notify the stakeholders. Do not rely solely on an announcement in the *Federal Register.*

Transmitting the EIS

Five copies (including appendixes) of the EIS are filed with the EPA office in Washington, D.C. Material incorporated by reference is not required to be filed. Addresses and contacts for filing the EIS with EPA are indicated in Table 2.17. The mailing address depends on the method used for transmitting the document. The contacts and details of this procedure should be checked, as they are subject to change.

Publishing the Notice of Availability

The EIS cannot be filed with the EPA before it has been transmitted to commenting agencies and made available to the public (§1506.9). A notice of availability (NOA), also commonly referred to as *the Notice,* is prepared to announce that the EIS is available for public review. The EPA is responsible for publishing the NOA in the *Federal Register.* The filing date for an EIS is defined as the date the NOA is published in the *Federal Register* and not by the date on which the statement is transmitted or received by EPA (§1506.10[a]).

Inviting Public Comments on the Draft EIS

In addition to EPA's publication of the NOA in the *Federal Register,* the lead agency may also be responsible for informing or contacting interested or

Table 2.17 EIS Filing locations based on method of delivery

EPA Contacts: Marilyn Henderson at (202)564-7153 or Pearl Young at (202)564-7167

Via United States Postal Service	Via Special Delivery (e.g., Federal Express, United Parcel Service)
U.S. Environmental Protection Agency Office of Federal Activities NEPA Compliance Division EIS Filing Section Mail Code 2252-A 401 M Street SW Washington, D.C. 20460	U.S. Environmental Protection Agency Office of Federal Activities NEPA Compliance Division Filing Section Ariel Rios Building (South Oval Lobby) Mail Code 2252-A, Room 7241 1200 Pennsylvania Avenue NW Washington, D.C. 20044

affected parties by other means. This may be done via vehicles including, but not limited to, personal letters, public advertisements and media announcements, and through the Internet (§1506.6[b]). Table 2.18 lists public and private parties from which an agency must request and, in some instances, obtain comments prior to preparing the final EIS (§1503.1[a]).

Environmental Protection Agency's Review

EPA's goal is to assist agencies in achieving decisionmaking that fully considers reasonable alternatives and environmental impacts, and facilitates public involvement. EPA's Web site is located at http://es.epa.gov/oeca/ofa/index.html.

Primary reviewers (i.e., associate reviewers) are assigned to review the EIS based on their expertise with the environmental issues addressed in the statement. A separate specialist is assigned for each principal area covered by the EIS (e.g., air quality, water quality, sensitive species). To facilitate EPA's review, the EIS staff may find it advantageous to contact and establish rapport with their counterparts in the EPA.

Based on this review, EPA assigns a rating to the EIS. The reviewers prepare a summary of their results and a cover letter to be sent to the agency. Under EPA's policy, the EIS rating is not negotiable. However, EPA does notify the agency prior to publishing its findings. A summary paragraph of each EIS reviewed by EPA, including the EIS rating, is published weekly in the *Federal Register*. The summary and ratings are also placed on the Internet.

2.6 Preparing and Issuing the Final EIS

All comments received from circulating the draft EIS must be reviewed. Public comments and the agency's response to each comment should be

Table 2.18 Parties from which comments must be obtained or requested

Obtain comments from:

- Any federal agency that has jurisdiction by law or special expertise with respect to any environmental impact involved, or that is authorized to develop and enforce environmental standards

Request comments from:

- Appropriate state and local agencies authorized to develop and enforce environmental standards
- Native American tribes (when a reservation may be affected)
- Any agency that requested a copy of an EIS on actions of such scope
- Applicant
- The public, especially persons and organizations who may be interested or affected

attached to the final EIS. If the draft EIS is found to be "so inadequate as to preclude meaningful analysis," the agency must recirculate a revised draft of the appropriate section(s) (§1502.9[a]).

The reader is reminded that under §1502.9(c)(4), CEQ has authority to approve alternative procedures for preparing, circulating, and filing supplemental and final EISs. The CEQ notifies the EPA of alternative procedures that are granted.

Managing Comment Resolution

The majority of public comments are often received from citizens or groups opposed to the proposal. It is not unusual for a draft EIS to receive hundreds of comments. Complex or controversial EISs may receive thousands. All comments should be submitted through a single contact responsible for coordinating and addressing them. Similarly, a single point-of-contact should be assigned to coordinate preparation of responses.

If the comments are voluminous, the EIS manager may find it desirable to capture public comments in a data base (see description of the COMTRACK data base in chapter 1). A computer data base conveys the advantage of automating the tracking and management of comments. Parameters such as the commenter's name, date of comment, either the entire comment or a summary thereof, and the agency's response can be captured in such a data base. As part of the ADREC, a database is also useful in demonstrating that the agency performed a systematic process.

Responding to Comments

All comments must be reviewed and addressed as appropriate. A response to a comment may require that the environmental impact analysis be modified, expanded, or completely redone. In some cases, the agency may need to modify one or more existing alternatives. In other cases, alternatives not previously investigated may need to be analyzed.

Comments can be extremely long. Verbose comments can be summarized (§1503.4[b]).

All substantive comments received on the draft statement (or a summary, where comments are exceptionally voluminous) are attached to the final EIS. However, if the comments are minor, the final EIS might consist only of an errata sheet (i.e., public comments, agency's response to those comments, and corrections to the EIS). In such cases, only the comments, responses, and the changes, and not the entire EIS, need be circulated to the public.

On the first pass, it is best to make changes in redline and strikeout mode so that internal reviewers can see the change and its corresponding effect on the document. When a comment is accepted, the agency must explain how and where it has been incorporated into the EIS. If a comment is deemed inappropriate or incorrect, the agency must explain, citing specific sources or reasons, why the comment has not been accepted.

A Methodology for Managing Comments and Responses

Various methods are in common use for managing comments resolution. One that is specifically designed to provide traceability is described in this section. This system can also be efficiently managed using a computer data base. A similar system (with some modification) can be used in capturing and responding to the earlier public scoping comments.

Comments received from review of the draft EIS are first arranged in order of receipt. A unique designation is assigned to each comment. In this context, a commenter may be an individual, group or organization, or an entire agency. As many individual comments may be linked to a single commenter, an alphanumeric designation is useful. A number can be assigned to designate a specific commenter, while a letter identifies each specific comment (e.g., issue or topic) raised by this commenter (see first column, labeled "Alphanumeric Commenter Designation," in Table 2.19). As an example, if "1" is the number assigned to Mr. A. Smith, "1-A" indicates his first comment, "1-B" his second, and so on. Thus, a specific issue or topic raised by each individual commenter can now be tracked by a unique alphanumeric comment designation.

The example shown in Table 2.19 allows the EIS team to cross-reference the alphanumeric comment designation with the commenter's identity and

Table 2.19 Cross-referencing draft EIS commenters with the location of their comment in the final EIS

Alphanumeric Commenter Designation	Individual or Organization	EIS Page Number (Appendix)
1-A	Mr. A. Smith	B-3
1-B	Mr. A. Smith	B-4
2-A	Organization of Concerned Citizens	B-5
3-A	Mr. J. Doe	B-5
3-B	Mr. J. Doe	B-6
4-A	The NIMBY Society	B-6
5-A	The Locally Undesirable Land Use (LULU) Committee	B-7
•	•	•
•	•	•
•	•	•

the page (e.g., appendix) of the final EIS where the comment can be found. For example, the first comment received from Mr. A. Smith can be found in the EIS appendix on page B-3. Similarly, Mr. A. Smith's second comment can be found on page B-4. The third column is left blank until the page location in the final EIS is determined.

A second table is constructed that cross-references each alphanumeric comment designation against the commenter's corresponding comment and the agency's response; the commenter's specific remark can now be cross-referenced against the agency's response.

The comment is entered as shown in the center column of Table 2.20; if necessary, verbose comments or agency responses can be summarized. The third column is reserved until a response is formulated.

As the final EIS is prepared, a third table is constructed that cross-references the alphanumeric comment designations against the draft EIS comment and the location in the final EIS where the agency's response has been addressed (see Table 2.21). For example, the agency addressed Mr. A. Smith's comment (i.e., 1-A) regarding groundwater impacts by modifying section 5.3 of the final EIS. Thus, this table directs the reader to the particular section of the EIS where a given comment has been addressed.

Together, matrixes similar to the examples shown (Tables 2.19 through 2.21) allow the EIS staff to track and manage comments and their disposition by cross-referencing the commenter, specific comment, location in the draft where the comment can be found, agency's acceptance or rejection of the comment, and location in the final EIS where the comment is addressed. To assist the public in reviewing the agency's response, tables similar to these may be included in or accompany the final EIS. Such tables should also be placed in public reading rooms and made available to all commenters.

Issuing the Final EIS

Once comments are incorporated and the final EIS is complete, an internal review is conducted. The final EIS is issued in a manner similar to that for the draft, although the agency is not required to respond to public comments that may be submitted.

Producing the Final EIS

To reduce cost, the EIS can be printed using double-sided pages. Consistent with the goals of NEPA, the EIS should be printed on recycled paper. As has happened in the past, the cost of publishing should not be overlooked in preparing the EIS budget. Factors such as the size, number of copies, use of color figures, and other elements should be included in the original budget.

Table 2.20 Cross-referencing draft EIS comments against the agency's responses

Alphanumeric Comment Designation	Public Comments Received from Review of Draft EIS	Agency's Response to the Public Comments
1-A	The analysis of groundwater impacts neglected to include nitrate leaching from fertilizers.	*Accepted:* The analysis has been revised to evaluate impacts from nitrate leaching.
1-B	It is my opinion that the hydrological spatial boundary should be expanded to include the adjacent northwestern watershed.	*Accepted:* The spatial boundaries have been expanded to include the adjacent watershed.
2-A	The analysis of air quality impacts should have included gaseous released through the HEPA filtration system.	*Accepted:* The analysis of air emissions has been revised to include gaseous releases through the HEPA system.
3-A	The proposal will result in psychological stress. Impacts of such stress must also be considered.	*Rejected:* The courts have ruled that psychological stress is not an impact on the environment requiring analysis in an EIS.
3-B	The proposal conflicts with local land use restrictions and zoning.	*Accepted:* All actions have been modified to be consistent with local land use and zoning restrictions.
4-A	Only EPA-validated air dispersion models should have been used in computing SO_x emissions.	*Rejected.* The analysis was performed using EPA-validated air dispersion models.
5-A	The EIS alternatives did not consider alternative sites.	*Accepted:* The alternatives will be revised so as to include alternative sites.
•	•	•
•	•	•
•	•	•

Distributing an EIS on compact disc (CD) can result in both cost savings and environmental benefits. For example, Los Alamos National Laboratory distributed a NEPA and a habitat management plan (totaling approximately 1,850 pages) on CD, saving $40,000. In all, nearly a quarter of million sheets of paper (and associated printing chemicals) were saved—equivalent to 25 trees.[46] Increasingly, agencies are also placing the entire document on the Internet.

Circulating and Filing the EIS

On completing the internal review, copies of the final EIS are provided to agency officials, the decisionmaker, and, if appropriate, members of Con-

Table 2.21 Cross-referencing of the comment topic with the location of the response in the final EIS

Alphanumeric Comment Designation	Comment Description	Location of Agency's Response
1-A	Groundwater impacts	Section 5.3
1-B	Spatial boundaries	Section 4.1 and 5.3
2-A	Air quality impacts	Section 5.4
3-A	Socioeconomic impacts	Rejected
3-B	Land use conflicts and restrictions	Section 3.6
4-A	Air dispersion models	Rejected. See Section 5.4
5-A	Land use alternatives	Sections 3.6, 4.4, and 5.2
•	•	•
•	•	•
•	•	•

gress. Copies of the final EIS are transmitted to public reading rooms and to affected or interested parties. The entire EIS is then filed with the EPA. As a cost savings measure, appendixes are often sent only to entities that have requested them or have provided comments on the draft. If the appendixes are not circulated with the EIS, they must be made readily available on request (§1502.18[d]).

CIRCULATING A SUMMARY TO REDUCE COST. The Regulations identify the summary as an appropriate tool for saving resources by reducing paperwork. Specifically, agencies may reduce cost and paperwork by: "Summarizing the environmental impact statement and circulating the summary instead of the entire environmental impact statement if the latter is *unusually long*" (§1500.4[h], emphasis added). Section 1502.19, however, acknowledges that stakeholders possessing special status through the nature of their participation (e.g., cooperating agency, applicant, commenter) must receive the full document, as must any person or organization who requests a copy.

The agency may find it advantageous to query potential recipients in advance (e.g., by providing forms at public meetings, return postcards, and electronic media) to identify entities wishing to receive either a summary or the full EIS document. To support recipients in making this choice, the query medium should indicate the expected length of the two statement options.

EPA's EIS REPOSITORY. Filed EISs are retained at EPA's Office of Federal Activities for two years. Documents in this repository are available only to staff personnel. At the end of the two-year period, EISs are sent to the National Records Center.

Final EISs issued between 1970 and 1977 and all draft, final, and supplemental EISs filed from 1978 to the present are maintained on microfiche for public viewing in the EPA library (first column, Table 2.22). Copies of EISs can also be purchased either on microfiche or hardback from the address listed in the second column of the table. A comprehensive collection of EISs, available either for viewing or loan, can be obtained from the address listed in the third column.

2.7 The Final Decision and Implementation of the Action

Once the EIS is filed, the responsible decisionmaker begins what is often the arduous task of balancing pertinent factors (e.g., environmental, economic, schedules, and technical considerations) in reaching a final decision regarding the course of action to be taken. Figures I.2 and 5.10 depict the process for coordinating the review and decisionmaking process.

Reviewing the Final EIS

The EIS document (including comments and responses) is to *"accompany the proposal* through existing agency review processes so that agency officials *use the statement in making decisions"* (§1505.1[d], emphasis added). The responsible agency official must carefully review the results of the EIS before reaching a final decision. No decision regarding the proposal may be made or recorded until:

1. 90 days after the draft EIS is filed with the EPA, or
2. 30 days after the final EIS is filed with the EPA (§1506.10[b]).

The reader should note that the filing date is defined as the date in which the NOA is published in the *Federal Register* and not the date on which the EIS

Table 2.22 Locations of EIS copies

EPA Microfiche Library of EISs	Purchasing EISs	Comprehensive Collection of EISs
Environmental Protection Agency Room 2904 Waterside Mall 401 M Street SW Washington, D.C. 20460	Angela Hitti Cambridge Information Group (CIG) 7200 Wisconsin Avenue Bethesda, MD 20814 (301)961-6744	Connie Avildsen Northwestern University Transportation Library—NEPA 1935 North Sheridan Road Evanston, IL 60201 (708)491-5275

is transmitted to the EPA. As indicated in Table 1.4, an exception to the restrictions prescribed in §1506.10(b) may be made where an agency's decision is subject to a formal internal appeal. Also under §1506.10(d), EPA has authority to both extend and reduce EIS time periods based on a showing of "compelling reasons of national policy."

Reaching a Final Decision

The EIS *must* be used by agency officials "in conjunction with other relevant material" in reaching a final decision (§1502.1). More to the point, the EIS must be used by agency officials in making "decisions that are based on understanding of environmental consequences" and for considering "actions that protect, restore, and enhance the environment" (§1500.1[c]).

Decision Factors

Recall from the description of Schmidt's model of purpose and need (see section 2.3) that purposes are objectives the agency wishes to achieve. It is in the decisionmaking process that these purposes comes into play. The purpose (i.e., decision factors) for taking action provides the decisionmaker with important input (in conjunction with environmental considerations) that is useful in discriminating among the analyzed alternatives. The factors used in reaching a final decision must be disclosed in the record of decision (ROD) (§1505.2[b]). Table 2.23 indicates factors commonly used in reaching a final decision.

The decisionmaker need not choose the preferred alternative. The decisionmaker is, in fact, free to choose any course of action as long as it has been appropriately described and analyzed in the EIS.

Though not technically required, copies of the ROD should be distributed to organizations and individuals that received a copy of the final EIS. Once a ROD is issued, the agency may still change its mind and select another alternative; this can be done by supplementing the ROD or issuing a new one.

Bounded Alternatives

In some cases, agencies have chosen a course of action not specifically covered in any one of the alternatives but that is appropriately bounded by two or more of the analyzed alternatives. In other cases, agencies have taken a mix-and-match approach, choosing a portion of one alternative and a portion of another, both of which were analyzed in the EIS.

Such decisions should be made judiciously; it is conceivable that the impact of an action bounded between two analyzed alternatives might exceed projections cited for either one. Likewise, the impact of mixing and matching aspects of different alternatives might exceed EIS projections for any

Table 2.23 Factors commonly used in reaching a final decision

Decision Factor	Explanation
Environmental impact	Selection of an alternative that reduces or eliminates an impact
Mitigation	Selection of a mitigation measure that reduces or eliminates an adverse impact
Cumulative impact	Selection of an alternative that reduces cumulative impacts
Human health risk	Selection of an alternative that reduces risks to human health
Mission	Selection of an alternative that satisfies the agency's underlying need and mission
Technological capability	Selection of an alternative that is scientifically or technologically feasible
Cost	Selection of an alternative that reduces cost
Socioconomic dislocation	Selection of an alternative that reduces economic dislocation or job losses
Environmental justice	Selection of an alternative that equitably spreads impacts across economic and ethnic populations
Consistency	Selection of an alternative that is consistent with other plans or requirements
Regulatory compliance	Selection of an alternative that complies with regulatory requirements, orders, compliance agreements, and other requirements
Implementation flexibility	Selection of an alternative that provides flexibility to respond to changing circumstances

one of them. If such an approach is taken, some mechanism should be implemented to demonstrate that the agency's decision does not exceed the EIS forecasts.

Finally, lest we forget, an EIS is prepared to provide information sufficient to discriminate among alternative courses of action so as to foster informed decisionmaking; a bounding analysis may accurately bound potential impacts without providing the decisionmaker or the public with sufficient information for discriminating between alternatives and reaching an 'informed' decision. Prudence should be exercised, as an inappropriately bounded alternatives analysis may provide fertile ground for a legal challenge.

Monitoring and Mitigation

Consistent with NEPA's goal of environmental protection, a monitoring plan should be implemented to ensure that (1) all actions are implemented in conformity with the decision described in the ROD, and (2) mitigation measures committed to in the ROD are correctly carried out. The relationship of

Table 2.24 Checklist of steps and tasks that should be considered and, if appropriate, performed as part of the formal EIS planning process

- Prepare for public scoping effort:
  - –Assemble scoping information packet (see Table 2.2).
  - –Obtain services of public involvement specialists.
  - –Prepare posters, exhibits, public advertisement, fact sheets.
  - –Take steps to establish rapport with the public.
- Publish and distribute the Notice of Intent (NOI) in the *Federal Register* and through other media.
- Prepare for and hold public scoping meeting(s) (if appropriate): Obtain services of an independent moderator.
- Finalize scope (range of actions, alternatives, and impacts to be analyzed).
- Prepare and issue an implementation Plan (if applicable, see Table 2.4).
- Initiate the alternatives synthesis and assessment phase (see section 2.3), applying a modified systems engineering approach (if applicable):
  - –Identify and integrate all pertinent planning factors.
  - –Identify all laws, permits, and licences that may need to be obtained.
  - –Consult with other authorities and agencies, and integrate all applicable environmental requirements with the EIS planning effort (see Tables 2.5. and 2.6).
  - –Perform input and functional requirements assessment step:
    - *Identify and organize all requirements from all sources.
    - *Identify any constraints and limitations on potential alternatives.
  - –Perform alternatives identification step: Apply the statement of underlying purpose and need in identifying potential alternatives.
  - –Perform alternatives screening and evaluation step:
    - *Define screening criteria.
    - *Define a range of reasonable alternatives for analysis.
  - –Perform alternatives synthesis step: Prepare outline of proposed action and alternatives (OPAA). Briefly outline each alternative for later analysis.
  - –Perform action identification and planning integration step:
    - *Identify all actions that make up each alternative.
    - *Integrate all other planning factors and requirements.
  - –Perform optimization step:
    - *Identify methods for improving alternatives.
    - *Identify potential mitigation measures.
  - –Perform alternatives description step: Prepare description of proposed action and alternatives (DOPAA).
- Initiate the interdisciplinary analysis phase (see Section 2.4):
  - –Prepare analysis plan: Determine analytical methods.
  - –Document assumptions: Prepare and distribute record of assumption (ROA) forms to the IDT.
  - –Perform analysis:
    - **Step 1:* Characterize potentially affected environment.
    - **Step 2:* Identify potential environmental disturbances.
    - **Step 3:* Perform preliminary investigation and screening of all potentially significant impacts.
    - **Step 4:* Prepare detailed assessment of potentially significant impacts.

Table 2.24 (*Continued*)

**Step 5:* Interpret significance.
**Step 6:* Assess potential monitoring and mitigation measures.

- Prepare draft EIS document:
  - –Circulate draft EIS for internal review.
  - –Circulate draft EIS for public review.
  - –Prepare and issue NOA for the draft EIS.
- File draft EIS with the Environmental Protection Agency (EPA).
- Incorporate public comments into the final EIS:
  - –Prepare NOA.
  - –Circulate final EIS for internal review.
  - –Circulate final EIS for public review.
- File final EIS with EPA.
- Reach final decision.

the monitoring and mitigation step to the EIS process and the project implementation phase is depicted in Figure 2.9. Applicable monitoring and mitigation measures are implemented throughout the project implementation phase (design, permitting, construction, and operational phases) and employed as necessary. Aspects of this procedure are described in more detail in chapter 4.

Summary

The intent of this chapter is to provide the reader with a comprehensive guide detailing the all principal steps and tasks that may need to be considered and, if appropriate, performed as part of an agency's EIS planning process. Table 2.24 summarizes these. Professional judgment must be exercised in determining those items most applicable to the existing circumstance. Ultimately, factors as diverse as agency culture, funding and schedule constraints, existing data, amount of experience the agency has with the proposal, and

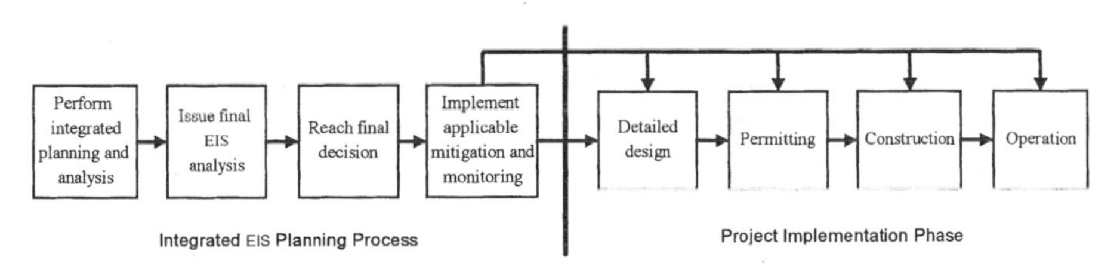

Figure 2.9 Integration of mitigation and monitoring with the project implementation phase

the size and complexity of the planning effort dictate which steps and tasks are most practical and appropriate.

This table can be used to assist practitioners in preparing a schedule and budget and as a road map for planning and managing the EIS effort. Depending on circumstances, items indicated in the table need not be performed in the order shown.

Notes

1. C.H. Eccleston, *The NEPA Planning Process: A Comprehensive Guide with Emphasis on Efficiency* (New York: John Wiley & Sons, 1999).
2. U.S. Department of Energy, *NEPA Lessons Learned* 18, 1 March 1999, 13.
3. Council on Environmental Quality (CEQ), *Guidance Regarding NEPA Regulations,* 48 FR 34263, 1983.
4. C.H. Eccleston, "Applying Value Engineering and Modern Assessment Tools in Managing NEPA: Improving Effectiveness of the NEPA Scoping and Planning Process," *Environmental Regulations and Permitting* 8, no. 2 (Winter 1998): 53–63.
5. CEQ, "Memorandum for General Counsels, NEPA Liaisons, and Participants in Scoping," Section II, Part A, no. 6, 30 April 1981; Also 46 FR 25461, "Notice of Availability of Memorandum to Agencies Containing Scoping Guidance," 7 May 1981.
6. Eccleston, *The NEPA Planning Process.*
7. 16 U.S.C. §470 et seq.
8. 36 CFR Part 800.
9. 64 FR 27044, 18 May 1999.
10. 42 U.S.C. §1996.
11. 25 U.S.C. §3001 et seq.
12. Executive Order 13112, *Invasive Species,* 3 February 1999; published in *Federal Register,* 64 FR 6183, 8 February 1999.
13. Eccleston, *The NEPA Planning Process,* section 11.3.
14. O.L. Schmidt, "The Statement of Underlying Need Determines the Range of Alternatives in an Environmental Document," The Scientific Challenges of NEPA: Future Directions Based on 20 Years of Experience, Session 13—The NEPA Process, Knoxville, Tennessee, 25–27 October 1989; also "The Statement of Underlying Need Defines the Range of Alternatives in Environmental Documents," 18 Environmental Law 371–81, 1988.
15. J.L. Lee, "The Power of Purpose and Need in Quality Planning: Three Case Studies," *Federal Facilities Environmental Journal* (Autumn 1997): 77–92.
16. *Webster's New Twentieth-Century Dictionary,* 2nd ed. (New York: Simon and Schuster, 1983).
17. *Webster's II New Riverside University Dictionary* (Boston: River Publishing, 1988).
18. O.L. Schmidt, personal communications (e-mail correspondence), 21 February 1999.
19. *Sierra Club v. U.S. Department of Transportation,* No 96 C 4768 (N.D. ILL. January 27, 1997).
20. *City of Carmel-by-the Sea v. United States Department of Transportation,* 95 F.3d 892 (9th Cir. 1996).
21. *Alaska v. Morrison,* 67 F.3d 723 (9th Cir. 1995).
22. CEQ, "Council on Environmental Quality—Forty Most Asked Questions Concerning CEQ's National Environmental Policy Act Regulations" (40 CFR 1500–1508), *Federal Register,* vol. 46, no. 55, 18026–18038, 23 March 1981, question 2a.

23. Ibid.

24. G.A. Miller, "The Magical Number Seven, Plus or Minus 2: Some Limits on Our Capacity for Processing Information," *Psychological Review* 63 (1956): 81–97.

25. Eccleston, *The NEPA Planning Process,* sections 8.5 and 11.2.

26. CEQ, "Forty Most Asked Questions," question 19a.

27. T.H. Lillie and H.E. Lindenhofen, "NEPA as a Tool for Reducing Risk to Programs and Program Managers," *Federal Facilities Environmental Journal* (Spring 1991).

28. *Fritiofson v. Alexander,* 772 F 2d 1225 (5th Cir. 1985).

29. L.R. Freeman, F. March, and J.W. Spensley, *NEPA Compliance Manual,* Government Institutes, 1992.

30. Eccleston, *The NEPA Planning Process,* section 12.1.

31. Ibid., section 8.6.

32. *Sierra Club v. U.S. Department of Agriculture,* 116 F.3d 1482 (7th Cir May 28, 1997).

33. Eccleston, *The NEPA Planning Process,* sections 11.4 and 12.4.

34. CEQ, *The National Environmental Policy Act—A Study of Its Effectiveness after Twenty-Five Years,* 1997.

35. C.H. Eccleston, "Continuous Spectrum Analysis: An Approach for Analyzing Alternatives and Impacts of Potential Projects," proceedings of the 18th annual conference on current and future priorities for environmental management, National Association of Environmental Professionals, Raleigh, North Carolina, 24–26 May 1993.

36. CEQ, *Memorandum: Guidance Regarding NEPA Regulations,* 48 FR 34263, 28 July 1983.

37. *California v. Block,* 690 f.2d 753 (9th Cir. 1982).

38. C.H. Eccleston, "Determining the Scope and Level of Detail Appropriate for a Programmatic Environmental Impact Statement," *Federal Facilities Environmental Journal* (Spring 1996): 59–69.

39. Eccleston, *The NEPA Planning Process.*

40. M.B. Gerrard, "Judicial Review of Scientific Evidence in Environmental Impact Statements," paper presented at the Ninth Oak Ridge National Laboratory Life Sciences Symposium, Knoxville, Tennessee, 1989; included as a chapter in *Environmental Analysis: The NEPA Experience,* eds. S.G. Hildebrand and J.B. Cannon, CRC Press, 1993).

41. Eccleston, *The NEPA Planning Process,* chapter 13.

42. CEQ, "Forty Most Asked Questions," question 4c.

43. Environmental Protection Agency, *Filing System Guidance for Implementing 1506.9 and 1506.10 of the CEQ Regulations, Federal Register* 54, no. 43, 7 March 1989.

44. DOE, *NEPA Lessons Learned,* 7.

45. *Sierra Club v. U.S. Department of Agriculture.*

46. DOE, *NEPA Lessons Learned,* 7.

3

The Environmental Impact Statement: Documentation Requirements

The essence of knowledge is, having it, to apply it; not having it, to confess your ignorance.

—Confucius

Be it war, commerce, or scientific analysis, the need to maintain a defensible posture is irrefutable. Perhaps, better than anyone, the legendary Chinese general Sun Tzu articulated this fact when he wrote:

> You can be sure of succeeding in your attacks if you only attack places that are undefended. You can ensure the safety of your defense if you only hold positions that cannot be attacked. That general is skillful in attack whose opponent does not know what to defend; and he is skillful in defense whose opponent does not know what to attack.[1]

The importance of performing a defensible Environmental Impact Statement (EIS) analysis cannot be overstated. As Sun Tzu would probably advise, adversaries seek to identify oversights and flaws that show a lack of adequate thought, planning, or adherence to regulatory requirements. As a defense, the astute EIS manager exercises vigilance in ensuring that all requirements are systematically identified and properly addressed. This chapter was prepared with this goal in mind.

Whereas chapters 1 and 2 addressed the question of how the EIS planning and analysis process is effectively implemented, this chapter focuses on providing the reader with a comprehensive description of the specific documentation requirements that the EIS must meet. Specifically, it centers on answering the question of what must be addressed in the EIS document. The reader is referred to the author's companion book, *The NEPA Planning Process: A Comprehensive Guide with Emphasis on Efficiency,* for additional details.[2]

The Council on Environmental Quality (CEQ) National Environmental Policy Act (NEPA) regulations (Regulations) specify strict documentation requirements that the EIS must meet. Unfortunately, these are scattered throughout the Regulations and across many guidance memoranda, making identifying and consolidating all pertinent documentation requirements a difficult task. Accordingly, a principal objective of this chapter is to integrate the relevant requirements, guidance, case law, and professionally accepted methods into a single, unified, and coherent source for use in preparing an EIS. Emphasis is placed on providing the reader with specific methods for reducing document size and compliance cost while rigorously complying with all regulatory requirements. For brevity's sake, this chapter summarizes many of these requirements; the reader is encouraged to read the actual regulatory provision for the details and precise wording.

Discussion of the documentation requirements is presented systematically, beginning with the notice of intent (NOI) in section 3.1. Section 3.2 introduces general direction and requirements governing preparation of the entire EIS document (e.g., page length guidance, requirements for disclosing opposing points of view). Section 3.2 sets the stage for an in-depth examination of the detailed EIS documentation requirements as presented in sections 3.3 through 3.11. Finally, section 3.12 presents the documentation requirements for preparing the record of decision (ROD). A comprehensive set of checklists for reviewing the adequacy of the EIS document is included in appendix A.

3.1 The Notice of Intent

As briefly described in chapter 1, an NOI, sometimes referred to as a *notice,* is prepared and issued prior to the public scoping phase. The NOI should be a short document, yet one NEPA practitioner reported reading one that was nearly 40 pages in length. The purpose of the NOI is to notify interested parties of the agency's intent to prepare an EIS. Chapter 5 describes certain procedures for effectively preparing and issuing the NOI.

The agency's proposed scoping process and any planned scoping meetings must be identified in the NOI. The name and address of an individual within the agency who may be contacted to answer questions about the proposal and the EIS process must be identified. Table 3.1 summarizes minimum documentation requirements that the notice must address (§1508.22).

To more effectively promote NEPA's public notification requirement, additional items not shown in Table 3.1 should also be considered; an expanded outline providing the public with a more comprehensive understanding of the proposal and the public scoping process is suggested in Table 3.2.

It is generally recommended that agencies avoid presenting an overly

Table 3.1 Items required to be addressed in the NOI

- Description of the proposed action and possible alternatives
- Description of the agency's proposed scoping process, including whether, when, and where scoping meeting(s) may be held
- Name and address of a person within the agency who can answer questions about the proposal and the EIS

detailed discussion of the actual proposal at this early stage, as it may substantially change during the scoping process. Instead, the agency should provide sufficient detail to inform the public on the nature and scope of potential actions. If the scoping process results in a significant change in either the proposal or its scope, the NOI must be revised and reissued (§1501.7[c]).

3.2 General Direction for Preparing the EIS

This section introduces general requirements governing preparation and content of the EIS. Table 3.3 presents direction and requirements that are generally applicable to the entire EIS. Direction summarized in Table 3.3 is elaborated in the following sections.

Differences Between the Draft and Final

An EIS is prepared in two stages. Consequently, there are differences, albeit minor ones, in the requirements for the *draft* and *final* statements. The

Table 3.2 Recommended outline for an NOI

- Identify any Cooperating agency(s)
- Brief description of the EIS process for unacquainted members of the public, including the purpose for publishing the NOI. Emphasize that no decision has been made and that the EIS will provide important input in reaching a final decision. Other discussion may include:
    - –The description of the agency's underlying purpose and need for taking action
    - –Background information, including historical context, of the proposal and why action is needed
- Brief description of proposal, including reasonable alternatives that may be evaluated
- Proposed schedule of the EIS
- Significant environmental issues and impacts that may be involved
- Description of the agency's proposed scoping process
- An invitation for the public to submit comments and attend scoping meetings
    - –Dates and locations of scoping hearings
    - –Location and availability of documents related to the proposal
    - –Name, address, and telephone number of a point-of-contact within the agency who can answer questions

Table 3.3 General requirements governing preparation of an EIS

- The EIS must provide "full and fair discussion of significant environmental impacts and shall inform decisionmakers and the public of the reasonable alternatives." (§1502.1)
- The draft EIS is expected to satisfy to the "fullest extent possible" requirements established for final EISs in Section 102(2)(C) of the Act. Moreover, the draft should be prepared in accordance with the scope determined during the scoping process. (§1502.9[a])
- An EIS is to be "clear," "to the point," and "written in plain language . . . so that decisionmakers and the public can readily understand" it. (§1500.2[b]; §1502.1; §1502.8)
- Every effort should be made to "disclose and discuss . . . all major points of view on the environmental impacts of the alternatives including the proposed action." (§1502.9[a])
- The EIS must "rigorously explore and objectively evaluate all reasonable alternatives." Where alternatives are eliminated from detailed study, the EIS must briefly explain why. (§1502.14[a])
- The EIS must encompass the "range of alternatives" that will be "considered by the ultimate agency decisionmaker." (§1502.2[e])
- The EIS must provide "the means of assessing the environmental impact of proposed agency actions, rather than justifying decisions already made." (§1502.2[g])
- Reasonable alternatives to proposed actions "that would avoid or minimize adverse impacts" must be objectively described. (§1500.2[e] and §1502.1)
- The EIS must "devote substantial treatment to each alternative considered in detail, including the proposed action, so that reviewers may evaluate their comparative merits." (§1502.14[b])

principal difference is that the final EIS includes comments received from circulation of the draft EIS and incorporates the agency's responses. Other minor differences in documentation requirements are described in appropriate sections. With respect to preparing the draft, the Regulations require that "to the fullest extent possible, the draft [EIS] must meet requirements established for final EISs" (§1502.9[a]).

Understandable Yet Rigorous

An EIS is to be "written in plain language," requiring writers "of clear prose" or editors to write, review, and edit the document. Statements should contain appropriate graphics so that decisionmakers and the public can readily understand the pertinent issues (§1502.8). Technical and scientific terms must be clearly defined and explained. For example, with respect to biological species, common names of organisms should be used in addition to the scientific names. However, scientific names should still be included to avoid ambiguity. To enhance readability, a glossary defining technical terms and a list of acronyms should be included.

Highly sophisticated methods are frequently employed in evaluating impacts. Analysts are habitually confronted with two opposing objectives. An

EIS is to be presented in a manner understandable by decisionmakers and the public, yet must provide an accurate, rigorous, scientific analysis of environmental impacts. Failure to comply with either of these competing objectives may provide sufficient grounds for successful litigation.

Using Plain Language

A presidential memorandum on plain language, dated June 1, 1998, expresses the objective of making the government more responsive and accessible in its communications with the public. The memorandum identifies helpful writing techniques and additional resources that may be found at the National Partnership for Reinventing Government's (NPR) Plain Language Action Network, located at http://208.204.35.97/ on the Internet.

TEXT. In implementing the objective of the presidential memorandum on plain language, common words should be used whenever feasible. Long sections of text should be broken up with informative headings and subheadings. Paragraphs should be kept concise and focused on one topic. It is suggested that short sentences be used. As illustrated in Table 3.4, the active voice is more informative and, therefore, preferable.

DEFINITIONS, ABBREVIATIONS, AND ACRONYMS. Technical terms should be defined. One useful technique involves highlighting definitions in boxes that are integrated with the text or provided in a list. Use of abbreviations and acronyms should be minimized. Although it is standard practice to spell abbreviated terms the first time they are used, it is burdensome to search for the first use of an unfamiliar abbreviation. This is why a list of abbreviations and acronyms should be provided. Some agencies also define an abbreviation or acronym each time it is used in a new section of the EIS.

USING GRAPHIC AIDS TO INCREASE COMPREHENSION. Graphic aids can greatly enhance the reader's comprehension of textual descriptions. Maps, tables, graphs, figures, and flowcharts can all enhance the effectiveness of the text. A programmatic EIS, for example, could contain a map showing the agency's pertinent facilities and sites and indicating interrelationships within the program.

ALTERNATIVES. Tables and graphs may be highly effective in comparing or summarizing other information. For example, a well-designed table or figure demonstrating how impacts are related to each of the alternatives can pro-

Table 3.4 Example of passive versus active voice

Passive voice does not identify the actor: "An air monitoring system was installed in 1993."
Active voice makes the actor clear: "The city installed an air monitoring system in 1993."

vide an excellent means of summarizing differences among alternatives. Remember that a graph normally shows an independent variable on the *x*-axis and dependent variable(s) on the *y*-axis. All tables and figures should have informative titles and headings. Where the alternatives are particularly complex (for example, multiple alternatives with multiple assumptions about a key variable), tables and other graphics may be particularly useful in identifying the particular case (e.g., alternative and issue) under discussion (see Figure 3.1).

The U.S. Printing and Binding Regulations recognize that color printing and graphics may add value, especially in increasing the clarity of maps and technical diagrams.[3] However, color printing often adds to the cost and complexity of printing; these regulations state that color printing must serve the end purpose of the document. Remember that if the document is photocopied, the color is normally lost.

UNITS OF MEASUREMENT. Consistent measurement units need to be used throughout the EIS. For example, if the cubic meter is chosen for expressing chemical volumes, the EIS should not also present some volume values in drums.

Metric units are sometimes followed by English units in parentheses (e.g., a 300-meter [984-foot] perimeter). A table of conversion factors should be included.

If scientific notation is used, it should be applied consistently throughout the text and tables (i.e., small numbers should not be expressed differently than larger numbers). For example, it is difficult to compare 0.001 and 5×10^{-4}.

A Full and Fair Discussion

General William Westmoreland, U.S. commander during the Vietnam War, once quipped, "Without censorship, things can get terribly confused in the public's mind." With respect to NEPA, censorship may be to a federal project what Vietnam was to General Westmoreland. Few mistakes are more likely to leave an agency open to challenge, or to breed public mistrust, discontent, and ill will than failure to fully and fairly investigate a controversial issue or concern.

An EIS must provide a "full and fair discussion of significant environmental impacts and . . . inform decisionmakers and the public of the reasonable alter-

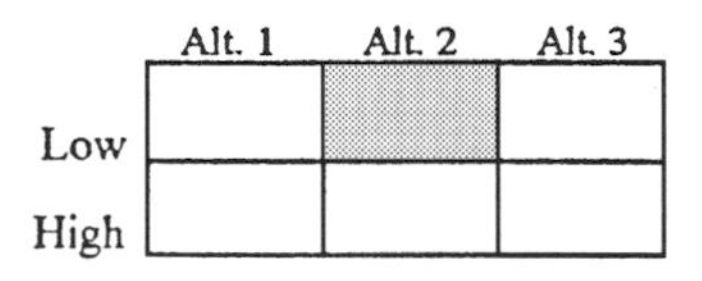

In this example, an EIS analyzes three alternatives, where values of "high" and "low" are assigned to a key environmental variable such as disruption to surrounding aesthetics. As each analytical case is presented (alternatives 1–3), a specific cell is shaded to indicate the analysis under discussion.

Figure 3.1 Example of using an icon as a visual aid

natives which would avoid or minimize adverse impacts or enhance the quality of the human environment" (§1502.1). Care must be exercised to ensure that the EIS properly encompasses the range of reasonable alternatives and not simply those that the agency favors. Agencies are generally much more vulnerable to challenge for failing to address an issue than for addressing it incorrectly.

Disclosing Opposing Points of View

As we have seen, agencies must make a concerted effort to "disclose and discuss . . . all major points of view" in the draft EIS (§1502.9[a]). Moreover, the final EIS must disclose "any responsible opposing view which was not adequately discussed in the draft statement." The agency's response to opposing views must be included in the final EIS (§1502.9[b]).

"Would" Versus "Will"

Every effort should be taken to avoid even the appearance of partiality. To this end, discussions of potential actions should be written using the conditional tense, as if the action *might* take place, to clearly delineate that no decision has yet been made. For example, this statement is inappropriate: "The action *will* affect an area of 5,000 square meters." A more impartial statement would read, "The *proposed* action *would* affect an area of 5,000 square meters." The words *proposed* and *would* clearly indicate to the reader that a final decision has not been made.

Dealing with Incomplete and Unavailable Information

James Matthew Barrie, author of Peter *Pan,* admitted, "I am not young enough to know everything." Because most EIS practitioners are beyond the age of knowing everything, a great deal of effort is normally devoted to obtaining data for the analysis. The Regulations require agencies to make a diligent effort to obtain data and evaluate impacts of potential actions. Notwithstanding, there are reasonable limitations on the cost and resources that an agency may expend on an analysis. In practice, cost constraints and limitations in the state of the art often prevent analysts from providing decision makers with a complete picture of environmental consequences.

The Regulations acknowledge such limitations by providing for circumstances that involve "incomplete" or "unavailable" information (§1502.22).[4] If the incomplete information concerning reasonably foreseeable significant adverse impacts is essential for making an informed choice among alternatives, and the overall costs are not "exorbitant," the information must be obtained and included in the EIS (§1502.22). If the information relevant to reasonably foreseeable significant adverse impacts cannot be obtained because (1) the overall costs of obtaining it are exorbitant, or (2) the means to obtain it are not known, the EIS must include:

1. A statement that such information is incomplete or unavailable
2. A statement of the relevance of the incomplete or unavailable information to evaluating reasonably foreseeable significant adverse impacts on the human environment
3. A summary of existing credible scientific evidence that is relevant to evaluating the reasonably foreseeable significant adverse impact on the human environment
4. An evaluation of such impacts based on theoretical approaches on research methods generally accepted in the scientific community. For the purposes of this requirement, "reasonably foreseeable" includes impacts that have catastrophic consequences, even if their probability of occurrence is low, provided that the analysis of the impacts (1) is supported by credible scientific evidence, (2) is not based on pure conjecture, and (3) within the rule of reason.

For more information on this requirement, the reader is referred to *The NEPA Planning Process*.[5]

Reducing Document Size

A substantial portion of the Regulations is devoted to streamlining preparation of an EIS. Experience indicates that much of the regulatory direction is either overlooked or, in some cases, blatantly disregarded. Table 3.5 provides specific direction that can be used in substantially reducing the size, duration, and level of effort expended in preparing the EIS.

As indicated in the first bullet of Table 3.5, the term *normally* was purposely chosen because the CEQ understood that agencies must be given a degree of latitude in responding to unusual circumstances.[6] With respect to this provision, the term *text* is considered to include the sections starting with "Purpose and Need" and continuing through the section on "Environmental Consequences" (see items 4–7, Table 3.10). One method for reducing the size of the main body of an EIS is to move material of less importance to the appendixes. While this may not result in a savings in effort, it will at least allow the reader to focus on material that is truly important.

The second bullet of Table 3.5 indicates that agencies should emphasize preparation of analytic analyses over encyclopedic ones. Adherence to this

Table 3.5 Specific direction for constraining the size of an EIS

- The text of a final EIS should normally be less than 150 pages in length. Proposals of "unusual scope or complexity" should "normally be less than 300 pages." (§1502.7)
- The draft is to be "analytic rather that encyclopedic," emphasizing material "useful to decision-makers and the public" while "reducing emphasis on background material." (§1500.4[f]) "Most important," the EIS "must concentrate on the issues that are truly significant to the action in question, rather than amassing needless detail." (§1500.1[b])
- An EIS "shall be kept concise and shall be no longer than absolutely necessary to comply with NEPA and with these regulations. Length should vary first with potential environmental problems and then with project size. (§1502.2[c])
- Efforts should focus on describing "significant environmental issues and alternatives" while reducing the "accumulation of extraneous background data" and "needless detail." (§1500.1[b], §1500.2[b], §1502.1)
- With respect to nonsignificant issues, only enough discussion should be presented to "demonstrate why more study is not warranted." (§1502.2[b])
- To reduce paperwork, agencies may, if changes are minor, attach and "circulate only the changes to a draft EIS rather than rewriting and circulating the entire statement." (§1500.4[m])

direction can substantially reduce EIS size because quantitative data can often be presented concisely while providing decision makers with a rational foundation on which to base decisions.

CEQ Page Limitations

Table 3.6 summarizes CEQ direction regarding page limits for various sections of an EIS. Note that no direction is propagated for more than half of the sections in a standard EIS outline. In practice, EISs commonly exceed CEQ's page limitations, occasionally running to several thousand pages. Analysts should strive, however, to comply with this direction, deviating only when necessary.

Documenting Assumptions

Uncertainty is an inherent aspect of predicting the future. With respect to preparing the EIS, uncertainties are most commonly dealt with by making reasonable assumptions. The balance between success and failure in predicting future outcomes often pivots on the ability to make rational and defensible assumptions. The EIS should clearly identify uncertainties and the assumptions used to bridge them. The credibility of an analysis hinges on the ability to substantiate such assumptions. Accordingly, the basis or rationale for each assumption should be clearly documented.

Table 3.6 Summary of CEQ direction on EIS page limits

Section of the EIS	Page Limit	Reference
Cover sheet	Not to exceed 1 page	§1502.11
Summary	Not to exceed 15 pages	§1502.12
Table of Contents	No direction	Not applicable
Purpose and Need	"briefly specify"	§1502.13
Alternatives section	No direction	Not applicable
Affected environment section	"succinctly describe the environment of the area(s)" Considerably smaller than the section on the environmental consequences	§1502.15 "Talking points on CEQ's Oversight of Agency compliance with the NEPA Regulations" (CEQ, 1980)
Environmental consequences section	No direction	Not applicable
List of preparers	"not to exceed 2 pages." "A line or two for each person's qualification"	§1502.17; CEQ's "Forty Most Asked Questions," Question 27c
List of agencies consulted	No direction	Not applicable
Index	No direction	Not applicable
Appendixes	No direction	Not applicable
Total text of final EIS	Normally less than 150 pages;less than 300 pages for proposals of unusual scope or complexity	§1502.7

Direction on Documenting the Analysis

This section presents general direction and requirements for performing the EIS analysis. As depicted in Table 3.7, analysts must conduct a rigorous and evaluation, which provides the decisionmaker with a full, fair, and objective treatment of all reasonable alternatives.

While trivial or nonsignificant issues should not be discussed in detail, neither should they be totally ignored (see last item, Table 3.7). Thus, the analysis should only briefly discuss impacts that were considered potentially significant but on closer examination were found to be nonsignificant. The idea is to clearly demonstrate that potentially significant impacts were indeed considered and found to be nonsignificant, and not simply overlooked or casually dismissed.

Table 3.7 General direction for preparing the EIS analysis

- "Information must be of high quality. Accurate scientific analysis . . . [is] essential to implementing NEPA." (§1500.1[b])
- An EIS is to be "analytic rather than encyclopedic." (§1500.4[b]; 1502.2[a])
- The EIS must "rigorously explore and objectively evaluate all reasonable alternatives." (§1502.14[a])
- Where alternatives are eliminated from detailed study, the EIS "must briefly explain the reasons for their having been eliminated." (§1502.14[a])
- The EIS must include "evidence that agencies have made the necessary environmental analyses." (§1500.2[b]; 1502.1])
- "Impacts shall be discussed in proportion to their significance. There shall be only a brief discussion of other than significant issues." With respect to nonsignificant issues, "there should be only enough discussion to show why more study is not warranted." (§1502.2[b])

The courts grant agencies a wide degree of latitude where there is an honest difference of opinion among experts. For example, the courts have repeatedly upheld statements such as "In our best professional opinion. . . ." The agency, of course, is responsible for ensuring that technically competent professionals have indeed reviewed the issue and firmly believe that such a statement is accurate.

Some agencies adopt a sliding-scale approach to preparing an EIS. This recognizes that the standard an EIS is expected to meet varies with the specific circumstances; thus, the amount of attention devoted to a specific issue or impact varies with the significance of the effect.[7]

Quantifying the Analysis

An EIS analysis must go beyond the obvious in explaining the significance of potential impacts. Consistent with a sliding-scale approach, the need to quantify an impact varies with its potential significance. As a rule, potentially significant impacts should be quantified wherever practical. Not only does a quantified analysis normally make for a smaller, more compact document, it also tends to provide information that is more useful in making a reasoned choice between alternatives. If it is not possible to quantify important issues, an explanation should be included explaining why.

INTENSITY AND DURATION. Where practical, both the intensity and duration of impacts should be quantified. For example, rather than stating that "sulfur dioxide emissions from the proposed action would be small and of short duration," the analysis might better indicate that "sulfur dioxide emissions from the proposed action would be less than 0.1 ton per month over a period of approximately 14 months."

Where practical, analysts should avoid describing impacts in terms of relative measurements. For instance, statements such as "radioactive emissions would increase by 7 percent" should be avoided when absolute metrics can be provided. Such a statement fails to provide the decisionmaker and the public with an absolute measure of the environmental consequences; the example statement fails to provide a direct means of determining actual environmental and health effects.

Comparison with Regulatory Standards

It is not uncommon to encounter statements in an EIS declaring that a particular discharge or activity "would be conducted in accordance with all applicable regulatory requirements," as the sole or primary source of evidence that an action would be innocuous. Such statements may provide supporting evidence, but they should not be relied upon as proof of nonsignificance. It is important to note that an action may comply with all applicable laws and regulations and still result in an impact, perhaps even one that is deemed to significantly affect environmental quality; conformance with laws and regulatory standards does not necessarily ensure that an action would not affect an environmental resource.

Witness, for example, the construction of a hydroelectric dam or nuclear reactor. Both of these actions may be permitted, constructed, and operated in accordance with all applicable laws and requirements, yet, no reasonable expert would argue that such projects do not, at least potentially, pose a significant threat to environmental quality.

Addressing Economic and Cost Considerations

In explaining why the company charged the Air Force nearly $1,000 for a pair of pliers, a spokesperson for a major defense contractor said, "They're multipurpose. Not only do they put the clips on, but they take them off." Fortunately, EIS practitioners generally provide the public with a more rational explanation of a project's cost versus derived benefit. The EIS should identify those considerations, including factors not related to environmental quality, that are likely to be relevant and important to a decision. Selected references related to incorporation of economic considerations, including preparation of a cost-benefit analysis, are provided in Table 3.8.

COST-BENEFIT ANALYSIS. As indicated in Table 3.8, the Regulations encourage incorporation of economic and other decisionmaking factors into the EIS analysis. A cost-benefit analysis may be incorporated by reference or appended to the EIS to support the evaluation of environmental conse-

Table 3.8 Selected references pertaining to the analysis and consideration of economic factors

- The analysis may incorporate use of a "cost-benefit analysis." (§1502.23)
- The analysis must "identify environmental effects and values in adequate detail so they can be compared to economic and technical analyses." (§1501.2[b])
- An agency may discuss preferences among alternatives based on relevant factors including economic and technical considerations and agency statutory missions. (§1505.2[b])
- Reasonable alternatives "include those that are practical or feasible from the technical and economic standpoint and using common sense, rather than simply desirable from the standpoint of the applicant."[8]
- The "agency's preferred alternative" is the one that the agency believes would fulfill its statutory mission and responsibilities, giving consideration to economic, environmental, technical and other factors.[9]
- The analysis of impacts includes "economic" effects. (§1508.8[b])

quences. If a cost-benefit analysis is included, the EIS must discuss the relationship between it and related analyses of unquantified environmental impacts, values, and amenities (§1502.23). A discussion and comparison of the merits and drawbacks of various alternatives need not be limited to a monetary cost-benefit analysis—and should not be, when there are important qualitative considerations.

COURT DIRECTION ON PERFORMING COST-BENEFIT ANALYSES. Prudence should be exercised in ensuring that costs versus derived benefits are objectively compared and evaluated. Witness a case where the Natural Resources Conservation Service was challenged for preparing an inadequate EIS, which was used by the U.S. Army Corps of Engineers in issuing a permit to construct a dam that could affect a number of species, two of which were endangered. Here, the court found that overly inflated estimates of economic benefits had been used in the analysis.[10] Specifically, the EIS cited gross rather than net economic benefits that would be derived from constructing the dam. The court concluded that the overestimated economic benefits, considered "crucial" in reaching the final decision, had impaired the ability to balance benefits against environmental damage.

How Much Is Enough? The Sufficiency Question

No two decisionmakers will likely agree completely on the amount of discussion necessary to provide coverage sufficient to allow the agency to make a decision. This observation is borne out, perhaps humorously, by what has become known as Cohn's Law:

> The more time you spend documenting what you do, the less time you have to do what you do. Equilibrium is reached when you do nothing but it's fully documented.

Consistent with Cohn's Law, the outcome of a NEPA review is often problematic; if two decisionmakers agree that a particular discussion does not provide adequate coverage, they may still disagree on the degree of analysis and discussion needed. Proponents of a project may believe that a NEPA analysis provides sufficient coverage for a proposed project, while critics may argue that the analysis needs to be taken to increasingly detailed levels. I call this the *sufficiency question:* How much information is enough?

Some aspects of a proposal may be considered sufficiently covered by merely mentioning that the specific action will take place. For others, an extensive analysis may be considered necessary. Since NEPA's inception, no definitive direction has been established for determining the amount of detail, discussion, and analysis that adequately covers a proposed action, yet agency decisionmakers are routinely called upon to do just that. Inevitably, such determinations are subjective.

Lacking definitive guidance, decisionmakers and critics may point to a universe of potential factors that can be used to defend a position that an action is or is not adequately covered. Assertions are often based on ambiguous opinions that can be neither proved nor disproved. While common sense is integral to decisionmaking, definitive guidance would greatly reduce the degree of ambiguity and subjectivity that currently exists. The author has developed a rigorous tool, referred to as the *Sufficiency-Test Tool,* for resolving this problem, which is described in the following sections. The reader is referred to the original publication for additional details on its use.[11]

The Sufficiency Test: A Tool for Determining When an Analysis Is Sufficient

As shown in Table 3.9, four criteria (or tests) are identified for resolving the sufficiency question. This tool is based on the premise that no useful purpose is served in providing detail if the information does not in some way contribute to the decisionmaking process. Each test also reflects specific provisions mandated in the Regulations. A flowchart describing the use of this tool is provided in Figure 3.2.

APPLYING THE SUFFICIENCY-TEST TOOL. Application of the Sufficiency-Test Tool begins by assessing a topic or issue that either is or will be discussed by asking, with respect each of the four criteria, "Would a more detailed analysis. . . ." A *no* answer to all four tests supports a decision that the issue or analysis is sufficiently covered.

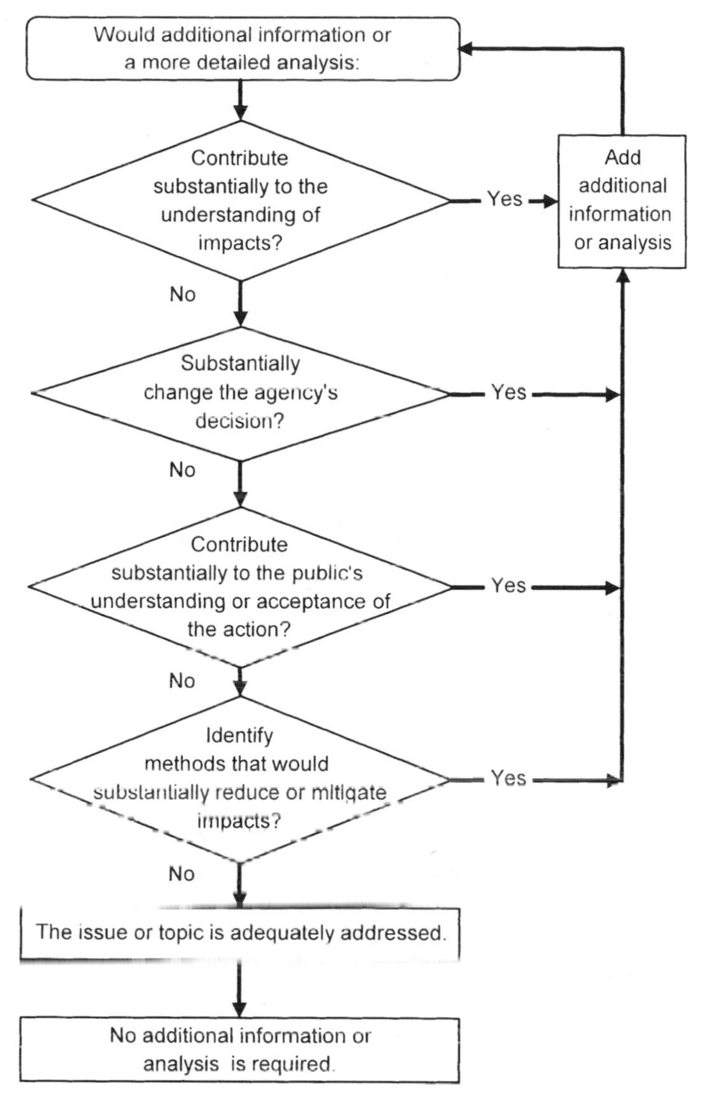

Figure 3.2 Sufficiency-Test Tool. The tool provides a systematic methodology for determining if the discussion of a topic or issue is sufficient.

A *yes* answer to any test supports a decision that more detailed discussion or analysis is warranted. Once the additional analysis is included, the process is repeated; that is, a review is again conducted (using this tool) to determine if the analysis is sufficient. This process is repeated until a *yes* response is no longer invoked.

The Sufficiency-Test Tool can be applied to an array of NEPA documentation requirements (EA and EIS). Specifically, it can be used in determining sufficient discussion of the proposed action and alternatives, analysis of environmental impacts, discussion of affected environment, and the appropriateness of introductory and background material. In all cases, however, the tool should be restricted to determining issues concerning "How much information is enough?"

This tool can be employed at various steps in the life cycle of the NEPA process. While preparation of a document is underway, for example, the tool may be used by analysts to determine whether or not existing discussion of a specific issue or topic is sufficient. If more discussion is deemed necessary, the tool can assist analysts in determining how much additional analysis is needed. Once the document is prepared, the tool can be used by reviewers in determining if particular topics or issues are adequately covered. During the comment incorporation process, the tool can assist analysts in determining a comment's relevance as well as the amount of discussion necessary. Moreover, it can assist decisionmakers in determining whether the amount of detail, discussion, and analysis is sufficient to support a final decision to pursue a proposed action; applied properly, it may help decisionmakers avoid the pitfalls of Cohn's Law. Finally, if the agency is challenged, the tool, as part of the administrative record (ADREC), provides evidence useful in demonstrating that the agency's analysis was sufficiently rigorous.

Table 3.9 Tests for determining if the discussion and analysis is sufficient

Would a more detailed analysis . . .

- *Contribute substantially to the understanding of environmental impacts?* A more detailed analysis is justified if it provides information that could substantially improve the agency's understanding of the impacts. (40 CFR 1500.1(c), 1502.1)
- *Substantially change the agency's decision?* Specifically, would the discussion allow the agency or public "to concentrate on the issues that are truly significant to the action in question," or would it simply amount to an exercise in "amassing needless detail"? A more detailed analysis is justified if, based on environmental factors, it could substantially affect the agency's decision. (40 CFR 1500.1(c), 1500.4(c), (f), and (l))
- *Contribute substantially to the public's understanding or acceptance or rejection of the proposed action or alternatives?* NEPA is a public process; a more detailed analysis is warranted if it could substantially affect the public's assessment (positive or negative) of the action or could substantially affect the public's ability to provide comments or input into the decision-making process. (40 CFR 1500.2(b) and (d), 1500.4(f), 1501.4(b), 1502.1, 1500.1(b))
- *Result in new measures that could substantially reduce the impacts associated with the project?* A more detailed analysis is justified if it may result in identifying mitigation measures that could substantially reduce adverse impacts. (40 CFR 1500.2(f), 1502.1, 1502.14(f), 1500.1(c))

CEQ's Recommended EIS Format

The final EIS should clearly demonstrate that all requirements are met, all significant environmental issues identified, and all important issues addressed. Rigorous adherence to such direction facilitates preparation of a defensible EIS while providing the decisionmaker with valuable information that can contribute significantly to decisionmaking.

Table 3.10 lists CEQ's recommended EIS format (§1502.10). As shown in this table, the standard EIS outline contains 11 sections, beginning with a cover sheet and ending with appendixes, which are optional. This outline should be followed unless there is a compelling reason to deviate. While other formats may be used, the EIS must, at a minimum, include sections 1, 2, 3, 8, 9, and 10 and the essence of the requirements depicted by sections 4, 5, 6, 7 and 11. Although it is generally best to minimize deviations from CEQ's recommended format, the most important consideration is that the agency satisfy its substantive content. Direction and requirements for preparing each of the sections shown in Table 3.10 are described in the following sections.

3.3 The Cover Sheet

As indicated in Table 3.10, the EIS must contain a cover sheet, not to exceed one page in length. Table 3.11 lists the specific items that the cover sheet must contain (§1502.11). A wrong title can misrepresent a proposal, confusing the public and possibly fueling controversy; care should be exercised in tailoring the title so that it accurately conveys the nature and scope of the proposal.

3.4 Summary and Table of Contents

As described in the following sections, an EIS must contain both a summary and a table of contents.

Table 3.10 CEQ's recommended EIS format

1. Cover sheet
2. Summary
3. Table of contents
4. Purpose of and need for the proposed action
5. Alternatives, including the proposed action
6. Affected environment
7. Environmental consequences
8. List of preparers
9. List of agencies, organizations, and persons to whom copies of the EIS are sent
10. Index
11. Appendixes (if any)

Table 3.11 Requirements for preparing the EIS cover sheet

1. List of the responsible agencies, including the lead agency and cooperating agencies
2. Title of the proposed action, including the state(s) and county(ies) in which the proposal would be located
3. Name, address, and telephone number of an agency contact who can supply additional information
4. Designation of the statement as either a draft, final, or supplemental EIS
5. One-paragraph abstract of the EIS
6. Date by which comments must be received

Summary

As specified in the Regulations, each EIS "shall contain a summary which *adequately* and *accurately* summarizes the statement" (§1502.12, emphasis added). To adequately represent the EIS, the summary must contain key information from each of the principal elements that make up the statement: purpose and need, affected environment, alternatives, the significant environmental issues and impacts that were analyzed, and the results. The summary can be viewed as a condensed version of the EIS, where each section is written in proportion to its importance. To accurately portray the EIS, the summary must not introduce ideas, information, or conclusions that are not discussed in the body of the statement.

The EIS summary provides a succinct mechanism for informing agency officials and the public about potential actions and their resulting environmental impacts. For many readers, the summary forms their first and lasting impression of the proposal. Because it allows the reader to quickly focus on issues of greatest concern, a skillfully composed summary is essential to the success of a planning document, especially as it is often the only section actually read. With respect to decisionmaking, the summary allows the reader to quickly assess and balance environmental implications of the decision against technical, economic, and other factors. Thus, the summary bears a great obligation to clearly communicate the essence of the EIS to the reader.

The following sections describe how to prepare a summary, required contents, how to use the summary to increase efficiency by reducing paperwork, and recommendations for writing an effective summary. Although this guidance specifically addresses the EIS summary, principles of good expository writing apply equally to the body of the EIS.

Preparing the Summary

The summary should be prepared as a standalone document; that is, technical terms and analyses need to be defined and described so that they can be readily understood by the average reader without referring to the main body of the EIS. Information should be extracted from the body of the EIS and packaged into a coherent narrative. This does not simply involve copying and pasting paragraphs from the main document; its preparation normally involves sifting through a large volume of material in an effort to identify that succinct information that will capture an important topic of interest.

The EIS manager may find it advantageous to assign the specialists involved in preparing each principal section of the EIS responsibility for preparing the corresponding section of the summary. It is essential to have a professional writer or editor play a key role in the coordination and editing of the summary. Allowing members unfamiliar with the subject matter to critique the summary provides a check on how well information is summarized for persons unfamiliar with the proposal—precisely the type of individual who is most likely to read the final summary.

REDUCING PAPERWORK. The Regulations identify cases where the summary may provide an appropriate vehicle for saving resources by reducing paperwork. Specifically, agencies are instructed to reduce excessive paperwork by "summarizing the environmental impact statement and circulating the summary instead of the entire environmental impact statement if the latter is unusually long" (§1500.4[h]). Important caveats on the use of this procedure are provided in §1502.19.

DRAFT VERSUS FINAL EIS. As necessary, the summary of the final EIS should be modified to describe significant or controversial comments. The agency's responses to such comments should also be indicated. The final summary should also identify any major changes to the draft EIS.

Contents of a Summary

The summary should sharply define the differences among environmental consequences of the analyzed alternatives. It should also clearly underscore the environmental implications of alternatives, controversies, and significant unresolved issues. The summary should indicate the decision that needs to be considered and eventually made, and the relationship of this decision with the purpose and need for taking action. An effective summary accurately emphasizes issues and impacts of greatest concern to the public and the decisionmaker. The significance of potential impacts must be clearly and accurately conveyed to the reader.

It is essential that the summary is informative, concise, and readable. Three elements must be emphasized in preparing the EIS summary—specifically, "The summary shall stress the *major conclusions, areas of controversy* (including issues raised by agencies and the public), and the *issues to be resolved* (including the choice among alternatives)" (Section 1502.12 emphasis added).

- *Major conclusions:* Emphasizes the environmental implications of choosing among alternatives. As the analysis and comparison of alternatives is the "heart" of an EIS, it must also be the focus of the summary.
- *Areas of controversy:* Clearly delineates areas of controversy. Describing controversial issues is important in the event of later EIS litigation. Acknowledging areas of controversy provides the agency with a means of demonstrating that it considered all relevant information, including views contrary to its position.
- *Issues to be resolved:* Identifies unresolved issues, which may include scientific and technical uncertainties, particularly those that may need to be resolved in lower-tier or supplemental EIS.

The summary should normally not exceed 15 pages. While a summary for a complex or programmatic EIS may exceed this target, it must nevertheless be concise. Note that brevity is not tantamount to conciseness. A brief summary is simply short, while a concise summary contains the information deemed essential to the decisionmaker and the public and does not include superfluous material or needless detail.

Suggestions for Enhancing Effectiveness of the Summary

The summary briefly explains the who, what, where, when, why, and how the proposed action or alternatives would be implemented. A question and answer format is sometimes used and may be particularly engaging. Here are proven methods for enhancing effectiveness of the summary:

- Provide a brief description of the intent and elements of the EIS process so that an unfamiliar reader understands why the statement has been prepared, what it will be used for, and how to participate in the process.
- To more succinctly emphasize impacts that are truly significant (e.g., changes in health and environmental resources), do not describe intermediate steps in the causal chain of events (e.g., changes in the amount of pollutants released).
- Exercise discretion in discussing trivial impacts, as such discussion often tends to blur or obscure the pivotal decisionmaking issues. The objective is to convey both the absolute and the relative importance

of each impact. For example, if all alternatives involve small or non-significant impacts of a particular type, it means little that the impact for one alternative is twice that of another; all of the alternatives have essentially no impact on that resource.

- Employ graphics and tables to summarize voluminous or important data. The summary might also contain one or more maps illustrating the location of the proposal, showing relevant political boundaries, roads, nearby facilities, population centers, surface waters, and other significant geographic features.

Table of Contents

The Regulations do not provide specific requirements for preparing the table of contents. The level of detail suitable for this table is left to the discretion of the EIS team. Various nomenclature systems for organizing the EIS are in common usage. Many agencies use the scientific system of headings (i.e., 1.0, 1.1, 1.1.1). In recent years, the trend has been away from a scientific nomenclature system toward use of different fonts and type styles in organizing headings and subheadings. Still other systems are hybrids that combine the two approaches.

3.5 Purpose and Need Section

The section on purpose and need briefly explains the "*underlying* purpose and need to which the agency is responding in proposing the alternatives including the proposed action" (§1502.13, emphasis added). The importance of crafting an accurate and concise description of the underlying need is often not fully appreciated, which can lead to serious problems. A need that is vague or inaccurately specified may be difficult to publicly defend. Defined correctly, the statement of purpose and need provides a rationale for distinguishing reasonable alternatives from those that are not, providing the agency with a defensible basis for dismissing courses of action. For additional information, the reader is referred to chapters 1 and 2, where this subject is discussed in more detail.

A recurring problem has involved mistakenly discussing the purpose and need for preparing the EIS document rather than that of the proposal.[12] For example, some EIS have mistakenly explained that the purpose and need is to comply with NEPA. Instead, the EIS must explain the purpose of and need for taking action. Although determining the underlying purpose and need may appear to be simple, it can be quite complicated. To help determine the purpose and need for a proposed action, ask, "Why are we considering the proposed action?"

3.6 Proposed Action and Alternatives Section

Agencies are not required to choose an alternative based solely on environmental considerations, nor is there a substantive requirement to mitigate environmental impacts. The EIS, however, must present the decisionmaker with information that will assist him in making an informed decision. Great importance is placed on flushing out alternative courses of action. For this reason, the centerpiece of the EIS is the section describing the proposed action and reasonable alternatives (alternatives section)—so much so that this section is described as the "heart" of the EIS (§1502.14). As used in this context, the terms *proposal* and *alternatives* should be interpreted as denoting the agency's proposed action (if one is defined) and the range of reasonable alternatives.

Alternatives versus Environmental Consequences

Confusion sometimes arises over the difference between the alternatives section and the section on environmental consequences (environmental consequences section). The alternatives section provides the actual description of each alternative. In contrast, the environmental consequences section provides the analysis of each alternative. Thus, the alternatives section provides information used by analysts evaluating impacts in the environmental consequences section.

To avoid duplication, the alternatives section should concentrate on describing and comparing the alternatives. Specific actions that would be conducted as part of each alternative should be clearly identified. This section summarizes but should not replicate the analysis presented in the environmental consequences section. The alternatives section allows the reader to compare options that are evaluated in detail in the environmental consequences section.[13]

Range of Alternatives

As described earlier, analysts must identify and flush out alternative approaches for meeting the purpose and need. The EIS should explain why the agency believes that the range of analyzed alternatives covers the full spectrum of reasonable alternatives. It is important to note that a reasonable alternative must be considered and, where appropriate, evaluated even if it lies outside the legal jurisdiction of the agency (§1502.14).[14] If such an alternative is desirable, the agency should clearly explain why it cannot be selected and why the law should be changed so that the agency may pursue it. Failure to adequately analyze alternatives outside the agency's jurisdiction has been a major problem in EISs.[15]

Dismissing Alternatives

The EIS must identify and briefly explain why potential alternatives were dismissed from detailed study as unreasonable (§1502.14[a]). Alternatives that were considered and then rejected are often placed under a section labeled "Alternatives Considered But Not Carried Forward." Discussion of such alternatives should be kept to a minimum—no more than is necessary to give the reader an adequate understanding of the alternative and why it was found unreasonable from the standpoint of economic, technical, or other considerations.[16]

Requirements for Describing Alternatives

Table 3.12 summarizes the principal regulatory requirements for describing the alternatives (§1502.14). Some of these requirements are explained in more detail in subsequent sections.

Comparing Alternatives

As specified in the Regulations, the alternatives section is to:

- Present the environmental impacts of the proposal and the alternatives in comparative form, thus sharply defining the issues and providing a clear basis for choice among options by the decisionmaker and the public (§1502.14).
- Devote substantial treatment to each alternative considered in detail . . . so that reviewers may evaluate their comparative merits (§1502.14[b]).

A crisp comparison of alternatives is among the most important functions of an EIS, yet, describing and comparing alternatives and complex scien-

Table 3.12 Regulatory requirements for describing alternatives

- Rigorously explore and objectively evaluate all reasonable alternatives. For alternatives eliminated from detailed study, briefly discuss the reasons for their elimination.
- Devote substantial treatment to each alternative considered in detail, including the proposed action, so that reviewers may evaluate their comparative merits.
- Include reasonable alternatives not within the jurisdiction of the lead agency.
- Include the alternative of no action.
- Identify the agency's preferred alternative(s), if it exist, at the draft EIS stage. Identify this alternative(s) in the final statement unless another law prohibits the expression of the preference.
- Include appropriate mitigation measures not already included in the proposed action or alternatives.

tific issues in a manner that can be clearly comprehended by nonspecialists can be arduous. Each of the alternatives must be described in detail sufficient to allow a reasonable comparison. Alternatives are to be compared with each other, not just with the no-action alternative.

As indicated by §1502.14[b], the analysis must devote "substantial treatment" to each of the reasonable alternatives.[17] The term *substantial treatment* was chosen in lieu of *equal treatment* because the degree of attention devoted to a particular alternative varies with the complexity of the alternative and the extent of its impacts. Unless technically justified, the level of consideration given to the proposed action should not differ substantially from that devoted to the other reasonable alternatives. Agencies are further instructed to identify environmental effects and values in adequate detail so they can be compared to economic and technical analyses (§1501.2[b]).

DISCRIMINATING BETWEEN ALTERNATIVES. The alternatives section draws on the scientific discussion presented in the environmental consequences section and is written so the reader can draw a sharp distinction among the impacts of each alternative (§1502.14 and §1502.16). The requirement to present environmental impacts in comparative form is considered so important that some agencies have placed this comparison in an individual chapter of the EIS.

A matrix of alternatives versus impacts and other important considerations provides a useful tool for illuminating the differences among alternatives. Table 3.13 depicts a simplified example of how such a matrix may be used.

Table 3.13 Example of how a matrix can be used to compare important characteristics and impacts among alternatives

Environmental Consequences	No Action	Proposed Action	Alternative A	Alternative B
Potentially affected population	8,000	1,350	1,750	6,550
Total land disturbed (acres)	160	600	340	1,120
Habitat disturbed (acres)	85	420	380	620
Water consumption (gal/day)	9,000	195,000	120,000	80,000
Peak noise level (dBA)	51	44	90	43
Number of deer killed	50	150	40	275
Miles of road built	1.1	1.1	1.8	2.7
Annual NO_2 impact (ug/m^3)	1.4	.11	.09	.8
Annual CO impact (ug/m^3)	44.6	12.6	5.2	18.9
Increase in respiratory illnesses over 10 years	180	65	30	110

Rigorous and Objective Alternatives Analysis

As noted earlier, the EIS must "rigorously explore and objectively evaluate all reasonable alternatives" (§1502.14[a]). The alternatives analysis must provide a "full and fair discussion of significant environmental impacts" (§1502.1).

Relationship to Related Actions

The alternatives section should clearly explain how the proposal may be related to any other action undergoing a NEPA review. When properly prepared, this explanation can assist the agency in demonstrating that any interim actions comply with CEQ's relevant requirements. Such discussion can also assist the agency in assessing the significance of potential impacts of connected and related actions. For example, this task is useful in determining if the proposal is "related to other actions with individually nonsignificant but cumulatively significant impacts" (§1508.27[b][7]). Finally, this discussion can help demonstrate that an action is not connected to another action in a way that would constitute illegal segmentation (§1508.25[a][1]).

Agency's Preferred Alternative

As described in chapter 2, the agency must identify the course of action it favors. This course of action is referred to as the agency's *preferred alternative* (§1502.14[e]). Identifying the preferred alternative in the draft EIS is advantageous because it provides an early signal of the direction in which the agency is leaning, allowing reviewers and the public to focus attention on the alternative that, in all likelihood, will be chosen. In addressing this requirement, the EIS should indicate "those considerations, including factors not related to environmental quality, which are likely to be relevant and important to a decision" (§1502.23).

If the agency does not have a preferred alternative, it must disclose this fact and explain that the preferred alternative will be identified in the final EIS. The only exception to this requirement occurs when another law prohibits identification of such an alternative (§1502.14[e]).[18]

Preferred versus Environmentally Preferable Alterative

Confusion sometimes arises between the use of the terms *preferred alternative* and *environmentally preferable alterative.* The agency's preferred alternative is the course of action the agency favors; it must be identified in the EIS.

In contrast, the environmentally preferable alterative is the one that, on balance, is most desirable from the standpoint of environmental quality. The agency's preferred alternative may or may not be the same alternative as the environmentally preferable alternative. The environmentally preferable alternative is not required to be identified in the EIS but must be identified in the ROD, along with any other alternatives that were considered in the EIS (§1505.2[b]).[19] The environmentally preferable alternative is discussed in more detail in section 3.12.

Mitigation Measures

Mitigation is a particularly important concept in NEPA—so much so that an EIS must discuss "appropriate *mitigation measures* not already included in the proposed action or alternatives (§1502.14[f], emphasis added).

NEPA does not require an agency to choose an alternative based solely on environmental factors, neither does it impose a substantive requirement to mitigate impacts. Agencies are, however, required to identify, consider, and evaluate mitigation measures. Until a final decision is reached, discussions of mitigation measures are to be treated as alternatives or options available to decisionmakers. Consistent with the methodology employed for describing alternatives, mitigation measures should be identified and described in the alternatives section of an EIS and their effectiveness evaluated in the environmental consequences section. CEQ direction for addressing mitigation is presented below.

Requirement to Address Mitigation Measures

Once a proposal is determined to result in significant impacts, mitigation measures *must* be considered, developed, and analyzed for all impacts, whether such impacts are significant or not (§1502.14[f], §1502.16[h], §1508.14).[20] Thus, the requirement to consider mitigation applies even to impacts that are not, by themselves, considered significant. Such direction is reasonable, especially from the standpoint of addressing cumulative impacts, where individual effects may be minor yet collectively significant.

Scope of Mitigation Measures

Consistent with common usage, the term *mitigation measures* does not normally include methods or technology considered standard engineering practice or required by law or regulations. Where applicable, mitigation measures must include methods such as land use controls and alternative designs for decreasing pollution emissions, construction impacts, and aesthetic intrusion.[21]

The scope of mitigation measures must address the range of potential impacts. Mitigation measures are described for all analyzed alternatives (including the proposed action, if one is defined). As with alternatives, all reasonable mitigation measures must be considered, even if they lie outside the jurisdiction of the agency (§1502.16[h], §1505.2[c]). This requirement extends to cases where the measures are unlikely to be adopted or enforced by the responsible agency. While the requirement might at first appear unjustified, it serves to alert officials and other agencies that possess the capability to implement these measures or to institute changes in existing rules or regulations.[22]

Evaluating Mitigation Measures

Some EISs identify mitigation measures without evaluating their effectiveness. Case law indicates that this practice is insufficient. The analysis must evaluate the effectiveness of mitigation measures in reducing potential impacts.[23] The description of mitigation measures must therefore be sufficiently detailed to allow analysts to evaluate their effectiveness.

For example, instead of simply stating that an environmentally sensitive area would be revegetated to mitigate construction disturbances, the discussion should describe specifically how and which areas would be revegetated, including the types of vegetation that would be used. The effectiveness of such measures is then assessed. Thus, the proposal might first be analyzed without mitigation measures and then reanalyzed with such measures to determine their effectiveness.

To ensure that environmental effects are fairly assessed, the ROD must address the likelihood that mitigation measures would be implemented. Where there is a history of nonenforcement or opposition to such measures, the EIS and ROD must acknowledge this. If the necessary mitigation measures will not be ready for a long time, this should also be indicated.[24]

Checklist for the Alternatives Section

A checklist useful in assuring that the alternatives section meets principal regulatory requirements is provided in Table 3.14.

3.7 Affected Environment Section

The significance of potential impacts is a function of both intensity and context. The affected environment section provides a baseline description of the existing environment against which both the intensity and context of potential impacts can be compared. This chapter is sometimes excessively long and detailed. Normally it should be considerably smaller than the environmental consequences section.[27] Guidance is provided in the following sections for reducing the detail often devoted to this section. Direction for collecting data and managing the task of preparing the affected environment section can be found in chapter 2.

Table 3.14 A partial checklist for reviewing the adequacy of the alternatives analysis

Does the EIS rigorously explore and objectively evaluate all reasonable alternatives? (§1502.1, §1502.14[a])
Does the EIS devote "substantial treatment" to each of the reasonable alternatives?[25] (§1502.14[b])])
Do the analyzed alternatives encompass the range of alternatives that will be considered by the ultimate agency decisionmaker? (§1502.2[e])
Have alternatives been excluded because they were not consistent with an action that the agency favors?
Do the analyzed alternatives include appropriate mitigation measures? (§1502.14[f])
Have reasonable alternatives been considered even if they lie outside the legal jurisdiction of the agency?[26] (§1502.14[c])
Does the EIS explain how each alternative, as well as any decision based on it, would or would not achieve the goals of NEPA and other environmental laws? (§1502.2[d])
Does the EIS indicate considerations, including factors not related to environmental quality, that are likely to be relevant and important in reaching a final decision? (§1502.23)

Limiting Range of Resources and Level of Detail

The EIS must "succinctly" describe potentially affected environmental resources. Discussions should be commensurate with the importance of the impact (e.g., sliding-scale approach). Unusual circumstances aside, each resource should be described in only enough detail to allow the reader to understand how it could be affected. Where applicable, data should be incorporated by reference (§1502.15).

It is not uncommon to find that an EIS provides an extensive discussion of environmental resources, even those that clearly have no potential for being affected. The EIS manager must exercise vigilance in ensuring that the range of environmental resources discussed is limited to those that are potentially threatened. As an example, local air quality may be dismissed if the proposal would not result in releases that could conceivably affect air quality. Alternatively, baseline contaminant levels (e.g., sulfur and nitrogen oxide levels) may need to be described if an action could significantly affect air quality. A cursory statement may be included that dismisses resources that have no potential for being affected and therefore are not addressed.

Describing the Affected Environment

Table 3.15 provides a generic outline for the affected environment section that should be tailored to meet specific circumstances. The introduction should describe the purpose that it is intended to satisfy (i.e., provide a baseline against which the potential impacts are measured). Discussion should be organized and presented on a resource-by-resource basis. Resources should be organized in the same order in which they are discussed in the environmental consequences section. If an action would be located near a minority or low-income community, the identity and proximity of this community should be discussed.

A commonly encountered problem involves mixing discussions of potential impacts with the description of the affected environment. The affected environment section simply presents the baseline environment as it presently exists—*not* as it would be affected if the proposal were implemented.

In lieu of conducting an adequate field survey, some EISs attempt to dismiss potentially significant environmental resource issues with a statement such as ". . . is not known to exist in the location of the proposed site." *Such responses are inappropriate* and may form the basis for a successful challenge. A site-specific survey is necessary when existing information regarding potentially significant resource issues is unavailable, insufficient, or questionable.

Environmentally Sensitive Resources

Particular emphasis is placed on describing environmentally sensitive resources. Table 3.16 lists categories of environmentally sensitive resources that may need to be addressed.

PRIME AND UNIQUE FARMLAND. As used in Table 3.16, the terms *prime* and *unique* with respect to farmlands deserve special mention, as confusion often arises with respect to their usage. *Prime farmland* refers to land possessing the best combination of characteristics for producing food, forage, fiber, and oilseed crops and that is available for use. In contrast, *unique farmland* is land other than prime farmland whose combination of characteristics can support production of high-value food or fiber crops. The U.S. Department of Agriculture and the Soil Conservation Service can provide additional information on addressing these resources.

RESTRICTION ON RELEASING SENSITIVE RESOURCE INFORMATION. In many instances, it is illegal to reveal locations of threatened and endangered species, archaeological sites, and other environmentally sensitive resources. References to such subjects, especially with regard to specific locations, may need to be restricted from public dissemination. Appropriate environmental specialists or legal counsel should be consulted before such material is publicly disseminated. The reader is directed to §1507.3(c) for procedures that may be helpful in addressing this issue.

Table 3.15 Generic outline for discussing the affected environment

3.0 Affected environment
- 3.1 Purpose that this section of the EIS satisfies
- 3.2 Local and regional environment
  - 3.2.1 Location of the proposal
  - 3.2.2 Physiography
  - 3.2.3 Nearby Native American reservations
- 3.3 Air resources
  - 3.3.1 Meteorology
  - 3.3.2 Air quality
  - 3.3.3 Air quality issues (e.g., acid rain, global warming)
- 3.4 Hydrology
  - 3.4.1 Surface water sources and quality
  - 3.4.2 Groundwater and hydrogeology sources and quality
- 3.5 Geological resources
  - 3.5.1 Soils
  - 3.5.2 Geomorphology
  - 3.5.3 Structure
  - 3.5.4 Geologic hazards
- 3.6 Ecology
  - 3.6.1 Plants
  - 3.6.2 Animals
  - 3.6.3 Aquatic life
  - 3.6.4 Habitats and fish and wildlife reserves
  - 3.6.5 Sensitive, threatened, and endangered species and habitats
  - 3.6.6 Biodiversity
- 3.7 Socioeconomic resources
  - 3.7.1 Population, local communities, employment, and income
  - 3.7.2 Local and regional industries and agriculture
  - 3.7.3 Traffic, transportation, public services, and utilities
  - 3.7.4 Archaeological, cultural, and historic resources
  - 3.7.5 Subsistence (e.g., hunting and gathering)
- 3.8 Land use
  - 3.8.1 General land use restrictions and policies
  - 3.8.2 Transportation and utility restrictions
  - 3.8.3 Residential, recreational restrictions
  - 3.8.4 Industrial, and agricultural restrictions
  - 3.8.5 Infrastructure (e.g., roads, sewage systems) restrictions
- 3.9 Other disciplines
  - 3.9.1 Ambient noise levels
  - 3.9.2 Ambient electromagnetic forces
  - 3.9.3 Background chemical contamination levels
  - 3.9.4 Background radiation levels
  - 3.9.5 Aesthetics

Table 3.16 Examples of environmentally sensitive resources

- Properties of historic, archaeological, or architectural significance
- Prime and unique agricultural lands
- Threatened and endangered, federal and state, listed and potentially listed species and critical habitats
- Special water resources, such as sole source aquifers, floodplains, and wetlands
- Protected natural resources, such as national and state forests and parks

3.8 The Environmental Consequences Section

The introduction to the environmental consequences section should explain that the purpose of the chapter is to evaluate potential impacts, thus providing the scientific and technical basis for comparing the alternatives presented in the alternatives section. The alternatives considered in this section should be organized and presented in the same order as they were introduced in the alternatives section.

General-Purpose Outline

Table 3.17 suggests a general-purpose outline for this section of the EIS. The actual format, of course, must be tailored to meet the agency's specific needs. Specific direction for preparing this chapter is presented in the following paragraphs.

Tools and Methodologies

Agencies are responsible for ensuring the professional and scientific integrity of the analyses. Scientific methodologies used in performing them must be identified, as must explicit reference to scientific and other sources relied upon in reaching conclusions (§1502.24).

Prudence must be exercised where there is scientific controversy concerning the selection of models and methodologies used in performing the analysis. Opponents may claim that a particular methodology was arbitrarily disregarded because it would lead to results unfavorable to the agency's objectives. The author recommends that the EIS briefly explain why particular methodologies were chosen. This discussion should also briefly describe methodologies that were considered and then rejected. In the event of a challenge, such explanations can help demonstrate that appropriate methodologies were not overlooked nor arbitrarily dismissed.

General Direction for Describing Impacts

As indicated in Table 3.17, analysis of impacts should be conducted resource by resource. The order used for describing environmental resource

Table 3.17 General-purpose outline for the section on environmental consequences

4.0 Environmental consequences by resources and disciplines
- –Explanation of the purpose of this section
- 4.1 Environmental consequences of taking no action (direct, indirect, and cumulative impacts)
  - 4.1.1 Air
  - 4.1.2 Surface water and groundwater
  - 4.1.3 Geology and soils
  - 4.1.4 Ecology
    - –Plants, animals, habitats
    - –Sensitive and endangered species and habitats
  - 4.1.5 Visual resources, land use, and land features
  - 4.1.6 Facilities and infrastructure
    - –Transportation and telecommunication systems
    - –Electric, water, and sewage systems
  - 4.1.7 Health and safety
  - 4.1.8 Socioeconomics
  - 4.1.9 Historical, archaeological, and cultural issues
  - 4.1.10 Amenities
    - –Local, state, and national parks and recreational areas
    - –Fish and wildlife reserves
  - 4.1.11 Other environmental disciplines and issues
    - –Radio and television interference
    - –Electromagnetic field levels
    - –Radiation levels
    - –Noise and odors
  - 4.1.12 Mitigation measures
- 4.2 Environmental consequences of the proposed action (similar to 4.1)
- 4.3 Environmental consequences of alternative A (similar to 4.1)
- 4.4 Environmental consequences of alternative B (similar to 4.1)
  - •
  - •
  - •
- 4.5 Comparison of alternatives and their consequences
  - –Environmental consequences, schedule, and cost. (This section is optional depending on how it is covered in the alternatives section.)
- 4.6 Discussion of agency's preferred alternative
  - –Discussion of factors leading to the selection. (This discussion can be included in this section or in the alternatives section.)
- 4.7 Adverse Impacts that cannot be avoided
  - 4.7.1 Air
  - 4.7.2 Surface water and groundwater
  - 4.7.3 Geology and soils
  - 4.7.4 Ecology
  - 4.7.5 Visual resources, land use, and land features
  - 4.7.6 Facilities and infrastructure
  - 4.7.7 Health and safety
  - 4.7.8 Socioeconomics

Table 3.17 *Continued*

4.7.9 Historical, archaeological, and cultural issues
4.7.10 Amenities
4.7.11 Other environmental disciplines
4.8 Irreversible and irretrievable commitment of resources
4.8.1 Air
4.8.2 Surface water and groundwater
4.8.3 Geology and soils
4.8.4 Ecology
4.8.5 Visual resources, land use, and land features
4.8.6 Facilities and infrastructure
4.8.7 Health and safety
4.8.8 Socioeconomics
4.8.9 Historical, archaeological, and cultural issues
4.8.10 Amenities
4.8.11 Other environmental disciplines
4.9 Energy requirements and conservation potential
4.10 Natural or depletable resource requirements and conservation potential
4.11 Relationship between short-term use of the environment and enhancement of long-term productivity
4.12 Relationship between the proposal and land use plans

impacts should follow the same sequence as the affected environment section. It is recommended that the analysis of a particular resource begin with an evaluation of the most significant impacts and end with those of least significance. The analysis may conclude that an impact originally considered significant was, on closer examination, found to be nonsignificant. In such cases, the analysis should clearly explain why.

While direct and indirect effects must both be evaluated, their discussion should not be segregated into separate sections but rather integrated into a common analysis. Because of its complexity and the special analytical methodologies that may be employed, the cumulative impact analysis is sometimes presented as an independent section.

One mistake that is occasionally made involves analyzing impacts on a particular resource without providing a corresponding discussion of the actual resource in the affected environment section. Lacking a corresponding baseline, it is difficult (if not impossible) to accurately gauge how a specific impact would change or affect an existing resource. A cross-check should always be conducted to verify that corresponding discussion appears in the affected environment section.

Site-Specific Analyses

The requirements of NEPA imply and the courts have mandated that a site-specific analysis be conducted. Consistent with this objective, the descriptions of each alternative should be site specific. If such an analysis is not practical, the agency may need to prepare a supplemental EIS or a more site-specific analysis that can be tiered from the EIS later. The location or area that would be affected by each alternative should be clearly delineated on maps.

Specific Impact Requirements

Table 3.18 summarizes environmental impact requirements that must be addressed (§1502.16). In addition to ecological effects, impacts also include but

Table 3.18 Specific requirements that must be addressed in the section on environmental consequences

- The analytic basis for the comparisons made under §1502.14 of the Regulations. (§1502.16)
- Adverse impacts that cannot be avoided, should the proposal be implemented. (§1502.16)
- Relationship between short-term use of man's environment and the maintenance and enhancement of long-term productivity. (§1502.16)
- Irreversible or irretrievable commitment of resources which would be involved in the proposal should it be implemented. (§1502.16)
- Direct effects and their significance. (§1502.16[a])
- Indirect effects and their significance. (§1502.16[b])
- Cumulative effects and their significance. [This requirement is sometimes addressed in an individual chapter.] (§1508.7, 1508.8, 1508.25[c][c])
- Possible conflicts between the proposal and objectives of federal, regional state, local, and tribal land use plans, policies, and controls for the area concerned. (§1502.16[c])
- Environmental effects of alternatives, including the proposed action. (§1502.16[d])
- Energy requirements, conservation potential of various alternatives, and mitigation measures. (§1502.16[e])
- Natural or depletable resource requirements, conservation potential of various alternatives, and mitigation measures. (§1502.16[f])
- Urban quality, historic and cultural resources, and the design of the built environment, including the reuse and conservation potential of various alternatives and mitigation measures. (§1502.16[g])
- Methods for mitigating adverse environmental impacts [if not already included in the alternatives section of the EIS]. (§1502.16[h])
- How alternatives considered in the EIS and decisions based on it will or will not achieve the requirements of sections 101 and 102 (1) of NEPA and other environmental laws and policies. (§1502.2[d]) [Note: This requirement is sometimes met in the alternatives section.]
- All federal permits, licenses, and other entitlements which must be obtained in implementing the proposal. If it is uncertain whether a federal permit, license, or other entitlement is necessary, the EIS must so indicate. (§1502.25[b]) [Note: This requirement is sometimes placed in the alternatives section or in an individual chapter.]
- A cost-benefit analysis, if appropriate. (§1502.23)

are not limited to aesthetics, historic, cultural, economic, social, and health effects (§1508.8[b]).

Short-Term Uses versus Long-Term Productivity

The discussion of environmental consequences is required to address "the relationship between short-term uses of man's environment and the maintenance and enhancement of long-term productivity" (§1502.16). The precise intent of this requirement is interpretive and has been the source of much confusion. It is desirable to explain the context in which the requirement has been interpreted. In the past, it has most often been interpreted to mean that the EIS must evaluate trade-offs between a short-term benefit (economic or otherwise) that would be derived from pursuing an action versus the long-term benefit or productivity that would be derived from not exploiting the environmental resources.

This discussion allows decisionmakers and the public to clearly grasp what would be gained or lost over both the short and the long term as a result of pursuing potential actions. This discussion should focus on a fair, open, and balanced assessment of such trade-offs. For example, the EIS might consider the short-term benefit derived from construction of a hydroelectric dam to generate electricity versus the long-term degradation of habitat and species and the scenic and recreational use of the river that would be lost.

Irreversible and Irretrievable Resources

The EIS must address "any irreversible or irretrievable commitments of resources which would be involved in the proposal should it be implemented" (§1502.16). This is sometimes referred to as the *I&I requirement.*

As these terms are not defined in the Regulations, the difference between *irreversible* and *irretrievable* has led to much confusion. *Webster's* offers these definitions:

> *Irreversible:* . . . incapable of being reversed
> *Irretrievable:* . . . that cannot be retrieved, restored, recovered, or repaired; irrecoverable.[28]

With respect to NEPA, an irretrievable commitment of resources can be interpreted as a use or consumption of a resource that cannot be replaced; however, given sufficient time, the effect may be reversible. In contrast, an irreversible commitment of resources may be interpreted as a use or consumption of a resource that can never be replaced and the effect of whose loss can never be reversed.

The reader is referred to the *Forest Service Handbook* for a working definition of these terms.[29] As used in the *Handbook,* an irreversible commitment of resources can be viewed as an action that results in a permanent use of a non-renewable resource. Such actions can never be reversed, at least in terms of human experience. While the exploitation of such resources may benefit the commercial sector, they also represent a permanent loss from the standpoint of the ecosystem. Examples of irreversible impacts are diverse and include extinction of a species, consumption of minerals or petroleum, and, possibly, the logging of an old-growth forest. A claim could be raised that, given sufficient time, such resources may replenish themselves. For instance, petroleum and mineral deposits are replenished over geological time, and old-growth forests may be reestablished over hundreds of years. In terms of human experience, however, such actions are not reversible and the resources are essentially lost forever.

In contrast, an irretrievable commitment of resources can be interpreted as the consumption of renewable resources that are not permanently lost but that may be exhausted for a long period of time. Examples include logging forest timber (excluding possibly old-growth forest) or the demise of a large number of salmon (which can eventually be recovered).

Where appropriate, accepted methods of practice may also dictate that the analysis consider *foregone resources,* a term that can be interpreted to include those environmental aspects that would not actually be used or consumed but that would be made inaccessible or unusable. An example is land inundated by water from a man-made lake.

In complying with the I&I requirement, analysts should discuss the commitment of resources expended as a result of both construction and operational activities. In the past, the I&I requirement has often been given only cursory treatment. As described in the following sections, careful adherence to this requirement may have important implications, as it can save untold sums as a result of natural resource damage assessments. Private applicants, in particular, should pay close attention to ensuring that the I&I requirement is adequately satisfied.

NATURAL RESOURCE DAMAGE ASSESSMENTS. Under Section 107 of the Superfund Act, responsible parties may be held liable for damages to publicly owned or managed natural resources.[30] Better known as the Natural Resource Damage Assessment (NRDA), this provision allows natural resource trustees to recover damages (i.e., money) for injury incurred to natural resources.[31] In 1994, the U.S. Department of the Interior published a final rule establishing administrative procedures for conducting NRDAs.[32]

An NRDA can be conducted to ascertain the results of a discharge of oil or release of a hazardous substance that damages an environmental resource.

NRDAs extend only to natural resource damages that occurred after the enactment of the Comprehensive Environmental Response Compensation and Liability Act (CERCLA). A claim may be brought regardless of whether or not a site is on the National Priorities List. Claims may also be brought against contractors managing federal facilities. State and tribal trustees also have authority to bring claims against the federal government for releases at federal facilities.

Damages to both use and non-use values are recoverable. Natural resource trustees may assess damages as high as $50 million per release of a hazardous substance plus standard cleanup costs assessed under CERCLA, the cost of assessing such injury, and any prejudgment interest. To date, settlements have ranged from $5,000 to over $24 million. Three sites have been identified where pending or future claims may exceed $500 million.[33]

As used in this context, the term *natural resources* is expansive and includes land, fish, wildlife, biota, air, surface water, groundwater, and other such resources controlled, managed, held in trust, or belonging to the United States or any Native American tribe. Natural resources are also generally interpreted to include all lakes, rivers, streams, and coastal waters bounded or unbounded by public lands. Liability most clearly extends to lands owned or managed by federal or state agencies or Native American tribes.

PROTECTION FROM NATURAL RESOURCE DAMAGE CLAIMS. Under limited circumstances, natural resource damages may be excluded from future liability. With respect to NEPA's I&I requirement, Section 107 of the Superfund Act provides an important exemption from NRDA claims if the following two conditions are met:

1. Potential damages to natural resources are specifically identified as an irreversible and irretrievable commitment of natural resources in an EIS or other comparable environmental analysis.
2. A decision to grant a permit or license authorizes the commitment of resources; and the action must be operated within the terms of any permit or license that is granted.[34]

In writing this provision, the Senate Committee explained the congressional intent.

> Federal officials make decisions in which resource trade-offs must necessarily be made, and in such cases liability for resource damage . . . should be limited. . . . In such a case, where the specific trade-offs are understood and anticipated in issuing the permit for such releases and the agency takes into account this knowledge and allows

the trade-off, then no liability will accrue for resources damaged pursuant to those permitted releases.[35]

Consistent with NEPA's procedural requirements, this exemption essentially allows natural resources to be irreversibly or irretrievably damaged, as long as the long-term consequences have been carefully assessed by the decisionmaker (i.e., described in the I&I section of the EIS) and the actions are undertaken in accordance with applicable permits and licenses.

Prudence should be exercised in identifying and describing any potential loss or damage to a natural resource that may be viewed as an I&I commitment of resources. For example, if remediation of a contaminated soil site could contaminate an adjacent stream, the potential degradation of the stream should be clearly described as a potential I&I commitment of resources.

Adverse Effects That Cannot Be Avoided

The intent of NEPA is to provide decisionmakers with pertinent information so that environmental consequences can be factored into their decisions. To this end, the EIS must discuss "any adverse environmental effects which cannot be avoided should the proposal be implemented" (§1502.16).

Essentially, this requirement constitutes a disclosure statement regarding adverse impacts. The reader should note that this requirement is not limited to those impacts deemed significant.

Land Use Conflicts

One of the purposes of NEPA is to promote integration of federal proposals with other planning and land use constraints, including state and local planning processes. Accordingly, an EIS must explore and address "possible conflicts between the proposed action and the objectives of federal, regional, state, and local (and in the case of a reservation, Indian tribe) land use plans, policies and controls for the area concerned" (§1502.16[c]).

With respect to this requirement, the phrase *land use plans* includes any formally adopted documents for land use planning or zoning. This term also includes plans that have been formally proposed by a government body and are under active consideration. Similarly, *land use policies* includes formally adopted statements of land use policy embodied in laws or regulations; it also covers land use policies that have been formally proposed but have not yet been adopted.[36]

The EIS evaluates the degree to which potential actions would affect land use plans and how such actions might impair the effectiveness of land use control mechanisms. The EIS preparer should investigate any possibility of resolving the conflict. Where an inconsistency is identified, the EIS must discuss the discrepancy and describe the extent to which the agency would reconcile its proposal with the plan or law (§1506.2[d]). Comments from officials with responsibility over the affected area should be solicited early and adequately addressed in the EIS.[37]

Inconsistencies with Other Laws and Plans

As an environmental planning process, an EIS provides a comprehensive umbrella under which developing plans may be coordinated with other laws and regulations. To this end, the EIS must discuss:

> any inconsistency of a proposed action with any approved state or local plan and laws (whether or not federally sanctioned). Where an inconsistency exists, the statement should describe the extent to which the agency would reconcile its proposed action with the plan or law. (§1506.2[d])

In complying with this requirement, the agency may find it useful to develop an environmental regulatory compliance plan to assist in demonstrating that important environmental laws and regulations have been identified, plans are properly coordinated and scheduled, and inconsistencies have been addressed. Such a plan may be incorporated by reference into the EIS. For example, if a programmatic memorandum of agreement is prepared, a discussion should summarize the agreement, describing how it is coordinated with other requirements. Wetland assessments and other environmental studies should also be described, especially in terms of how they are coordinated with the planning process at large. Such a plan can reduce duplication and inefficiencies and is useful in demonstrating that all laws and regulations have been adequately considered before a decision is made to pursue a particular course of action.

Energy and Natural Resources Consumption

Consuming upward of one third of the world's usage, the United States has a voracious energy appetite. Satisfying this demand is not without a price. Energy consumption extracts a heavy toll on the environment. To this end, the environmental consequences section must evaluate "energy requirements and conservation potential of various alternatives and mitigation measures" (§1502.16[e]).

This provision is interpreted to mean that agencies must analyze the total energy costs and benefits, including hidden and indirect costs (economic and otherwise), associated with implementing a proposal.[38] Consistent with this requirement, the EIS must also evaluate "natural or depletable resource requirements and conservation potential of various alternatives and mitigation measures" (§1502.16[f]).

Federal actions often involve consumption of large amounts of nonrenewable resources (e.g., land, water, wood, steel, and geological and mineral resources). In many cases, use of such materials represents a long-term commitment of natural resources. Consistent with this requirement, analysts should compute, evaluate, and list any significant consumption of important resources.

Urban, Historic, and Cultural Resource Impacts

The environmental consequences section must discuss "urban quality, historic and cultural resources, and the design of the built environment, including the reuse and conservation potential of various alternatives and mitigation measures" (§1502.16[g]).

It is interesting to note that this provision was not included in the Regulations, as originally drafted; it was added based on public comments received during the review period for adopting the Regulations, recommending that this issue also be addressed.[39]

In fulfilling this requirement, analysts should evaluate impacts to both the prehistoric and historic settings and how these impacts can be avoided or mitigated. Information should be presented with the goal of allowing the reader to judge whether the merits of the project outweigh the potential adverse impacts to cultural resources.

Emphasis is placed on evaluating facilities and sites listed on the National Register of Historic Places (NRHP), as well as potential candidates for such listings.[40] Determining the significance of impacts on historic places is normally based on evaluation criteria established by the NRHP and guidelines established by the Secretary of the Interior.[41] Impacts to Native American resources are often evaluated against the factors established by the NRHP and the American Indian Religious Freedom Act.[42]

Where possible, the discussion should describe how potential work would be coordinated with the State Historic Preservation Officer (SHPO). Areas that have yet to be surveyed for historic resources but are considered to have such potential should be clearly delineated. If a programmatic memorandum of agreement will be prepared, the discussion should describe how this agreement will be coordinated with the EIS planning process.

How Alternatives Achieve NEPA's Goals

The EIS must "state how alternatives considered in it and decisions based on it will or will not achieve the requirements of sections 101 and 102(1) of the Act and other environmental laws and policies" (§1502.2[d]).

As Sections 101 and 102(1) of NEPA establish a national policy for preserving the environment for future generations of Americans, the EIS must explain how analyzed alternatives and decisions that will be made would or would not achieve this goal. The reader is referred to Sections 101 and 102(1) for details on the specific policy goals that should be addressed in this section of the EIS.

The provision is one of many ways in which the viability of NEPA's national environmental policy goals can be elevated. The reader should note that, according to the CEQ, a common problem with EISs in the past has been failure to adequately explain how the proposal would or would not achieve NEPA's requirements.[43]

Listing Permits, Licenses, and Other Entitlements

An EIS must also list:

> all federal permits, licenses, and other entitlements which must be obtained in implementing the proposal. If it is uncertain whether a federal permit, license, or other entitlement is necessary, the draft environmental impact statement shall so indicate. (§1502.25[b])

This provision does not indicate where this requirement is to be addressed in the EIS. Some address it in the alternatives section, while others address it in the environmental consequences section. Still others incorporate a separate chapter devoted solely to this subject, particularly in circumstances where such listings may be lengthy.

A common problem has been that EISs fail to adequately identify all required federal permits and approvals.[44] Permits and other requirements are often identified through agency consultation. Work on preparing this list should begin early, as it often provides a basis for identifying other federal agencies that may need to be consulted or brought into the scoping process. This effort also seeks to reduce project surprises by alerting managers of pending requirements that may need to be addressed.

This section of the EISs should also specify the entity responsible for granting and obtaining each permit or license. Each alternative may require a unique set of permits, licenses, and other entitlements. If there is considerable variation in the requirements, a regulatory compliance matrix may prove invaluable in allowing readers to compare the permitting and licensing requirements to the corresponding alternatives. Table 3.19 provides a simplified representation of such a matrix.

Special Analytical Considerations

Guidance for addressing impacts and issues that require special consideration are described in the following sections.

Impacts on Human Health and Safety

Where applicable, the EIS analysis evaluates potentially significant impacts to public health. Human health impacts should be evaluated against recognized standards such as regulatory limits established by the Safe Drinking Water Act and the Occupational Safety and Health Act.

While the Regulations state that impacts to human health must be considered, they are silent about corresponding requirements to consider safety. Factors established for determining significance, however, specify that the

Table 3.19 Simplified example of a regulatory matrix of alternatives versus required permits, licenses, and entitlements

Permit, License, or Entitlement	No Action	Alternative 1	Alternative 2
RCRA Permitting	RCRA Part A & B Permit	RCRA Part A & B Permit	RCRA Part B Permit
State Hazardous Materials Permitting	State Materials License No. 20171-1	State Hazardous Materials Facility Siting Permit	State Radioactive Materials License
State Hazardous Waste Management Permitting	State Hazardous Waste Management Act Permit	State Hazardous Waste Management Act Permit	State Air Pollution Control Permit
State Hazardous Waste Generation Permitting	State Generator's User Permit No. B-1043/A	State Health and Welfare Construction and Operation Permit	New Contaminant Source Construction Permit
Environmental Protection Agency (EPA) Permitting	EPA approval for PCB thermal treatment under TSCA	EPA approval for PCB thermal treatment under TSCA	EPA approval for PCB thermal treatment under TSCA
Nuclear Regulatory Commission (NRC) Approvals	Nuclear Regulatory License (NRC)	Radioactive materials license required by the state Radiation Control Regulations and NRC	Nuclear Regulatory License
Transportation Approvals	U.S. Department of Transportation certificate	State transportation permit	N/A

assessment of significance may depend on "the degree to which the proposed action affects public *health or safety*" (1508.27[b]]2], emphasis added). This requirement implies that, as appropriate, the EIS must consider potential safety issues, because an EIS must analyze significance in terms of the degree to which public safety may be compromised.

POTENTIAL ACCIDENTS. Some agencies include accident scenarios in their EISs. For example, as applicable, the U.S. Corps of Engineers and the Bureau of Reclamation include scenarios such as the overtopping of dams and dam failures. Such investigations may be a necessary component of an analysis, where the results could have a bearing on determining significance, mitigation, or agency decisionmaking. This is particularly true where an accident scenario may pose grave consequences. The analysis may need to consider potential consequences of low-probability–high-consequence incidents as well as high-probability–low-consequence events. The author has proposed a methodology for determining how and when an accident analysis should be addressed in an EIS analysis.[45] Additional information on addressing accident analyses can be found in the *The NEPA Planning Process.*[46]

NATURAL DISASTERS. Natural phenomena may profoundly influence the impacts of certain types of actions (e.g., dams, nuclear facilities, and chemical processing plants). An analysis may need to consider how geological or atmospheric hazards could affect safety issues and potential environmental impacts, for example. Analysis of geological hazards may involve evaluating how phenomena such as earthquakes, floods, volcanic eruptions, or landslides might affect federal actions. Similarly, atmospheric hazards may involve an analysis of events such as high wind storms and tornados.

Socioeconomic Impacts

In assessing cause-and-effect relationships between economic downturns and their sociological ramifications, President Calvin Coolidge is alleged to have confidently concluded, "When more and more people are thrown out of work—unemployment results." Analysis of such esoteric relationships is no less important in preparing the environmental analysis. Agencies are routinely obligated to make decisions that affect employment, to say nothing of the prosperity and infrastructure of communities. This aspect of the EIS explains how social and economic resources, both local and regional, could be affected, especially with respect to induced population growth. An analysis of impacts on employment patterns and income levels is often relevant, as are impacts on public and private institutions such as housing, schools, public utilities, recreational resources, and public and emergency services.

For example, impacts resulting from increased usage of sewage treatment and solid waste disposal facilities, electricity, water, and other services may be relevant. If appropriate, the analysis may need to estimate the number of new dwelling units that may be required as a result of primary and secondary population growth and impacts on the existing housing market. The analysis may also need to consider how the capacity of the existing transportation system (i.e., roads, railroads, airports, and port facilities) could be affected and if the system would require expansion. For more information on addressing socioeconomic impacts, see *The NEPA Planning Process.*[47]

ENVIRONMENTAL JUSTICE. In promoting the goal of environmental justice, the Clinton administration issued Executive Order 12898, which focuses federal attention on human health and environmental conditions within low-income and minority neighborhoods.[48] A presidential memorandum accompanying this order instructs federal agencies to:

> analyze the environmental effects, including human health, economic and social effects, of Federal actions, including effects on minority communities and low-income communities, when such analysis is required by the National Environmental Policy Act.[49]

The EPA has issued draft guidance for incorporating environmental justice concerns in NEPA compliance analyses.[50] The CEQ has also issued guidelines for addressing environmental justice.[51]

To this end, an EIS may need to consider factors that may unfairly burden or place disproportionate adverse environmental and socioeconomic impacts on minority and low-income segments of a community. Evaluating disproportionately high and adverse impacts should be considered in terms of both environmental and health effects.

Environmental Effects: The following factors should be considered in determining the appropriateness of evaluating disproportionately adverse environmental impacts on minority and low-income populations:

- Could an impact on the natural or physical environment occur that significantly and adversely affects a minority or low-income population? Such impacts include ecological, cultural, economic, and social impacts when they are related to impacts on the natural or physical environment.
- Could a significant adverse impact occur to a minority or low-income population that is likely to appreciably exceed that to the general population or other appropriate comparison group?

- Could the environmental impact occur to a minority or low-income population that would be affected by multiple or cumulative adverse exposures from other environmental hazards?

Health Effects: The following factors should be considered in determining the appropriateness of evaluating disproportionately high and adverse health impacts on minority and low-income populations:

- Could the health effect be significant or above generally accepted norms? Adverse health effects include death, illness, and bodily impairment.
- Could the risk of an impact to a minority or low-income population be significant and appreciably exceed that to the general population or other appropriate comparison group?
- Could the health effect occur to a minority or low-income population that would be affected by multiple or cumulative adverse exposures from other environmental hazards?

For example, a preliminary screening analysis might indicate that effects associated with a specific action that is a component of a larger proposal would be significant in terms of both environmental and human health impacts. However, the closest identified low-income community is located 15 kilometers away from the proposed site. The environmental and human health effects on this community would not be high nor would there be a disproportionately adverse impact on this community when compared to the general population or other appropriate comparison groups. Based on this evaluation, an analysts would probably be justified in concluding that impacts to minority or low-income populations would not be disproportionately high or adverse.

Mitigation measures for reducing or eliminating disproportionately significant adverse impacts to minority and low-income populations should be evaluated. For example, an analysis of a hazardous waste transportation proposal might need to consider if the transportation route would unfairly jeopardize the health or safety of such a population. Potential alternatives and mitigation measures might involve spreading potential risk across a number of port cities, docks, and railroad and highway corridors.

It is recommended that potential impacts to minority and low-income communities be addressed in a subsection devoted to analysis of socioeconomic impacts. Because of the expense and complexity of this analysis, agencies may find it desirable to apply a sliding-scale approach in which the level of analysis is proportional to the significance of the disproportionate impacts. Where appropriate, the ROD should specifically mention the conclusions reached regarding the potential significance of environmental justice impacts on minority or low-income populations. See *The NEPA Planning Process* for additional information on this topic.[52]

PROTECTION OF CHILDREN FROM ENVIRONMENTAL HEALTH RISKS. Where federal actions may disproportionately affect children, Executive Order 13045 dictates that agencies are to ensure that their policies, activities, and standards address such risks; agencies are to assign a high priority in assessing such environmental, health, and safety risks.[53]

To this end, an EIS should consider any disproportionate impacts on children. For example, the analysis should focus on products or substances that a child is likely to come into contact with or ingest.

Conformity

State implementation plans provide a mechanism for enforcing criteria pollutant standards (i.e., carbon monoxide, lead, nitrogen dioxide, ozone, particulate matter, and sulfur dioxide) under the National Ambient Air Quality Standards (NAAQS). Recently, a new dictate issued by the EPA directs that federal actions must comply with state implementation plan conformity criteria[54]—that is, they must not cause or contribute to any new air quality violation. EPA's requirement pertains only to criteria pollutant levels that could be affected by direct and indirect emissions resulting from a federal action within a nonattainment or maintenance area. Pursuant to the Clean Air Act, indirect emissions are those that are reasonably foreseeable and within the control of a responsible federal agency. Conformity determinations are not required for any major stationary source that is already covered by Prevention of Significant Deterioration (PSD) programs. Numerous types of activities are also exempted from the conformity determination.

Under the new direction, conformity determinations are to be integrated into the NEPA process. Where applicable, an EIS analysis should consider direct and indirect emissions in evaluating conformity with state implementation plan criteria. As necessary, mitigation measures must be identified.

3.9 List of Preparers and Entities to Whom the EIS is Sent

The EIS must include a section listing the individuals who prepared the EIS. There are three reasons for including such a section. First, agencies are required to use an interdisciplinary approach in analyzing impacts. A section listing preparers provides a basis for determining if an interdisciplinary approach has truly been used. Second, the list instills accountability, which tends to enhance professional competency. Finally, the list promotes the professional standing of EIS staff members by recognizing their contribution to the analysis.[55]

The list of preparers must indicate both the name and qualifications of the individuals who were primarily responsible for preparing the EIS. Qualification material should briefly cover expertise, experience, and professional discipline (§1502.17).[56] The author suggests that this section also include, at a minimum, an individual's academic degrees and years of experience. The CEQ suggests that one or two lines of text is sufficient to cover an individual's qualifications. As a rule, the list should not exceed two pages in length. For this reason, individuals who have had only minor input or responsibilities need not be included.

If a consulting firm is hired to prepare the EIS, those members who made substantial contributions to the effort should be listed.[57] Individuals responsible for preparing important background papers should also be included (§1502.17). If the agency uses information submitted by the applicant, names of the persons responsible for independently evaluating and verifying the accuracy of these data must be included (§1506.5[a]). Individuals responsible for reviewing or editing the document should also be included.[58]

In the past, EISs have sometimes failed to provide adequate information on the preparer's education and experience, making it difficult to identify individual areas of responsibility. This information is important, as it can assist the reader in determining if a systematic, interdisciplinary approach has been followed.[59]

List of Entities to Whom the EIS Is Sent

The Regulations do not provide specific direction regarding the content or preparation of the list of agencies, organizations, and persons to whom copies of the statement are sent. They do, however, specify parties to which a copy of the EIS must be furnished (§1502.19). Section 2.5 in chapter 2 elaborates on this requirement.

3.10 Index, Glossary, and Bibliography

An EIS must contain an index. The well-constructed statement also contains a glossary and a bibliography. Presented below is guidance on preparing these sections of the EIS.

Index

Although an index is required under the standard EIS format (§1502.10), the Regulations are silent as to its preparation or content. CEQ's "Forty Most

Asked Questions," however, provides supplemental guidance. The index should reference more than obvious topics and issues but does not need to cover every conceivable term or phrase. As a rule of thumb, the index should reference any topic believed to be of reasonable interest to a reader.[60]

Key-Word Indexes Facilitate Compliance

While not required, it is highly recommended that a key-word index be used, consisting of descriptive terms for identifying relevant concepts and issues.[61] This approach enables readers to quickly locate areas of interest such as types of alternatives, affected resources, impacts and issues, and so on. Moreover, it facilitates incorporation of information into computer data bases.

The most frequent user of a key-word index is often the agency itself. After a ROD is issued, a comprehensive key-word index can assist practitioners in efficiently addressing future NEPA compliance issues by determining if a given activity has been adequately addressed in the EIS. The index can also assist agencies in quickly locating material that can be used or incorporated by reference in preparing lower-level or further NEPA documents.

Glossary and List of References

The Regulations do not require the EIS to contain either a glossary or list of references; however, it is considered good practice to include both. At a minimum, a glossary should contain standard NEPA, scientific, and technical terms used in the EIS. A list of references allows the reader to quickly locate documents referenced in the EIS. As applicable, lists of symbols, acronyms, and abbreviations and a table of measurement conversions should also be included.

3.11 Appendixes

The inclusion of appendixes in an EIS is optional. Used properly, appendixes provide the EIS team with an efficient mechanism for preparing a succinct analysis of reasonable alternatives and impacts, which can markedly improve both the readability and usefulness of the EIS. The body of the EIS should be designed to present the decisionmaker with information necessary to evaluate potential impacts and reach an informed decision. Consistent with this objective, material that is highly technical in nature should normally be presented in the appendixes. Table 3.20 summarizes CEQ direction regarding the appropriate use and content of the appendixes (§1502.18).

Where feasible, detailed descriptions of scientific methodology used in the analysis should be placed in the appendixes so that they are available for public review and comment (§1502.24). (Other material appropriate for inclusion in the appendixes includes lists of affected species and related stud-

Table 3.20 Direction regarding the appropriate use of appendixes

- Appendixes contain material prepared in connection with the EIS. Material not prepared directly in connection with the EIS should be incorporated by reference. (§1502.18[a])
- Appendixes normally contain material that substantiates any analysis fundamental to the EIS. (§1502.18[b], §1502.18[c])
- Appendixes should be circulated with the EIS. If they are not, they must be made readily available upon request. (§1502.18[d])
- Lengthy or detailed descriptions of the scientific methodology used in the analysis should be placed in the appendix.[62] (§1502.24)
- Comments and responses on the draft EIS may be placed in the appendixes.[63]

ies.) When this is done, the main body of the EIS need provide only a summary discussion of the analytical methodologies and models used in the analysis. The reader can then be directed to the appropriate appendix for details.[64] To allow decisionmakers to focus on issues of greatest concern, it is also recommended that comments received from circulation of the draft EIS and the agency's responses be placed either in the appendixes or a separate section of the EIS.[65]

Appendixes versus Incorporation by Reference

As indicated by the first item in Table 3.20, CEQ draws a distinction between the ways material prepared directly and indirectly in connection with the EIS is to be presented. Material prepared directly in support of the EIS is to be placed in the appendix. In contrast, material not directly prepared in connection with the EIS should be incorporated by reference (§1502.18[a]).

Appendixes must be completed and available for review at the time the EIS is filed with the EPA. The appendix should accompany the EIS whenever practical. If the appendixes are not circulated with the EIS, they must be placed in locations accessible by the general public or furnished to commenters on request.[66] In contrast, material incorporated by reference must be accessible to the general public but does not need to accompany the EIS. This can be done by citing publicly available information, furnishing copies to central locations such as public reading rooms, or sending copies to commenters on request. Such material must be publicly available for the full length of the minimum public comment period.[67]

3.12 The Record of Decision

As part of the final decisionmaking process, the agency is responsible for preparing a ROD, the purpose of which is to publicly record the agency's final decision. The ROD is a concise statement describing the agency's final choice among the alternatives considered and may be integrated into any other record prepared by the agency. It is important to note that a decisionmaker may not choose a course of action unless it has been adequately described and analyzed (§1505.1[e]).

The decisionmaker is legally mandated to consider and balance the environmental consequences against other decisionmaking factors in reaching a final decision regarding the course of action to be taken. The decision may also be based on factors including economic and technical considerations as well as the agency's statutory missions. The agency must also identify and discuss all factors considered and how they were weighed before reaching its final decision (§1505.2[b]). Table 3.21 lists the specific items that the ROD must address.

Table 3.21 Summary of items addressed in the ROD

- Provide a statement of the agency's decision. (§1505.2[a]) This statement must explain what the decision was, how it was made, and what mitigation measures are being imposed to lessen adverse environmental impacts of the proposal.[68]
- Identify the alternatives considered by the agency in reaching its decision. Preferences among alternatives based on relevant factors, including economic and technical considerations as well as the agency's statutory missions, may be discussed. The agency must also identify and discuss all factors considered and how they were weighed by the agency before reaching its final decision. (§1505.2[b])
- Specify the alternative(s) considered environmentally preferable. (§1505.2[b])
- State whether all practicable means have been adopted to avoid or minimize environmental consequences associated with the selected alternative, and if not, why they were not. A monitoring and enforcement program must be adopted and summarized, where applicable, for any mitigation. (§1505.2[c])
- Provide a concise summary of the mitigation measures that the agency has committed itself to adopt. The ROD must identify the mitigation measures and monitoring and enforcement programs that have been selected and clearly indicate that they are being adopted as part of the agency's decision. Discussion of mitigation and monitoring must be more detailed than a general statement that mitigation will be adopted, but not so detailed as to duplicate discussion of mitigation in the EIS.[69]
- If the proposal involves issuance of a permit or other approval, the specific details of the mitigation measures must be included as appropriate conditions in whatever grants, permits, funding, or other approvals are being made by the federal agency. If the proposal is to be carried out by the federal agency itself, the ROD should delineate the mitigation and monitoring measures in sufficient detail to constitute an enforceable commitment, or incorporate by reference the portions of the EIS that do so.[70]
- The ROD should indicate the likelihood that mitigation measures will be adopted or enforced by the responsible agency(s). If there is a history of nonenforcement or opposition to such measures, the ROD should acknowledge it. If the necessary mitigation measures will not be ready for a long time, this fact should also be recognized.[71]

Preparing the ROD

In the past, RODs have sometimes not clearly identified the environmentally preferable alternative. The CEQ has also indicated that RODs sometimes fall short in describing the considerations that led to a decision not to adopt the environmentally preferable alternative. Finally, in adopting a mitigation and monitoring plan, RODs have sometimes been weak in describing whether all practical means are to be employed in mitigating environmental impacts.[72]

Plaintiffs sometime focus a legal challenge on the ROD rather than the EIS itself and, as a result, prudence should be exercised in preparing the ROD. It should clearly demonstrate that the responsible official(s) understand the potential actions (i.e., whom, what, where, when, why, and how) and the resulting environmental consequences. Emphasis should be placed on explaining the rationale used in choosing the alternative.

Conclusions regarding environmental impacts must be specifically tied to the analysis contained in the EIS. Under no circumstances should the ROD record a course of action that has not been evaluated by the alternatives analysis presented in the EIS. The rationale used in choosing a final course of action should be carefully documented; if the agency is later challenged, a soundly documented rationale can strengthen the agency's defense, particularly in cases where it is accused of making an arbitrary or capricious decision. For this reason, among others, the trend in recent years has been towards lengthier RODs.

Occasionally, an alternative that meets the goal of one agency may not meet that of a cooperating agency. In such cases, each agency may identify its own preferred alternative within the EIS. Each agency may then prepare and issue a separate ROD identifying the alternative that it will pursue. In this case, the environmentally preferable alternative (see next section) identified in the ROD by one agency doesn't necessarily need to be the same as that identified in a second agency's ROD.[73]

Environmentally Preferable Alternative

While the environmentally preferable alternative must be identified in the ROD, it need not be in the EIS (§1505.2[b]). The reader should note that failure to choose a practical environmentally preferable alternative is one factor that can be used in determining if an action should be referred to the CEQ (§1504.2[f]).

The environmentally preferable alternative is the one that, on balance, is considered to best promote the goals expressed in Section 101 of NEPA. The CEQ has interpreted this as the alternative that "causes the least damage to the biological and physical environment" and that "best protects, preserves, and enhances historic, cultural, and natural resources." Thus, at least three distinct factors are to be considered in identifying the environmentally preferable alternative:

1. Biological resources
2. Physical environment
3. Historic, cultural, and natural resources[74]

As many competing factors may need to be balanced in identifying the environmentally preferable alternative, this determination is not always straightforward. This observation was noted during the public comment period for adoption of the Regulations, where commenters expressed the belief that it might be difficult to determine which alternative is the most environmentally preferable. For instance, is the protection of prime farmland in one area preferable to protection of a wildlife habitat in another area? Is protection of an endangered species more important than protection of a historic building? In response to this public comment, the draft Regulations were revised to allow an agency to choose more than one environmentally preferable alternative, regardless of whether the alternatives are "equally" preferable.[75]

Mitigation and Monitoring Plans

A monitoring and enforcement program for ensuring that decisions are appropriately implemented should be adopted and summarized in the ROD, especially in important cases. Such a plan must be adopted and summarized, where applicable, for any mitigation measures that are chosen (§1505.2[c]; §1505.3). Any mitigation measures that are adopted must be adequately evaluated in the EIS. To reduce paperwork, the ROD may incorporate discussion of mitigation measures by reference to the EIS.

It is important to note that under federal administrative law, commitments made in the ROD are legally binding and agencies are held accountable for their implementation. Such commitments are enforceable by other agencies and private entities alike.[76] Partly for this reason, the ROD should discuss the likelihood that the mitigation measures will actually be enforced by the responsible agency(ies). Where there is a history of nonenforcement or opposition to such measures, this fact should be acknowledged in the EIS and the ROD.[77]

References

1. Sun Tzu, *The Art of War,* James Clavell (New York: Dell Publishing, 1983), p. 26.
2. C. H. Eccleston, *The NEPA Planning Process: A Comprehensive Guide with Emphasis on Efficiency* (New York: John Wiley & Sons, 1999).
3. Senate Publication 101-9, No. 26, February 1990.

4. Eccleston, *The NEPA Planning Process,* 85–88.
5. Ibid., Section 3.4.
6. CEQ, Preamble to Final CEQ NEPA Regulations, 43 *Federal Register,* 55978, Section 3, November 29, 1978.
7. Eccleston, *The NEPA Planning Process,* Introduction.
8. CEQ, "Council on Environmental Quality—Forty Most Asked Questions Concerning CEQ's National Environmental Policy Act Regulations" (40 CFR 1500–1508), *Federal Register,* vol. 46, no. 55, 18026–18038, (March 23, 1981) question 2a.
9. Ibid., question 4a.
10. *Hughes River Watershed Conservancy v. Glickman,* 81 F.3d 437 (4th Cir. April 12, 1996).
11. C. H. Eccleston, "NEPA: Determining When an Analysis Contains Sufficient Detail to Provide Adequate NEPA Coverage for a Proposed Action," *Federal Facilities Environmental Journal* (Summer 1995): 37–50.
12. CEQ, public memorandum, 1980, "Talking Points on CEQ's Oversight of Agency Compliance with the NEPA Regulations."
13. CEQ, "Forty Most Asked Questions," question 7.
14. Ibid., question 2b.
15. CEQ, Talking Points.
16. CEQ, "Forty Most Asked Questions," question 1a.
17. N. C. Yost and J. W. Rubin, The National Environmental Policy Act, unpublished.
18. CEQ, "Forty Most Asked Questions," question 4b.
19. Ibid., question 6a.
20. Ibid., question 19a.
21. Ibid.
22. Ibid., question 19b.
23. *The Steamboaters v. Federal Energy Regulatory Commission,* 759 F. 2d 1382 (9th Cir. 1985); *Northwest Indian Cemetery Protective Association v. Peterson,* 795 F. 2d 688 (9th Cir. 1986).
24. CEQ, "Forty Most Asked Questions," question 19b.
25. Yost and Rubin, National Environmental Policy Act.
26. CEQ, "Forty Most Asked Questions," question 2b.
27. CEQ, "Talking Points."
28. *Webster's New Twentieth-Century Dictionary* (unabridged), 2nd ed., (New York: Simon and Schuster, 1983).
29. Forest Service Handbook H1909.15, FR 57:182, p 43189, September 18, 1992.
30. Superfund Act, 43 CFR 11, Section 107(a)(4)(C).
31. Superfund Act, 43 CFR 11, Section 107.
32. Ibid.
33. U.S. General Accounting Office, *Superfund: Status of Natural Resource Damage Claims,* GAO/T-RCED-95-239, 1995.
34. Superfund Act, 43 CFR 11, Section 107 (f)(1).
35. U.S. Senate, Senate Report Number 848, 96th Congress, 2nd session, 1980.
36. CEQ, "Forty Most Asked Questions," question 23b.

37. Ibid., question 23a.
38. CEQ, Preamble to Final CEQ NEPA Regulations, Section 4.
39. Ibid.
40. 36 CFR 60.2.
41. 36 CFR 60.2 and 48 FR 44723.
42. 43 CFR 7.
43. CEQ, Talking Points.
44. Ibid.
45. Forthcoming in *Environmental Practice,* National Association of Environmental Professionals, 1999.
46. Eccleston, *The NEPA Planning Process,* sections 8.6 and 12.2.
47. Ibid., section 12.3.
48. Executive Order No. 12898, Federal Actions to Address Environmental Justice in Minority and Low-Income Populations, February 11, 1994.
49. Presidential Memorandum for the Heads of all Departments and Agencies, released concurrently with Executive Order No. 12898, 1994.
50. EPA, *Draft Guidance for Incorporating Environmental Justice Concerns in EPA's NEPA Compliance Analyses* (Washington, D.C., July 1996).
51. CEQ, *Guidelines for Addressing Environmental Justice under the National Environmental Policy Act,* March 1998.
52. Eccleston, *The NEPA Planning Process,* Section 9.1.
53. Executive Order 13045, Protection of Children from Environmental Health Risks and Safety Risks, 62 *Federal Register.*
54. Final rule, "Determining Conformity of General Federal Actions to State or Federal Implementation Plans," 58 *Federal Register* 63214, No. 228, November 30, 1993; took effect on January 31, 1944 (40 CFR parts 6, 51, and 93).
55. CEQ, Preamble to Final CEQ NEPA Regulations, Section 4.
56. CEQ, "Forty Most Asked Questions," question 27c.
57. Ibid., question 27a.
58. Ibid., question 27b.
59. CEQ, Talking Points.
60. CEQ, "Forty Most Asked Questions," question 26a.
61. Ibid., question 26b.
62. Ibid., question 25a.
63. Ibid.
64. Ibid., question 25a.
65. Ibid., questions 25a and 29a.
66. Ibid., question 25b.
67. Ibid.
68. Ibid., question 23c.
69. Ibid., question 34c.
70. Ibid.

71. Ibid., question 19b.

72. CEQ, Talking Points.

73. CEQ, "Forty Most Asked Questions," question 14b. National Environmental Policy Act Regulations (40 CFR 1500–1508), Federal Register, Vol. 46, No. 55, 18026–18038, March 23, 1981.

74. Ibid., question 6a.

75. CEQ, Preamble to Final CEQ NEPA Regulations, Section 4.

76. CEQ, "Forty Most Asked Questions," question 34d.

77. Ibid., question 19b.

4

Implementing the Agency's Decision

Had I been present at the creation, I would given some useful hints for the better ordering of the universe.

—Alfonso X

The focus of this chapter is on providing the reader with modern tools, techniques, and approaches for ensuring that agency decisions are appropriately implemented. *Isos,* the Greek word for *equal,* is also commonly used in referring to the International Organization for Standardization (ISO). This chapter places emphasis on describing how the National Environmental Policy Act (NEPA) can be effectively integrated with an ISO 14000–consistent environmental management system (EMS).

4.1 Challenging the Agency's Decision

Agencies have frequently been successfully challenged for failing to fulfill commitments made in the environmental impact statement (EIS) or record of decision (ROD), or for undertaking actions not adequately evaluated in the chosen alterative. This section discusses the process of challenging an agency's action. For an in-depth description of the court system and litigation process, the reader is directed to the companion book, *The NEPA Planning Process: A Comprehensive Guide with Emphasis on Efficiency.*[1]

Initiating Legal Action

Once an agency complies with NEPA's procedural requirements, it may proceed with an action even though it may be environmentally destructive. For this reason, advisories tend to focus their efforts on challenging procedural violations. Although advisories may find this limitation disheartening, procedural challenges still offer a formidable weapon for challenging an agency's action.

A trivial violation does not normally, by itself, give rise to a lawsuit (§1500.3). Experts intimately familiar with all aspects of NEPA should be consulted to identify procedural grounds for challenging the agency's action.

Plaintiff versus Defendant

With respect to NEPA litigation, the plaintiff is the party challenging an agency's action; the defendant is the agency against which the challenge is lodged. The mere threat of a legal challenge is often sufficient to pressure an agency into reconsidering its course of action. If the plaintiff successfully argues its case before the court, a project may be halted and the agency may be compelled to initiate or repeat the NEPA process. Politically, this can be quite embarrassing, to say nothing of diminishing the agency's public credibility.

PLAINTIFF'S STRATEGY AND RESPONSIBILITY. The plaintiff needs to demonstrate that it is the proper party to challenge the federal action and that the issues being raised are those that the court is authorized to consider. This is known as *standing.*

A case can be strengthened by evidence that an action could harm a large number of people. A class-action suit can provide a mechanism for the man on the street to exploit such strength. The plaintiff's position is strengthened when it is able to demonstrate that environmental interests affect the public, or a large segment thereof, in addition to themselves. Likewise, the plaintiff should focus on showing that the action would result in environmental degradation beyond simple economic or social loss.

PRELIMINARY INJUNCTION. The plaintiff normally asks for a preliminary injunction. In reaching a decision to grant an injunction, the court considers the specific circumstances of the case.

Challenging the Adequacy of the EIS Analysis

A large number of NEPA lawsuits involve claims that an EIS is inadequate. The plaintiff is normally required only to provide a reasonable argument, not undisputable evidence, that the EIS is inadequate. The agency bears the burden of proof in demonstrating that the document is adequate.

Challenging the ROD

NEPA's mandate is not fulfilled by merely preparing an EIS to satisfy a procedural requirement and then completely disregarding the analysis in the subsequent decisionmaking process. The decisionmaker must review and consider the EIS analysis before reaching a final decision, which is recorded in the ROD. The agency must be able to demonstrate that results of the NEPA process were seriously considered and balanced against competing factors before the final decision was reached.

The Court's Role

The court's role is to determine if the agency has conducted an adequate and objective effort, based on the rule of reason, to present all significant environmental factors and alternatives for the decisionmaker's consideration.

It is not uncommon for scientific experts to disagree on technical issues. In cases where the agency's expertise and judgment are questioned, courts have typically sided with the agency.

The Administrative Record

The court's role is generally limited to reviewing the administrative record (ADREC) to determine if the agency made a sound decision based on available information. The court's role is normally not to second-guess the professional judgment of an agency; instead, its job is to determine if the agency can show that it carefully reviewed the facts before reaching a decision. Thus, the importance of establishing a thorough ADREC demonstrating that the agency properly weighed all pertinent factors in reaching a final decision cannot be overemphasized.

Occasionally, the court must look beyond the ADREC to verify that the agency considered all relevant factors. Allegations that an EIS did not investigate reasonable alternatives, overlooked significant impacts, or swept stubborn problems or serious criticism under the rug may raise questions sufficient to justify introduction of evidence outside the ADREC.

Remedy

The plaintiff may obtain a remedy if the court determines that the agency has violated NEPA's procedural requirements. Sometimes this is an order for the agency to give further consideration to the issue before taking action. In other cases, a remedy may be in the form of an injunction to take or not to take some sort of action by a certain time or after certain other actions have been undertaken.

4.2 Mitigation and Postmonitoring

As described in this section, mitigation and postmonitoring efforts are elements critical to the goal of environmental protection.

Mitigation

If an EIS involves issuance of a permit or approval, mitigation measures adopted in the ROD must also be included as a condition in whatever approvals are later made by the agency.[2] Mitigation and other conditions committed to in a ROD are enforceable through litigation. Such commitments may be enforced by other agencies and private entities alike; for this reason, the responsible agency official must ensure that sufficient funds and resources are available and that such measures are technically achievable.

As applicable, a mitigation action plan should be prepared in conjunction with the ROD. Any such plan should contain the following five essential elements:

1. Specific and detailed mitigation measures (how, where, and what measures will be taken)
2. Assignment of responsibility for successfully implementing such measures
3. A specific schedule
4. Appropriate funding authority
5. Measurable performance criteria

Monitoring

Postmonitoring is an essential step in ensuring that agency commitments are not lost in the haste of implementing a project. A monitoring program should be adopted and summarized in the ROD, especially in important cases (§1505.3). As applicable, a monitoring and enforcement plan should also be developed, summarized in the ROD, and adopted for mitigation measures that are chosen (§1505.2[c]).

At a minimum, a monitoring plan should be prepared for circumstances:

- That are controversial
- Where the effectiveness of the mitigation measures is problematic
- Where the impacts would occur over a protracted period of time

Such a plan may be particularly important where a proposal involves changing circumstances, from either an environmental or a developmental perspective.

An agency official should be designated to make results of relevant monitoring available to the public. Additional NEPA analysis may be required if the monitoring program reveals significant new circumstances or information, unanticipated impacts, or impacts significantly different from those originally addressed. A tool is provided in the next section for assisting the reader in determining if additional NEPA analysis must be prepared.

4.3 The Smithsonian Solution: Determining When a Proposed Change Requires Additional NEPA Analysis

Once the EIS is completed, the agency is at risk for subsequent changes that may not be adequately described or for complete invalidation of the EIS. Neither NEPA nor its implementing regulations (Regulations) provide detailed direction for determining the degree to which an action may vary before preparation of new or supplemental documentation is necessary. Notwithstanding, decisionmakers are routinely called upon to determine if a change to an action described in an EIS departs to such an extent from the description presented in the statement that additional documentation must be prepared; no two decisionmakers are likely to completely agree on this issue. Specifically, this problem involves answering the question, "To what degree can an action be changed before the corresponding description in the NEPA documentation must be supplemented to provide sufficient coverage for the action?"

Without a definitive methodology on which to base such decisions, decisionmakers and critics alike may point to a universe of potential considerations in defending their claim that a change in an action does or does not require new or additional NEPA documentation. Assertions are often based on equivocal opinions that can neither be proved nor disproved. This can result in a prolonged decisionmaking process (i.e., project delays), inconsistencies in decisionmaking, and increased risk of a legal challenge.

The author has developed a tool for streamlining such determinations by reducing the degree of subjectivity. I began considering this problem while strolling along the Potomac River in Washington, D.C., so I refer to it as the *Potomac Paradox*. Similarly, I formulated the solution at a table in the Smithsonian Institute's cafeteria; thus, I call the proposed decisionmaking tool for resolving the Potomac Paradox the *Smithsonian Solution*.*

Basis for the Tool

Under the Smithsonian Solution, a proposed change to an action is first reviewed against the description presented in the existing NEPA document. If the proposed change is a new action—one outside the scope of what was previously considered—the requirement to prepare additional NEPA documentation (categorical exclusion, environmental assessment, or EIS) is automatically triggered. The reader is referred to the author's paper on the subject for a detailed explanation of the tool.[3]

If the issue in question is not a new action but represents a change in the description of the previously considered action or impacts (including new information or circumstances), the decisionmaker complies with requirements prescribed in §1502.9[c][1]. Specifically, this provision states that an EIS must be supplemented if one of the following occurs:

* The concept is strictly that of the author and does not reflect opinions of the Smithsonian Institution.

1. The agency makes substantial changes in the proposed action that are relevant to environmental concerns.
2. Significant new circumstances or information arise relevant to environmental concerns and bearing on the proposed action or its impacts.

Because any action is potentially subject to requirements of an EIS, until proven otherwise, this provision can be considered equally applicable to actions that do not necessarily involve an EIS (i.e., environmental assessment). While §1502.9[c][1] provides general factors for determining if a change requires additional review, it does not provide specific criteria for assessing and making an actual determination. Criteria are proposed in the following section for assisting the reader in evaluating the two factors specified in §1502.9[c][1].

Evaluation Criteria for Assessing Changes

A thorough legal review of over 200 NEPA cases involving supplemental EISs was performed in constructing a practical tool for resolving the Potomac Paradox. Five discrete criteria (Table 4.1) were identified for determining if a change to a proposed action requires additional NEPA documentation. All five criteria were found to be consistent with existing case law. The reader is referred to the author's paper on the subject for a detailed justification of the criteria cited in Table 4.1.

Table 4.1 Criteria for determining when a change to an action requires preparation of additional NEPA documentation

1. There is a change in a previously described action that might result in a significant new impact not investigated in the earlier NEPA document.
2. There is a change in a previously described action that might cause an analyzed impact to deviate significantly from projections described in the existing NEPA document (i.e., there is a reasonable possibility that the proposed change could significantly alter impact projections investigated earlier).
3. There is a reasonable expectation that new alternatives could be identified for achieving the purpose and need of the proposed change that were not considered in the existing NEPA document and might affect the environment in a manner substantially different from the environmental effects of the proposed change.
4. Significant new circumstances or information relevant to environmental concerns has been obtained that could substantially change the agency's decision or could allow the public to contribute comments that could substantially improve or affect the manner in which the proposed change to an action is implemented.
5. Significant new circumstances or information relevant to environmental concerns has been obtained that could substantially change the public's understanding (or acceptance/rejection) of the proposed change in a manner substantially different from that which existed when the NEPA document was prepared. The public would benefit from an additional NEPA review.

The Smithsonian Solution Tool

Based on the criteria listed in Table 4.1, a general purpose tool (i.e., the Smithsonian Solution Tool), for determining when a proposed change requires additional NEPA documentation is presented in Figure 4.1. This tool provides decisionmakers with a systematic, rigorous, consistent, and defensible set of tests for performing what is otherwise a relatively subjective procedure.

Using the Tool

The tool is initiated with the first rectangle at the top of Figure 4.1. The proposed change is reviewed against the existing NEPA document. If the proposed change is a new action (first decision diamond shown in Figure 4.1) not previously considered, additional NEPA documentation must be prepared. If the proposed change involves a previously considered action, the action is reviewed in terms of the five remaining tests.

The next three tests are considered with respect to the question, "What could additional NEPA analysis of the previously described action reveal?" All three tests (second, third, and fourth decision diamonds in Figure 4.1) are considered in determining the outcome of this question. A *yes* to any one of the three tests is sufficient to reach a determination that additional NEPA analysis must be prepared (e.g., supplemental EIS). In a similar fashion, the remaining two tests are examined with respect to the question, "Could significant new circumstances/information (not previously considered) that is relevant to the potentially significant impacts, substantially change?" A *yes* to either of these two tests is sufficient to reach a determination that additional NEPA analysis must be prepared.

A response of *no* to all six tests supports a decision that the proposed change could be implemented without preparing additional NEPA documentation. A *yes* answer to any one of the six tests provides a basis for concluding that additional NEPA documentation is necessary. Where the answer to any test is not obvious, the user should err on the side of conservatism. Each test is evaluated according to the decisionmaker's best professional judgment. In applying this tool (Figure 4.1), the user is encouraged to refer to the more detailed criteria in Table 4.1.

Advantages and Restrictions

Whereas the Smithsonian Solution does not completely eliminate subjectivity, the vast array of potential considerations is essentially reduced to six narrowly defined tests. One must be prepared to justify why the analysis does or does not meet one or more of the tests. Views based on vague or ambiguous arguments are not justified.

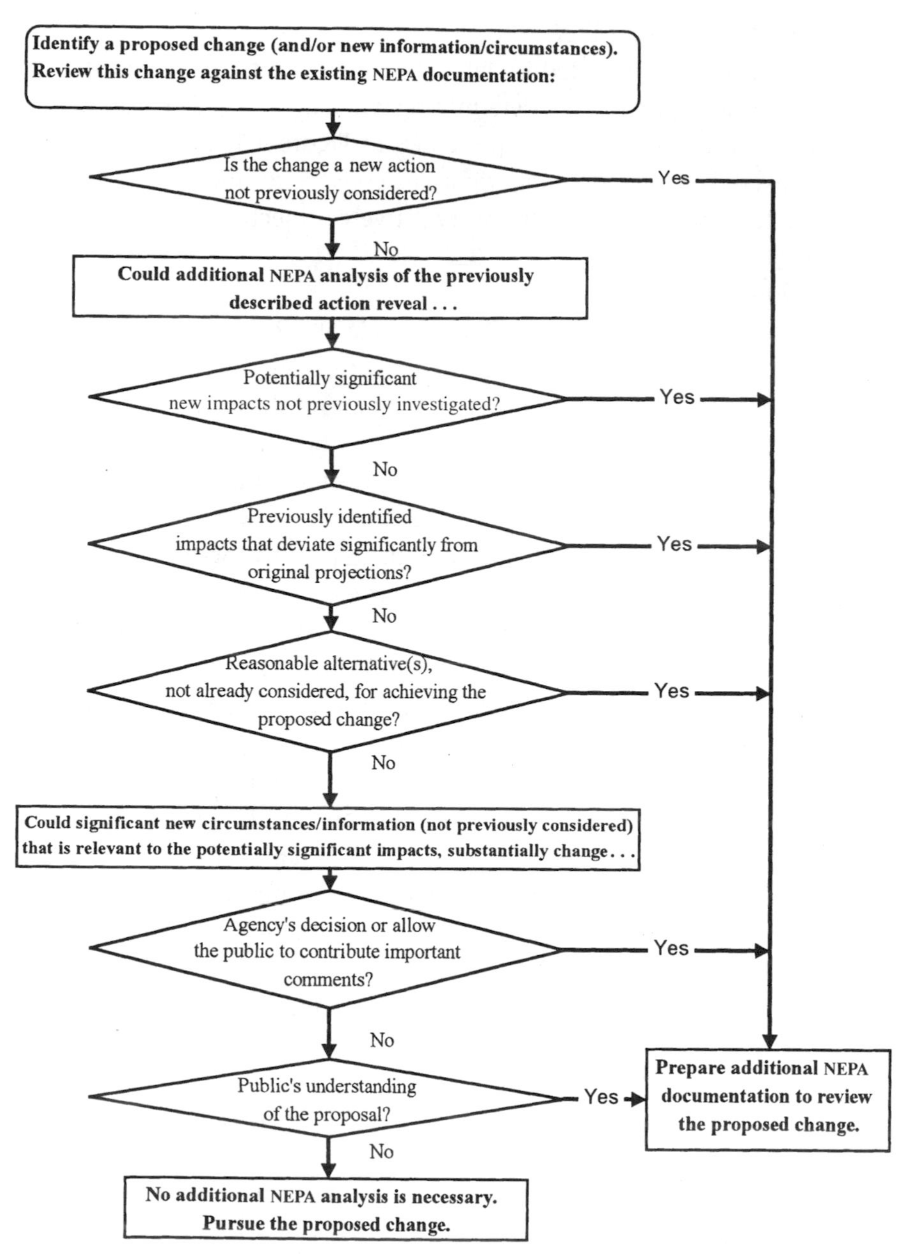

Figure 4.1 Smithsonian Solution. The tool provides a systematic methodology for determining when a proposed change requires additional NEPA documentation.

Exceptions might occasionally arise that cannot be completely addressed with this tool. In such instances, additional factors might need to be considered in reaching a final decision. Where an exception occurs, decisionmakers and critics alike are expected to provide a rational argument and specific evidence justifying why additional factors need to be considered.

4.4 Integrating NEPA with an ISO 14000 Environmental Management System

While NEPA establishes a national policy for protecting the environment and requires agencies to comply with certain action-forcing mechanisms such as preparing an EIS, it generally lacks a substantive mandate requiring agencies to make decisions or take actions protecting the environment. It is imperative that federal managers consider new approaches for enhancing environmental protection while reducing redundancies and cost. The diverse array of confusing and sometimes inappropriate or conflicting regulatory requirements compounds compliance complexities and increases the need to seek resourceful solutions. Faced with mounting environmental issues, compliance requirements, competing resources, and tightened budget constraints, agencies must seek innovative approaches for doing more with less.

A New Paradigm

As we have seen, an EIS provides a proven and powerful tool for formulating policy and planning federal actions. Strong parallels exist between the goals and requirements of NEPA and those of an ISO 14000–consistent EMS. The combination of NEPA with an EMS promises to infuse NEPA's substantive national policy goals into federal decisionmaking. Such a strategy could lead to more effective planning and enhanced environmental protection while streamlining compliance.

An integrated NEPA/EMS paradigm provides a key for increasing the effectiveness and uniformity of implementing NEPA at the early planning stage while reducing cost, delays, and redundancies. Effectively integrated, NEPA satisfies perhaps the most important one of the five principles of an EMS: environmental planning. NEPA's regulatory requirements are not only consistent with the objective of an EMS but actually enhance its effectiveness. An integrated approach adds the benefits of increased environmental coordination and heightened communications, which translate into cost reduction and fewer delays.

To promote efficiency, minimize redundancy, and enhance environmental protection, a strategy is described for effectively integrating NEPA with an EMS. This strategy is designed to balance the rigors of an international standard with the need to efficiently implement an integrated NEPA/EMS system, given a diverse set of challenging circumstances and constraints. To this end, this strategy is specifically designed with adaptability in mind, such that it can be adopted by any agency and applied at virtually any level of federal program or project implementation. The reader is referred to the author's paper for additional details on the integrated NEPA/EMS strategy.[4]

Why NEPA and EMS Complement One Another

As stated earlier, strong parallels exist between the goals and requirements of NEPA and those of ISO 14001 EMS. While the following strategy is directed at integrating NEPA with an ISO 14001–certified EMS, it could be applied to integration with any EMS that is consistent with the ISO 14001 standards. Hence, the term *EMS* is interpreted here to mean an ISO 14001–consistent EMS.

Not only are the goals and requirements of NEPA and ISO 14001 internally consistent, their strengths and weaknesses tend to complement one another. Table 4.2 summarizes the similarities and synergistic strengths of an integrated NEPA/EMS. These similarities and strengths are outlined in the following sections.

Policies and Plans Are Actions Subject to NEPA

The Regulations identify categories of federal activities that are subject to NEPA (Table 4.3). It is important to note that NEPA provides an inclusive framework for integrating and unifying early federal planning requirements and processes. Establishment of federal policies and plans is an action subject to the requirements of NEPA. Accordingly, federal policies and plans established as part of an EMS are potentially subject to the requirements of NEPA.

NEPA Promotes Integration of Environmental Requirements

As expressed in the Regulations, federal agencies are instructed to integrate NEPA with other environmental reviews (e.g., regulatory requirements, permits, agreements, project planning, and policies) so that procedures run concurrently rather than consecutively; this requirement reduces duplication of effort and delays in compliance and minimizes the overall cost of environmental protection:

- Integrate the requirements of NEPA with other planning and environmental review procedures. (40 CFR §1500.2[c])

Table 4.2 How NEPA and ISO 14001 Complement One Another

Comparison	NEPA	ISO 14001
Scope	All federal actions are subject to NEPA's requirements.	Optional. A federal agency may choose to adopt an ISO 14001 EMS.
Goal	To protect the environment by ensuring that environmental factors areconsidered during the early planning process.	To protect the environment by identifying impacts and, using a system of continual improvement, to reduce these impacts.
Mandate	Lacks a substantive mandate to protect the environment.	Requires that substantive actions be taken thatlead to continual improvement in environmental protection.
Planning function	Mandates that a comprehensive environmental planning process be conducted but lacks an environmental quality system for ensuring the decisions are properly implemented.	Requires a planning function and provides a system for ensuring that decisions are appropriately implemented, but does not prescribe a detailed process for performing the planning function.
External input	Defines a detailed formal public scoping process for identifying significant impact and eliminating nonsignificant issues.	Requires that a procedure (not public) be used to record and respond to external parties, but does not prescribe a detailed process.
Other environmental requirements	Executive orders and CEQ guidance direct federal agencies to integrate pollution prevention measures, environmental justice, biodiversity, and a host of other considerations with NEPA (see Appendix B).	Requires a top-level environmental policy, including a commitment to prevention of pollution that is broadly defined.
Life-cycle assessment	Requires analysis of reasonably foreseeable impacts over the life cycle of the action.	The ISO 14040 series describes in detail how to perform a life-cycle assessment.
Impact assessment requirements	Provides detailed specifications for analyzing direct, indirect, and cumulative impacts.	Requires an investigation of environmental aspects. Little specificity is provided on the requirements of this investigation.
Accumulated environmental experience	Nearly 30 years of experience has been accumulated in the planning and analysis of significant environmental impacts.	A relatively new requirement that has accumulated only limited experience in the planning and analysis of significant environmental issues.
Significance	Defines specific factors for determining the significance of environmental impacts.	Provides no detailed direction for interpreting or determining the meaning of *significance.*
Mitigation	Requires that mitigation measures are identified and analyzed as part of the planning processes.	Provides a system for ensuring that mitigation measures are implemented.

(continues)

Table 4.2 *Continued*

Comparison	NEPA	ISO 14001
Monitoring	Encourages (and sometimes requires) postmonitoring measures.	Mandates monitoring as part of the continual improvement cycle.
Continual improvement	CEQ is promoting a cyclical process based on adaptive management.	A continual improvement process is a basic concept inherent in an EMS.

- Identify other environmental review and consultation requirements. . . . [P]repare other required analyses and studies concurrently with, and integrated with, the environmental impact statement. (40 CFR §1501.7[a][6])
- Any environmental document in compliance with NEPA may be combined with any other agency document. (40 CFR §1506.4)[5]

The NEPA Regulations define detailed analytical and documentation requirements for performing the NEPA analysis. With little modification, a NEPA analysis can be used to advance and streamline compliance with requirement of an ISO 14001–consistent EMS.

PLANNING VERSUS IMPLEMENTATION. Environmental planning is a mandatory element under ISO 14001. Unfortunately, ISO 14001 provides only limited specifications for conducting the planning function. Procedures and requirements for scoping, investigating environmental aspects, defining tem-

Table 4.3 Categories of federal activities subject to NEPA

Federal actions tend to fall within one of the following categories (40 1508.18[b]).

1. "Adoption of official policy . . . formal documents establishing an agency's policies which will result in or substantially alter agency programs."
2. "Adoption of formal plans, such as official documents prepared or approved by federal agencies which guide or prescribe alternative uses of federal resources, upon which future agency actions will be based."
3. "Adoption of programs, such as a group of concerted actions to implement a specific policy or plan."
4. "Approval of specific projects:"
   - "projects and programs . . . new or revised agency rules, regulations, plans, policies, or procedures." (40 1508.18[a])
   - "Actions include new and continuing activities, including projects and programs entirely or partly financed, assisted, conducted, regulated, or approved by federal agencies; new or revised agency rules, regulations, plans, policies, or procedures; and legislative proposals." (1508.18[a])

poral and spacial bounds, interpreting significance, and executing many other requirements are, at present, only vaguely defined or implied.

In contrast, NEPA's regulations provide highly prescriptive direction and requirements for ensuring that an accurate and scientifically defensible analysis is prepared, providing decisionmakers with information sufficient to make an informed choice. These requirements are reinforced by nearly 30 years of experience gained by agencies engaged in diverse missions and environmental issues. Properly combined, a NEPA/EMS system provides a synergistic process for planning actions and implementing decisions in a manner that protects and enhances environmental quality while minimizing cost.

Substantive versus Procedural

Viewed by the courts, NEPA is largely a procedural requirement—that is, an agency must comply with the procedural aspects of NEPA but is not obligated to select an environmentally beneficial alternative or to demonstrate that its decision conforms to the environmental goals established in Section 101 of the Act.

NEPA's contribution derives not from a substantive mandate to choose an environmentally beneficial alternative but from its procedural requirement forcing decisionmakers to rigorously evaluate and consider the effects of potential actions on the environment, just as they would balance more traditional factors, such as cost and schedules. In contrast, an ISO 14001–consistent EMS involves a commitment to take substantive actions to improve environmental quality. Not only must environmentally beneficial actions be taken, they must be undertaken in a cyclical process of continual environmental improvement. Thus, an EMS provides a mechanism for enforcing the substantive environmental mandate that NEPA lacks.

Similarly, NEPA requires analysis of mitigation measures but places no substantive mandate on decisionmakers to enact such measures. In contrast, ISO 14001 requires organizations to establish target objectives for improving environmental performance. Obtaining such targets necessitates implementing actions similar to that of NEPA's mitigation measures. Again, NEPA prescribes rigorous requirements for planning and investigating mitigation measures, while ISO 14001 provides the teeth for implementing them.

Analytical Similarities

The Regulations provide highly prescriptive requirements for ensuring that an accurate and defensible analysis is performed that provides a decisionmaker with information sufficient to support informed decisionmaking.

NEPA is more demanding in requiring a comprehensive analysis of direct, indirect, and cumulative impacts. In contrast, ISO 14001 requires investigation of significant "environmental aspects," which are the specific activities that affect the environment. While the environmental aspects must be determined, their environmental consequences or impacts on environmental resources are not required to be evaluated.

The NEPA process is reinforced by nearly three decades of experience accumulated by a diverse range of federal agencies, each faced with a unique, organic mission and a wide spectrum of environment issues. From a planning perspective, NEPA provides a rich and rigorous platform for ensuring that environmental impacts are identified, evaluated, and considered before a decision is made to pursue an action.

LIFE-CYCLE ANALYSIS. To the extent practical, NEPA requires that an analysis be performed over the entire life cycle of an action, including connected actions. Both short- and long-term effects must be considered. The reasonably foreseeable impacts of future actions must be identified and evaluated. The ISO 14040 series describes in detail how a life-cycle analysis should be performed. Integrating such requirements reduces cost and paperwork. A tool for determining the appropriate integration of NEPA with a life-cycle assessment (LCA) is described shortly.

Significance

Significance of environmental impacts is a theme central to both NEPA and ISO 14001. NEPA requires analysis of potentially significant impacts of federal actions. The concept of significance permeates NEPA's regulatory provisions, which include a definition and specific factors to be used by decisionmakers in reaching determinations regarding significance. Not so with ISO 14001. Here, *significance* is vaguely defined and connotes no specific factors for use in reaching a determination. Again, NEPA brings three decades of experience to bear on the problem of determining significance. NEPA's regulations provide specific, publicly reviewed factors, reinforced by case law, for assisting decisionmakers in reaching such determinations.

Integrating Pollution Prevention

CEQ has issued guidance indicating that, where appropriate, pollution prevention (P2) measures are to be coordinated with and included in the scope of a NEPA analysis.[6] A number of federal agencies have also issued similar directives. ISO 14001 speaks to the merits of P2, but mainly from the standpoint of establishing a top-level P2 policy. Under an integrated process, NEPA provides an ideal framework for evaluating and integrating a comprehensive

P2 strategy while ISO 14001 provides a top-down policy for ensuring that P2 is actually incorporated at the operational level.

Public Participation

Public participation is essential to the NEPA process. Decisions regarding significance and the choice of alternatives are highly dependent on the concerns of stakeholders. In contrast, the ISO 14001 series requires only that a plan for external communications and inquiries be developed. This weakness is apparent in almost all parts of ISO 14001. This is another case where NEPA's three decades of experience with public scoping and participation balances the weaknesses of the ISO 14001 EMS.

Monitoring and Continuous Improvement

As described earlier in this chapter, the Regulations strongly encourage and, in some instances, mandate incorporation of monitoring. The courts, however, generally have not insisted that agencies incorporate monitoring as part of the NEPA process. In contrast, monitoring is inherent to an EMS. A properly integrated NEPA/EMS ensures that monitoring is correctly executed.

ADAPTIVE MANAGEMENT AND STRATEGIC PLANNING. Recently, CEQ has begun advocating two new paradigms.[7]

The first, referred to as *adaptive management,* consists of five steps: predict, mitigate, implement, monitor, and adapt. The intent of this new approach is to allow midcourse corrections based on the findings of environmental monitoring. Under an EMS, the monitoring step is a basic element used in ensuring that the organization's environmental policy and plan are implemented properly. As appropriate, a plan is developed to correct deficiencies and improve environmental performance. The cycle is repeated. The EMS plan and policy are revised and reimplemented. What appears to have gone largely unnoticed is that adaptive management is not only consistent with but is, in fact, surprisingly similar to the continual improvement cycle underlying an EMS (see Figure 4.2).

The second paradigm, referred to as *strategic planning,* incorporates a collaborative approach for identifying and solving environmental problems within the agency's internal planning process at the early planning stages. An EMS could provide a crucial mechanism for integrating strategic planning into agency operations.

Strategy for Integrating an EMS with NEPA

A strategy for integrating an EMS with NEPA is depicted in Figure 4.3. Conceptually, Figure 4.3 is composed of three discrete functions or phases: (1)

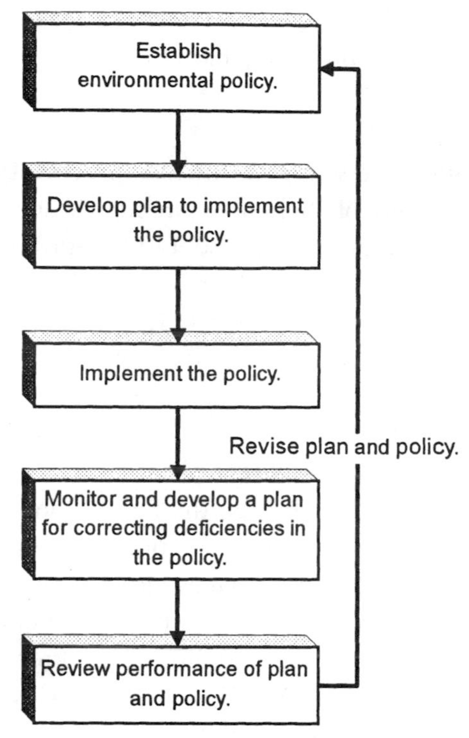

Figure 4.2 Overview of a typical environmental management system

centralized planning, analysis, and decisionmaking, (2) implementation, and (3) environmental monitoring and enforcement. The approach is designed to be implemented either in a standalone mode or, preferably, integrated with the total federal planning (TFP) strategy described in chapter 5.

Centralized Planning, Analysis, and Decisionmaking Phase

The proposed scheme is initiated with the establishment of a high-level organizational environmental policy and a commitment to environmental quality (block #1, Figure 4.3). Next, an effort is mounted as part of the NEPA process to develop a specific plan for implementing the environmental policy (block #2, Figure 4.3). Proposals for federal actions are also evaluated in this phase.

PUBLIC SCOPING. NEPA's public scoping process (described in detail in chapters 1 and 2) is used to obtain public input and sort significant from non-significant issues (block #3, Figure 4.3). Specifically, the scoping process is used to define the range of actions, alternatives, and impacts that will be evaluated.

As described earlier, NEPA's definition of significance and its 10 factors are used in determining the significant impacts (attributes) that will be evalu-

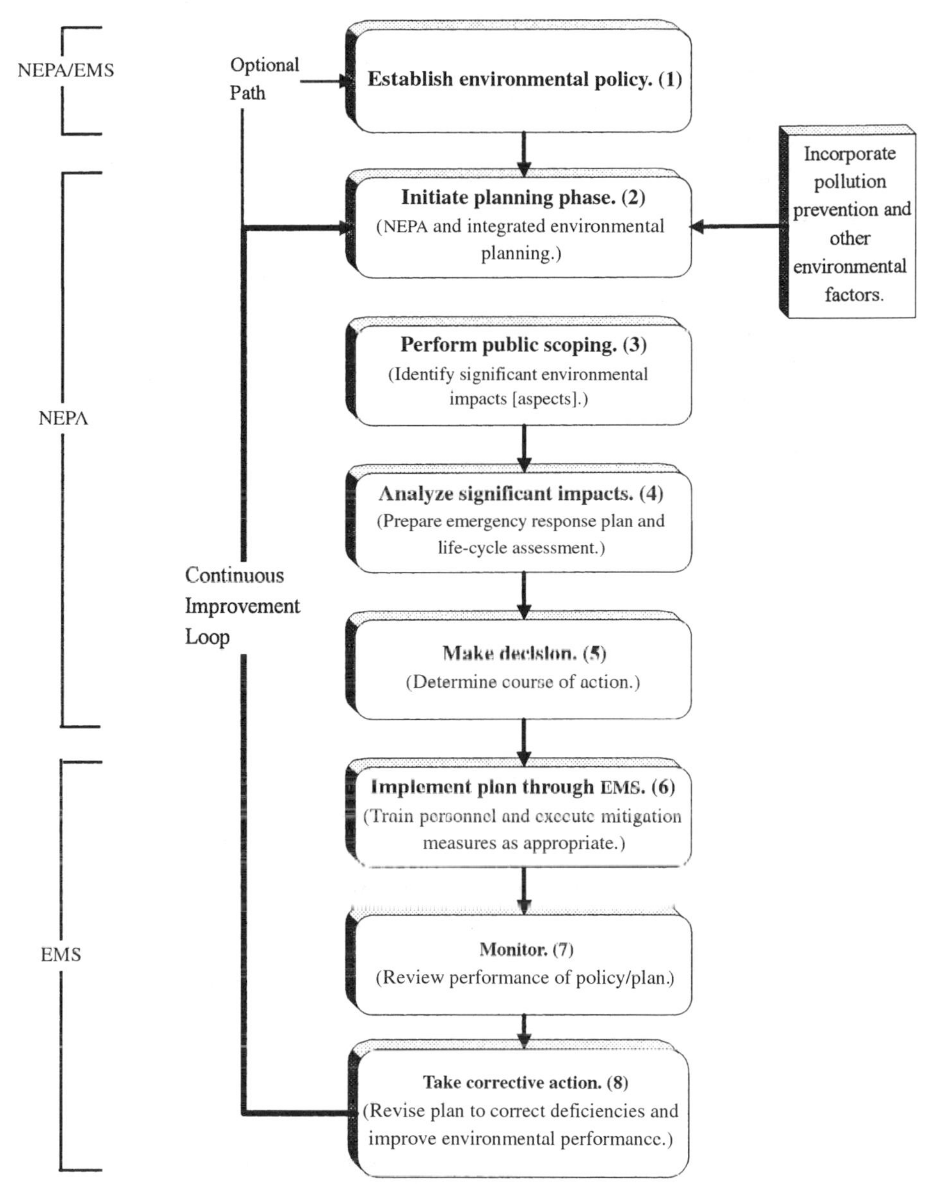

Figure 4.3 Conceptual process for integrating an EMS with NEPA

ated. *The* NEPA *Planning Process* provides an in-depth discussion of significance and its determination.

ANALYSIS, SIGNIFICANCE, AND DECISIONMAKING. Drawing on a diverse array of expertise and experience, an interdisciplinary effort is used in developing an environmental implementation plan. This plan may be prepared for a major federal program or installation, a particular facility or operation, or a project-specific action. NEPA provides the interdisciplinary framework for integrating and coordinating all early environmental planning, reviews, and analyses necessary to support formulation of the plan; thus, the environmental implementation plan might incorporate an array of related environmental issues such as P2, safety procedures, habitat management practices, environmental justice, sustainable development, and related goals or disciplines. Chapters 1 and 2 offer a description of tools and approaches that can be used in performing this task.

Consistent with NEPA, scenarios and alternative approaches are investigated as part of the process for developing the environmental implementation plan (see chapter 2). Actions, alternatives, and impacts are evaluated pursuant to the requirements of NEPA and additional requirements that might be promulgated as part of the EMS (block #4, Figure 4.3). The NEPA analysis can also be used to prepare an emergency response plan, thus satisfying an important EMS requirement.

The NEPA analysis, combined with other pertinent planning studies, is used by the agency in reaching a final decision (block #5, Figure 4.3).

Implementation Phase

Once the environmental implementation plan is completed and a final decision reached, an implementation effort is begun. Pursuant to ISO 14001, personnel are trained and applicable mitigation measures are implemented (block #6, Figure 4.3).

A centralized planning function (see chapter 5) is useful in coordinating implementation of the plan among the relevant federal facilities and operations. At the lower facility or operational level, environmental compliance officers or equivalents can be assigned responsibility for preparing a more detailed site-specific facility implementation plan for carrying out the Environmental Implementation Plan.

Under this scheme, the environmental implementation plan provides high-level direction and constraints that the facility implementation plan must meet. Thus, individual facility implementation plans could be tiered from the environmental implementation plan, providing facility- and project-specific direction for instituting results of the planning process by way of the EMS. The centralized planning office could be assigned responsibility for approving each facility implementation plan to ensure continuity and consistency. Implementation problems and cross-cutting issues could also be elevated to the centralized planning office for resolution.

Further, the EMS requires extensive job-appropriate training of all affected employees to ensure that the facility implementation plans are implemented correctly. Defining and tracking the appropriate training requirements could be a centralized function and performing training an operations-specific function.

Monitoring and Enforcement Phase

As depicted in block #7 of Figure 4.3, a centralized oversight office (consistent with a TFP approach) could be assigned responsibility for performing reviews and monitoring facility and operational compliance. An environmental compliance officer (or equivalent) could be assigned responsibility for preparing and transmitting input and status reports to the oversight office. Audits could be performed periodically by the oversight office to verify compliance.

The monitoring data are evaluated to verify compliance and effectiveness of the EMS in meeting the established policy and plan. As appropriate, the organizational policy and plan are revised to correct deficiencies (block #8, Figure 4.3). Substantial changes could be made at the centralized planning level, while less significant changes might be implemented at the project or facility level by revising the facility implementation plan. Ultimately, the concept behind the EMS is that impacts will eventually dissipate such that the revised plan would address impacts different from those in the existing plan. Such a process ensures a continuous improvement cycle, which is the hallmark of an EMS; it also promotes NEPA goals as well as CEQ's paradigm of adaptive management.

Integrating Life-Cycle Assessment with a NEPA Analysis

The LCA is a valuable cradle-to-grave accounting tool for use in identifying and mitigating negative environmental and economic impacts of products and processes. In many circumstances, an LCA can provide an effective methodology for detecting resource inefficiencies and major sources of waste generation. Direction for performing an LCA is contained in ISO 14040.[8]

Many questions have been raised regarding the applicability of an ISO 14040–consistent LCA with a NEPA analysis. Specifically, under what conditions does it make sense to perform an integrated NEPA/LCA?

Preparation of a LCA can be challenging, time-consuming, and costly. Clearly, prudence must be exercised in reaching a decision to perform an integrated NEPA/LCA analysis. The author has proposed a decisionmaking tool consisting of six discrete factors that can assist practitioners in reaching a decision regarding the appropriate integration of a NEPA analysis with an LCA. Properly applied, this tool reduces the risk that an LCA may be inappropriately prepared and integrated with a NEPA analysis. The reader is referred to the

Table 4.4 Factors considered in determining if a federal action is a candidate for a combined NEPA/LCA analysis

1. *Significant impacts.* As an LCA requires extensive analytical effort, the environmental assessment does not normally provide a practical mechanism for performing this analysis. In contrast, the EIS is prepared on federal proposals that may significantly affect the environment and does provide an effective mechanism for incorporating an LCA, as it is a large undertaking requiring a rigorous analysis of potential actions, alternatives, and their significant impacts.
2. *Reasonably foreseeable impacts.* The courts have generally concluded that the analysis of environmental impacts under NEPA is governed by the rule of reason, impacts need be evaluated only to the extent they are deemed reasonably foreseeable. The requirement to investigate impacts does not extend to impacts deemed remote or speculative.
3. *Ripeness for decision.* While the scope of an EIS inquiry is normally much broader than that of an LCA, the latter normally necessitates a level of assessment that, while narrower, is also more detailed than the former. For this reason, an LCA should be performed only if the proposal has matured to a stage where sufficiently detailed information is available to support an analysis. Because it requires detailed information, an LCA may need to be performed at a later stage of project development than an EIS.

   The NEPA regulations recognize such limitations as they encourage federal agencies to tier EISs to focus on actual issues "ripe for decision at each level of environmental review." (§1502.4[b], §1502.20, §1508.28) If a particular action or process has not advanced to the stage where it is ripe for decision (e.g., analysis), it may be deferred to a later stage when more information is available. If an action has not matured to the point where sufficient information is available to prepare an LCA, it may be appropriate to defer preparation until such time as any lower-tier NEPA analysis may be prepared and tiered from the EIS.
4. *Planning and decision-making process.* NEPA is a planning and decision-making process. As indicated in the Regulations:

   "The NEPA process is intended to help public officials make decisions that are based on understanding of environmental consequences, and take actions that protect, restore, and enhance the environment." (§1500.1[c])

   Use the NEPA process to identify and assess the reasonable alternatives to proposed actions that will avoid or minimize adverse effects of these actions upon the quality of the human environment. (§1500.2[e])

   Use all practical means . . . to restore and enhance the quality of the human environment and avoid or minimize any possible adverse effects of their actions upon the quality of the human environment. (§1500.2[f])

   Consistent with this direction, an LCA might be justified if it has potential to improve the agency's final decision. (§1500.1(c), §1500.4(f), §1501.2, §1502.1, §1502.5, §1502.14)
5. *Unresolved conflicts.* In complying with NEPA, agencies are required to "study, develop, and describe appropriate alternatives to recommended courses of action in any proposal which involves unresolved conflicts concerning alternative uses of available resources."[10] Many LCA studies are conducted to evaluate unresolved conflicts in the use of environmental resources. A LCA, therefore, might be justified if it would help settle or clarify unresolved conflicts in the commitment of environmental resources.
6. *Clear basis for choice among options.* The section on alternatives is the "heart" of the EIS. In describing the proposal and alternatives, this section supports the ultimate goal of "sharply defining the issues and providing a clear basis for choice among options." (§1502.14) An LCA might be justified if it could help to define the issues, thus providing the decisionmakers with a clear basis for making a reasoned choice among alternatives.

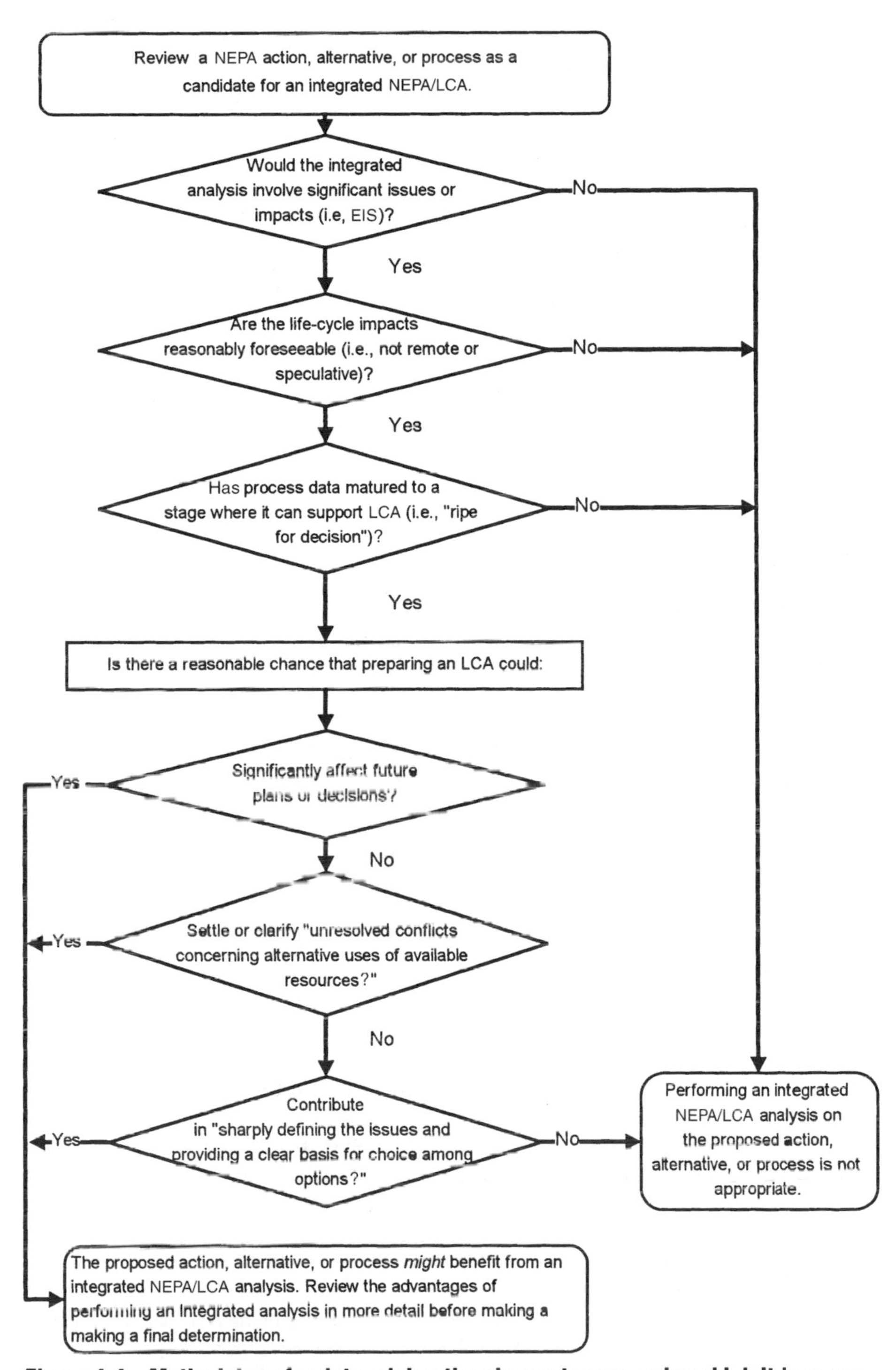

Figure 4.4 Methodology for determining the circumstances under which it is appropriate to combine an ISO 14040 LCA with a NEPA analysis

author's publication on the subject for additional details on the basis and use of this tool.[9]

Criteria Useful in Making a Determination

Based on the purpose and regulatory requirements of NEPA, factors are proposed in Table 4.4 for determining if a federal action should be the subject of a combined NEPA/LCA.

A Decisionmaking Tool

A decisionmaking tool consisting of six distinct tests, based on the factors presented in Table 4.4, is presented in Figure 4.4. Undeniably, this tool does not eliminate subjectivity inherent in such determinations. However, it does provide a consistent methodology for reducing the vast number of factors that may inundate practitioners and decisionmakers as they ponder this issue.

APPLYING THE TOOL. Application of this tool is initiated by the rectangle at the top of Figure 4.4. The user reviews a proposal (e.g. alternative, subprocess, mitigation measure) to determine if an integrated NEPA/LCA analysis is warranted. A series of six tests is used to assist the user in reaching a decision.

The first three tests are screening criteria. A *no* response to any one of these is sufficient to reach a determination that an integrated NEPA/LCA analysis is not warranted. If the response to the first three tests is *yes,* the user proceeds to the second rectangle, labeled with the following question: "Is there a reasonable chance that preparing an LCA could?" This question is followed by three additional tests. A response of *yes* to any one these is sufficient to support a determination that some aspect of the proposal might benefit from an integrated NEPA/LCA analysis.

References

1. C. H. Eccleston, *The NEPA Planning Process: A Comprehensive Guide with Emphasis on Efficiency* (New York: John Wiley & Sons, 1999), ch. 13.
2. CEQ, "Council on Environmental Quality—Forty Most Asked Questions Concerning CEQ's National Environmental Policy Act Regulations (40 CFR 1500–1508)," *Federal Register* vol. 46, no. 55, 18026–18038, March 23, 1981, question 34c.
3. C. H. Eccleston, "Determining If a Change to a Proposal Requires Additional NEPA Documentation—The Smithsonian Solution," *Federal Facilities Environmental Journal* 10, no. 1 (Spring 1999): 119–132.
4. C. H. Eccleston, "A Strategy for Integrating NEPA with an EMS and ISO 14000," *Environmental Quality Management* 7, no. 3 (Spring 1998): 9–17.
5. CEQ, Regulations for Implementing the Procedural Provisions of the National Environmental Policy Act, 40 Code of Federal Regulations, Pts. 1500–1508, 1978.
6. CEQ, Guidance on Pollution Prevention and the National Environmental Policy Act, published at 58 FR 6478, January 29, 1993.

7. CEQ, *The National Environmental Policy Act: A Study of its Effectiveness after Twenty-Five Years,* 1997.
8. ISO 14040: Environmental Management—Life Cycle Assessment—Principals and Framework
9. C. H. Eccleston, "Integrating a Life-Cycle Assessment with NEPA: Does it Make Sense?" *Environmental Quality Management Journal* 8, no. 3 (Spring 1999): 43–48.
10. Section 102(2)(E) of NEPA.

5

Total Federal Planning: A Unifying Strategy for Integrating Federal Planning

We trained hard, but every time we began to form up teams we would be reorganized. I was to learn later in life that we tend to meet any new situation by reorganizing; and a wonderful method it can be for creating the illusion of progress while producing confusion, inefficiency, and demoralization.

—Petronius Arbiter
210 B.C.

The introduction to this book described traditional problems that have hampered the effectiveness of federal planning in general and the National Environmental Policy Act (NEPA) of 1969 in particular. A repertoire of state-of-the art approaches and methodologies for addressing these problems and enhancing the effectiveness of federal planning was also introduced. These methodologies form the basis of a modern approach to federal planning, referred to in this book as *total federal planning* (TFP). The TFP strategy is designed to address problems depicted in Table I.2. Specifically, the strategy is advanced with the expressed purpose of unifying large-scale federal planning efforts under a single, systematic, structured, and holistic framework.[1] The TFP strategy provides an innovative approach to planning that has potentially profound implications for the way planning is managed at the federal level.

TFP considers planning comprehensively, such that NEPA is only one of many elements essential in forging an effective and all-encompassing planning process. Thus, the scope of this chapter is much broader than the traditional bounds of NEPA. Its intention is nothing less than arming the reader with a strategy and tools for rethinking and, if warranted, challenging the traditional framework, problems, and inefficiencies that have frequently plagued federal planning. At a time when Congress is striving to reengineer the governmental framework, a TFP strategy offers a sound and systematic approach to unifying the disjointed and convoluted planning processes often encountered by planners and project managers alike. Potentially, this approach has widespread implications in the way federal planning is approached. While scattered cases exist where certain features of the TFP strategy were used, documented cases have not been identified where it was implemented comprehensively.

The TFP strategy is specifically designed for large or complex programs and projects, particularly where preparation of an environmental impact statement (EIS) is required. The strategy complies with all NEPA regulatory requirements for preparing an EIS. Agencies can also benefit from applying certain aspects of this strategy to small-scale project planning as well. TFP is designed such that planners and decision makers can customize and adapt the approach to their particular needs. Thus, TFP can be modified and tailored case by case to accommodate existing circumstances. While the TFP strategy is specifically designed for federal program and project planning, it could with proper modifications, be applied to private or nonfederal planning processes.[2] Professional judgment must be exercised in the application of this model.

As this chapter is intended to provide only a high-level overview, the reader will find more detailed guidance for coordinating and preparing an EIS in chapters 1 and 2. Chapter 3 provides a detailed description of the documentation requirements that the EIS must meet. Chapter 4 describes additional tools and procedures that can be integrated with TFP.

This chapter assumes that the reader is acquainted with or has access to a text describing EIS process requirements. For this reason, detailed EIS regulatory procedures and requirements are not described. Instead, the reader is referred to the author's companion book, *The NEPA Planning Process: A Comprehensive Guide with Emphasis on Efficiency,* for a detailed discussion of the NEPA process and the specific regulatory requirements governing preparation of an EIS.[3]

5.1 NEPA Provides a Framework for Integrating All Federal Planning

Concinnity (kun SIN i tee) is the act of blending distinct elements so skillfully as to form an elegant and harmonious whole. Concinnity, or the blending of all relevant decisionmaking factors into a single integrated process, is a component essential to the success of any comprehensive and holistic planning process. Planning processes that fail to blend pertinent decisionmaking factors are destined to experience disconnects or redundancies, ultimately leading to costly or ineffective actions. Conceptually unpretentious and alluring in its simplicity, concinnity is often not realized in the real-world setting of federal planning and policy.

The Council of Environmental Quality's (CEQ) NEPA regulations (Regulations) mandate explicit requirements governing the preparation of an EIS.[4] Yet, there is a distinct difference between specifying what procedural requirements must be satisfied and how such requirements are best implemented. The NEPA regulations are quite explicit in specifying what procedural requirements must be met. In promoting a flexible process that can be tailored to a diverse array of federal projects, agencies are granted unusual latitude in determining how such requirements actually will be implemented. This chapter focuses on describing a modern general-purpose approach to how NEPA should be integrated with the federal planning process.

No rigorous and structured general-purpose approach has been accepted for determining how NEPA's regulatory requirements are most effectively implemented. At best, minimal progress has been witnessed in developing or integrating new tools, methods, and procedures for effectively implementing the EIS process.

New approaches that incorporate innovative methodologies for harnessing NEPA's untapped potential to facilitate federal planning are needed if NEPA is to truly achieve the goals Congress laid out over a quarter-century ago. Interestingly, NEPA's regulations provide the regulatory basis for constructing a general-purpose process for combining all pertinent decisionmaking factors into a single, integrated planning process.

NEPA is Unique

In creating NEPA, Congress laid the foundation for a comprehensive planning framework. In the broad array of federal planning and compliance requirements, NEPA is unique. It is the only federal planning process that is both mandated and applicable to all major federal actions. From its inception, NEPA was envisioned by its authors to provide a framework so flexible that it could be molded and adapted to meet the diverse missions and specific circumstances of any federal agency. Echoing this philosophy, the Regulations were specifically written to provide agencies with such flexibility. Three decades after NEPA's enactment, the Act and its implementing regulations continue to provide the basic elements, flexibility, and framework essential for welding a diverse array of planning considerations and requirements into a single, comprehensive, and integrated planning process.

NEPA Provides All Elements Necessary for General-Purpose Planning

As depicted in the left-hand column of Table 1.2, NEPA contains the essential requirements and elements requisite to the success of any general-purpose planning process. For example, agencies are instructed to integrate NEPA with other environmental reviews so that they "run concurrently rather than consecutively," thereby avoiding duplication of effort and delays in compliance and minimizing cost.[5] The regulations direct agencies to coordinate and "integrate the NEPA process with other planning at the earliest possible time."[6] Moreover, the "heart" of the NEPA planning process is the requirement to provide an open and unbiased evaluation of "all reasonable alternatives."[7]

As the only general-purpose planning process required for all major federal actions, NEPA provides the underlying *Tao* (path) for achieving concinnity. The flexibility that NEPA offers as a comprehensive process for integrating all early planning factors (environmental and nonenvironmental) into a single framework has gone largely unrecognized. Environmental considerations are only one of many factors evaluated. As described below, the TFP strategy provides an integrated approach under which all relevant decisionmaking factors (e.g., cost, safety, schedules, risk, technological constraints, and regulatory requirements) are considered together within a single interdisciplinary framework.

5.2 A New Approach to Federal Planning

TFP offers a structured framework for integrating all early environmental and nonenvironmental decisionmaking factors into a single, comprehensive process for planning major federal actions. To promote effectiveness and efficiency, modern tools and principles from the disciplines of value engineering (VE), systems engineering (SE), and total quality management (TQM) (see Introduction) are intrinsic to the concept of TFP. Other related planning processes and methodologies that can also be integrated with TFP are described in chapter 4. Properly integrated, such tools and principals provide the rigorous, structured, and disciplined framework essential in achieving concinnity.

An overview of a general-purpose strategy incorporating all pertinent steps essential for addressing problems commonly encountered in federal planning is introduced in the following section; this section provides an overview of the TFP approach from the perspective of an organizational structure implementing all early planning under an autonomous interdisciplinary steering team (IST).

Overview of the Total Federal Planning Process

TFP can be viewed as consisting of three underlying concepts:

1. The NEPA process, which provides the unifying framework for integrating all pertinent planning factors
2. A structured and systemized approach for performing integrated planning that addresses traditional problems outlined in the introduction

3. Incorporation of management tools and principles that are used throughout this process to facilitate efficiency and effectiveness.

VE tools and principles are used to identify and manage important information and factors, while TQM principles are used to optimize alternatives and to ensure quality in the planning process. SE principles are used as a unifying tool for integrating and conducting the investigation of alternatives and various other aspects of the analysis. Table 5.1 summarizes succinctly the benefits and features of TFP. A detailed description of the TFP procedural process for integrating federal planning that complies with all applicable EIS regulatory requirements is described in section 5.3.

Centralizing the Planning Function

As is often the case in both federal and private organizations, elements essential to a well-orchestrated planning function are detached from one another; to address the problems resulting from such organization (see problems described in Table I.1 Introduction), the TFP approach organizes all elements critical to the objective of performing comprehensive integrated planning into a centralized planning function.

The small, nonshaded circles flanking the upper half of the circle denoting the IST (Figure 5.2) represent some of the functions essential to an effective planning function. Under the TFP approach, elements essential to the success of planning are centralized within an integrated planning framework. Centralizing such functions allows specialists from many interrelated disciplines to more effectively communicate and work together in developing integrated plans. Such a system must include more than just planners; all elements essential to the success of integrated planning must be included.

Organizational Structure

Under TFP, the planning process must be designed to cross functional boundaries. This necessitates creation of cross-functional team(s) with responsibility for developing a unified and optimum plan in lieu of simply satisfying members' individual functional responsibilities. The planning effort must involve every pertinent decisionmaking level, from management through project staffing. To this end, the IST must be structured and given independent authority so as to facilitate coordination of efforts, communications, and the collection and dissemination of information.

While a well-orchestrated planning function may benefit from physically consolidating its functions into a single department or even a single building, this is not necessarily a requirement under the TFP approach; at a

Table 5.1 Principal features underlying the TFP process

- The TFP process is holistic, combining and integrating early planning efforts into a single, interdisciplinary, unified planning process conducted under the NEPA umbrella. Specifically, this system enhances ability to determine an optimum course of action; increases understanding of interrelationships; reduces redundancies, miscommunications, and problems that may otherwise go unrecognized; and enhances understanding of infrastructure and resource requirements.
  - –Nonenvironmental studies are included as part of the EIS or in supporting documents integrated with the NEPA analysis.
  - –All planning factors relevant to making an informed decision are evaluated within a single, integrated interdisciplinary analysis (e.g., environmental, cost, legal, performance specifications, schedules, technical capability).
- Planning is both centralized and semi-autonomous of the project proponent; while project proponents provide input and are involved in planning, they are not given direct control or supervision over the planning function. An independent IST oversees the planning process, enhancing objectivity and the ability to identify courses of action that best optimize all factors.
  - –The TFP approach is designed to centralize critical planning and decisionmaking elements so as not to hinder an agency's operations or implementation functions. Thus, TFP allows the agency to centralize its planning function, independent of and without hindering other critical mission-oriented functions (construction, operations, and other activities), which tend to be performed or organized in a decentralized fashion.
  - –The IST identifies a set of alternatives that are developed in sufficient detail to support a detailed study by a project or engineering office and later analysis by an interdisciplinary team (IDT).
  - –The IST and IDT define data requirements that will be developed by the project or engineering office.
  - –A proposed action is not defined. Instead, a set of reasonable alternatives is identified for analysis. Emphasis is placed throughout the process on optimizing potential alternatives.
- The planning process is integrated under a NEPA umbrella using modern management tools and principles such as:
  - –SE, to provide a rigorous methodology useful in systematically managing and analyzing complex operations while minimizing schedule and cost constraints
  - –TQM, which promotes optimization of alternatives, accountability, data adequacy, and integrity of the analysis
  - –VE tools, which promote objectivity by identifying, reviewing, discriminating, and prioritizing various planning factors and options, and which helps in reaching group consensus.

minimum, however, all essential planning elements, including the NEPA group, must be organized or matrixed in such away that they can communicate and cooperate.

Applying TQM Principles in Reorganizing Agency Planning

In developing a centralized planning function, it is critical that the following TQM factors be addressed:

- The sphere of influence within an organization, and who controls it, must be reviewed by management and, if appropriate, reorganized.
- A candidate project should be carefully selected to spearhead the TFP initiative.
- The structure, staff, training, and techniques that will be required to successfully complete the task must be determined.
- The TFP process is conducted. Progress, problems, and successes are documented.
- Areas for improving the TFP process are identified and evaluated.
- Adjustments and improvements are incorporated as deemed necessary. The modified process is again performed on a new proposal. Failures or problems are studied to determine what can be learned and how future projects can be improved; successes are publicized throughout the organization.

This process is depicted in Figure 5.1. As a first step in implementing TFP, it may be advantageous to perform a quality audit to determine the state and effectiveness of the existing planning function. This may require input from all levels of the agency.

An Interdisciplinary Steering Team

Special measures are incorporated into TFP for ensuring that an open and impartial planning process is conducted and that all reasonable alternatives are considered and afforded fair treatment. Independence is of prime importance in achieving this goal. Accordingly, an independent body is given control over the planning process, rather than the project proponent who might have a vested interest in the outcome of the final decision.

To this end, an IST is created to oversee the planning process; its purpose is to ensure that an open and impartial planning process is conducted that explores all reasonable alternatives. In this book, the IST is viewed as a small nonpermanent team of specialists representing a diverse spectrum of disciplines (e.g., environmental, risk assessment, project engineering, safety, health, regulatory, infrastructure, legal, economic, and SE and VE) relevant to the planning process at hand. A specific IST could be formed to oversee each major project that arises. As warranted, an IST could also be assembled for smaller planning activities as well.

The IST oversees the entire planning process (with input and representation from the project proponent), from conception to the point where a final decision is made. On completing the planning process, the IST is dissolved

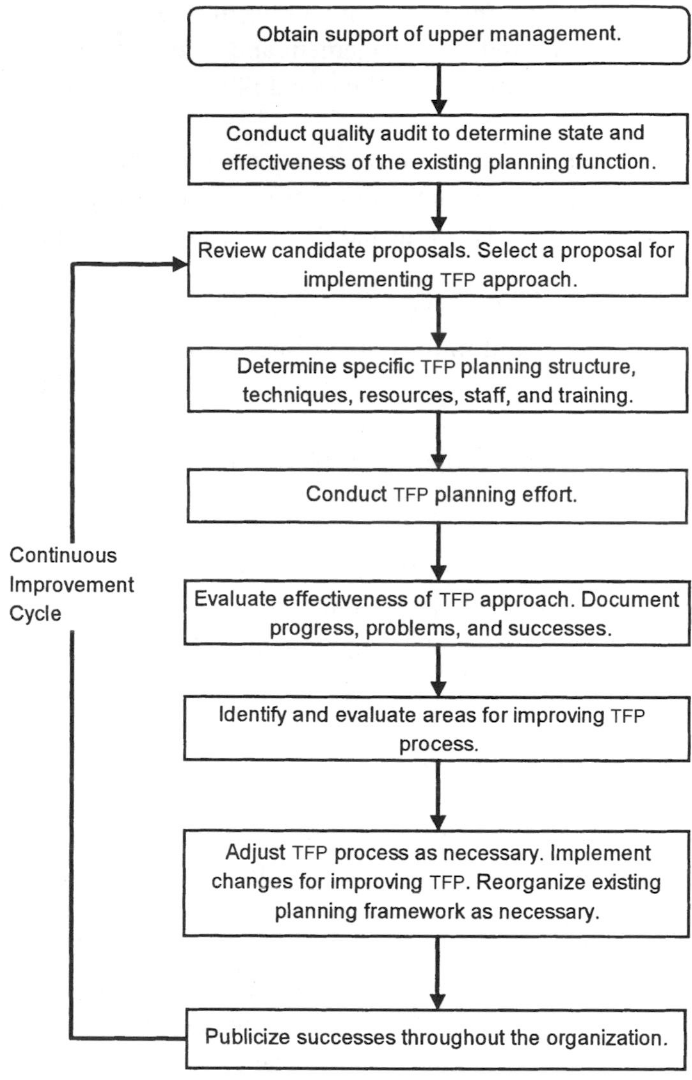

Figure 5.1 Using a TQM approach to initiate, conduct, and improve implementation of the TFP strategy within an agency's existing planning structure

and responsibility for implementing the final decision is turned over to the project office or project proponent. Given its temporary status and interdisciplinary nature, an IST is unlikely to become institutionalized or to be influenced by vested interests or organizational agendas. Thus, this approach is much more likely to uncover an optimum course of action. The reader should note that, based on specific organizational structure, internal cultural biases, and other constraints, a permanent IST body may actually be more appropriate. Regardless, either a permanent IST or a similar body has responsibility for monitoring implementation to ensure that the project conforms to the original plan and constraints.

The IST is responsible for overseeing, integrating, and coordinating project planning into a unified process; it is represented by the large shaded circle in the center of the Figure 5.2. The principle functional entities responsible for performing data collection and development, analysis, decisionmaking, and project implementation are represented by the smaller shaded circles located below the IST icon. Some but not necessarily all of the organizations, functions, and processes that are critical to the success of the IST and its planning function are represented by the small nonshaded circles surrounding the upper half of the IST circle; these functions are organized so that they provide critical information and input that support the IST's role in preparing an integrated plan.

The concept of an IST is a new feature distinguishing this approach from most other federal planning approaches and processes. While the IST is specifically designed to introduce objectivity and independence, project proponents and other vested interests are represented and integrally included. However, while project proponents provide input and are given representation, they do not control the planning process. Instead, engineering or project proponents are given responsibility for conducting design studies only after the IST obtains consensus from all representatives (including the project proponent) regarding the range of alternatives that will be evaluated.

Interdisciplinary Team

The IST should not be confused with the interdisciplinary team (IDT). The IDT comprises the technical specialists, scientists, and analysts responsible for conducting the actual planning analysis. In contrast, the IST does not prepare the actual planning analysis. Instead, it oversees the planning process, instilling an open, objective, and integrated atmosphere that fosters the objective of identifying and considering all reasonable alternatives. In reality, members of the IST might also participate in the analysis as part of the IDT, and vice versa.

Initiating the TFP Process

As described earlier, when a need for action is identified, an IST is formed with responsibility for guiding and providing independent oversight over the entire planning process. The IST is composed of representatives from pertinent planning and decisionmaking organizations, such that all planning functions are coordinated though the team. Thus, planning entities and functions that are often detached from one another are integrated into a centralized interdisciplinary planning format. Under this approach, NEPA provides an umbrella for integrating all pertinent planning efforts and requirements into a single unified framework, overseen by the IST. As described below, the IST also plays a prominent role in integrating TQM principles into the planning process, thus ensuring the excellence of the final product.

Using results from the public scoping process (not shown on Figure 5.2) in conjunction with other internal input (depicted by the nonshaded circles surrounding the upper half of the IST circle in Figure 5.2), the IST determines the final set of alternatives that will be evaluated in the EIS. The team defines the set of alternatives and decisionmaking factors (with input from the project proponent) that must be investigated to provide information sufficient to reach an informed decision. Under a similar approach, the project proponent might also develop a set of alternatives and data specifications that are submitted to the IST for review and formal approval. Alternatively, the project proponent could provide the IST with some minimal set of functional requirements and constraints; the team and the project proponent would then work in partnership to develop a set of reasonable alternatives that meet these requirements.

Holistic Planning

Figure 5.2 provides an overview of the TFP approach. The thin, straight arrows extending from the nonshaded circles surrounding the upper half of the IST icon indicate flow of information as well as how pertinent planning entities and functions are integrated into a centralized planning function. This information supplements input received from public scoping. Bold arrows (in the lower half of the figure) represent the integration, relationships, and decisionmaking flow paths between the IST, IDT, engineering and design office, project office, decisionmaker, and monitoring and mitigation function. Relationships depicted by the bold arrows are designed to inject objectivity, accountability, and quality into the final decisionmaking process.

Data Development

Once agreement is reached on the range of alternatives, the IST provides a formal set of alternatives and design requirements and specifications to the project proponent (arrow labeled "1. Provide design requirements," Figure 5.2). It should be emphasized that this set of alternatives and design specifications is generated using input from a range of sources including the project proponent.

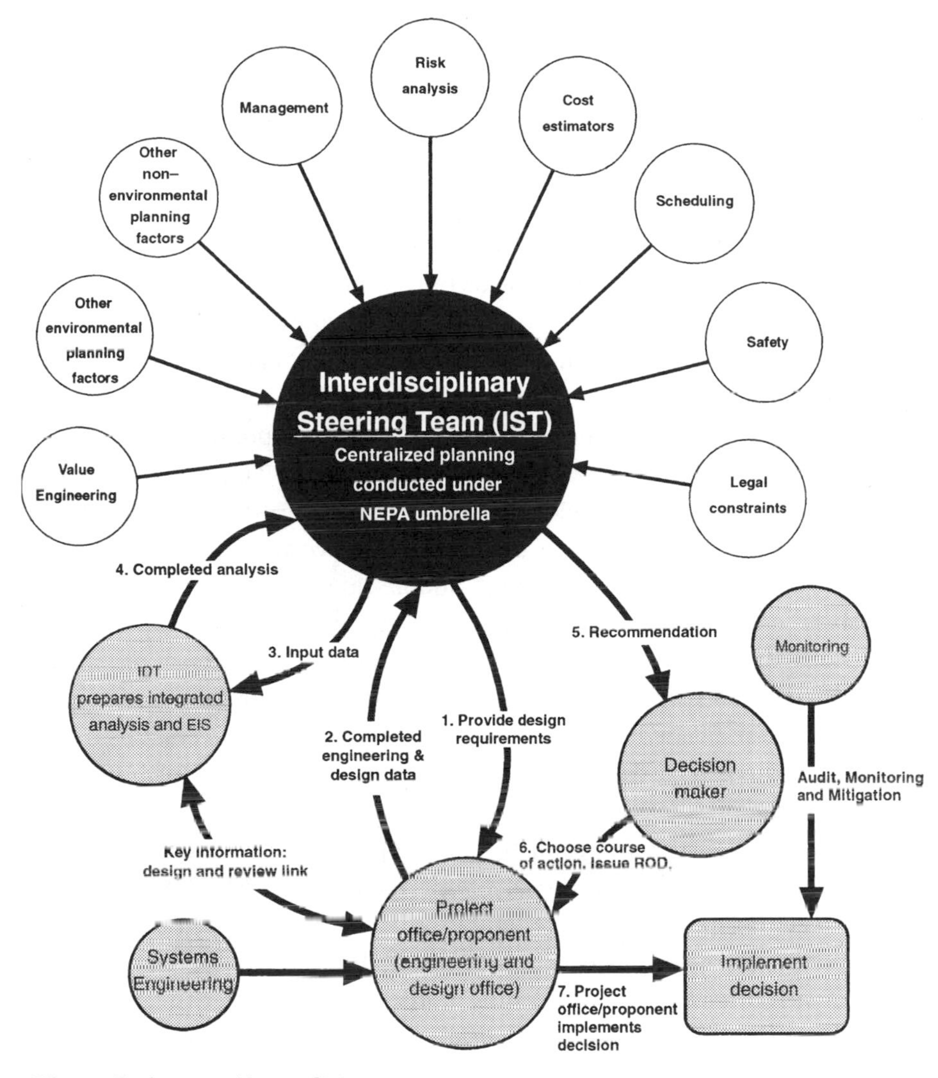

Figure 5.2 Overview of the organizational structure for implementing the TFP strategy. This organizational structure integrates all early planning under an independent IST.

Based on these requirements, the project proponent (e.g., engineering and design office) designs and develops the reasonable alternatives, including any engineering/design data, to a stage sufficient to support an analysis of the alternatives. As appropriate, an SE approach may be used for generating this data (see circle labeled "Systems Engineering"). The completed data package is transmitted back to the IST (arrow labeled "2. Completed engineering and design Package,"), which is responsible for checking the analysis for completeness and accuracy before the information is provided to the IDT. These individual steps are intrinsic to a well-orchestrated SE methodology.

Alternatives Analysis

Experience indicates that conflicts in interest may arise when an entity assigned the task of developing new technical and engineering input data is also responsible for performing the analysis. Separating these entities and functions is important in helping to ensure analytical quality, objectivity, and cost constraints; to facilitate a more open and objective evaluation, the actual analysis of alternatives is conducted by the IDT. The IST provides the IDT with technical information (arrow labeled "3. Input data," Figure 5.2) that was generated by the project proponent. However, the entities responsible for developing data and performing the analysis do not work in total isolation from one another. As indicated by the double-sided arrow in Figure 5.2, the analysis is coordinated and conducted with support from the project proponent (e.g., engineering and design office). For example, the project proponent may be requested to generate newly identified input data (consistent with specification established by the IST) needed for preparing the analysis by the IDT. Thus, as necessary, data may also be transferred back and forth between the IDT and the project proponent.

The IDT investigates and compares both environmental and nonenvironmental considerations critical to making a reasoned decision regarding the course of action to be taken. The final product is an integrated EIS that considers both environmental and nonenvironmental factors, or an EIS tightly coupled with other pertinent planning studies. The IDT issues the completed EIS to the IST for independent review (arrow labeled "4. Completed analysis," Figure 5.2).

Quest for an Optimum Alternative

A planning process that is skewed toward analysis of a proposed action is less likely to identify an optimum course of action. In promoting the objective of an open and unbiased planning process, TFP avoids the standard step of identifying a proposed action. This approach even avoids the term *proposed action.* Instead, a range of reasonable alternatives is developed, each given fair and equitable consideration, in an endeavor to identify an optimum alternative. This set of alternatives is simply referred to as *the proposal.* As defined herein, the term *optimum alternative* is the one that best meets all relevant environmental and nonenvironmental planning factors (e.g., public sentiments, cost, schedules, environmental impacts, performance factors, technical factors). Conducted properly, this optimum alternative will probably be the agency's preferred alternative, which must be identified in the EIS (§1502.14[e]).

This approach has many advantages, some of which could result in substantial cost savings. Because TFP promotes a more open, independent, and interdisciplinary process, the agency stands a greater chance of identifying innovative and perhaps more cost-effective alternatives that might otherwise be overlooked.

Data Verification and Implementation

The IST is responsible for assuring the objectivity, accuracy, and completeness of the analysis before the EIS (or integrated planning study) is provided to the decisionmaker. This step is essential in maintaining quality of the final product.

On completing the review of the planning studies, the IST may recommend an appropriate course of action to the responsible decisionmaker (arrow labeled "5. Recommendation," Figure 5.2), with whom the final decision lies (arrow labeled "6. Choose course of action. Issue ROD," Figure 5.2). The recommended action is recorded in the agency's ROD.

Once the final decision is recorded, control is passed to the project proponent who is responsible for conducting detailed engineering designs or studies necessary for implementing the decision (arrow labeled "7. Project office/proponent implements decision," Figure 5.2). The project proponent is responsible for implementing the final decision. As a final step, an independent office (possibly a centralized planning office) is responsible for periodically auditing and monitoring implementation of the action to ensure that it conforms with the agency's final decision. To reduce or avoid impacts, the project proponent is also monitored to ensure that mitigation measures are properly incorporated into the final design.

Advantages of TFP

Properly conceived and executed, a TFP approach reduces miscommunication and redundancy. Potential conflicts are easier to identify, evaluate, and rectify. Communication is enhanced because representatives from all relevant planning disciplines, including project proponents, are integrated into a unified planning process. Potential delays and design inconsistencies are minimized because associated infrastructure and regulatory problems and requirements are identified early in the process. Trade-offs and interrelationships are easier to identify and analyze. Finally, the chance that an important problem or issue is overlooked is greatly diminished. By providing a more comprehensive analysis, which may uncover a more optimal alternative, this approach may also result in resource and cost savings. Such benefits should be expected to increase with the size and complexity of the project.

Advantages to the Project Manager

The project proponent does not actually relinquish control over implementation of the project. Instead, the success of the planning process is shared among all pertinent planning entities. Correctly executed, an interdisciplinary planning approach can expedite project implementation and possibly reduce implementation costs. When diverse entities are given an opportunity to participate in the planning process, they tend to develop a sense of ownership of the outcome. Not only do the planning entities have greater appreciation for the need to take action, they also are better able to identify and address potential roadblocks. Consequently, once a final decision is reached, the project proponent may well find loyal supporters who can be counted on to resolve potential snags during implementation.

5.3 Total Federal Planning: A Detailed General-Purpose Strategy

Section 5.2 provided an overview of the TFP approach from an organizational perspective. This section describes TFP from the perspective of the detailed step-by-step procedural process, in which early planning is managed by an independent IST. Figure 5.3 provides a generalized overview of this procedural process. For comparison, Figure 5.4 indicates the principal steps followed in preparing a standard EIS. Figures 5.5 through 5.11 depict the detailed step-by-step procedure for implementing the planning strategy.

A general-purpose procedural process is described for implementing the TFP approach, which consists of eight distinct phases:

Phase 1: Pre-scoping
Phase 2: Formal Scoping
Phase 3: Data Development and Alternatives Design
Phase 4: Optimization and Data Validation
Phase 5: Analysis
Phase 6: Quality Control
Phase 7: Review and Decisionmaking
Phase 8: Auditing, Mitigation, and Monitoring

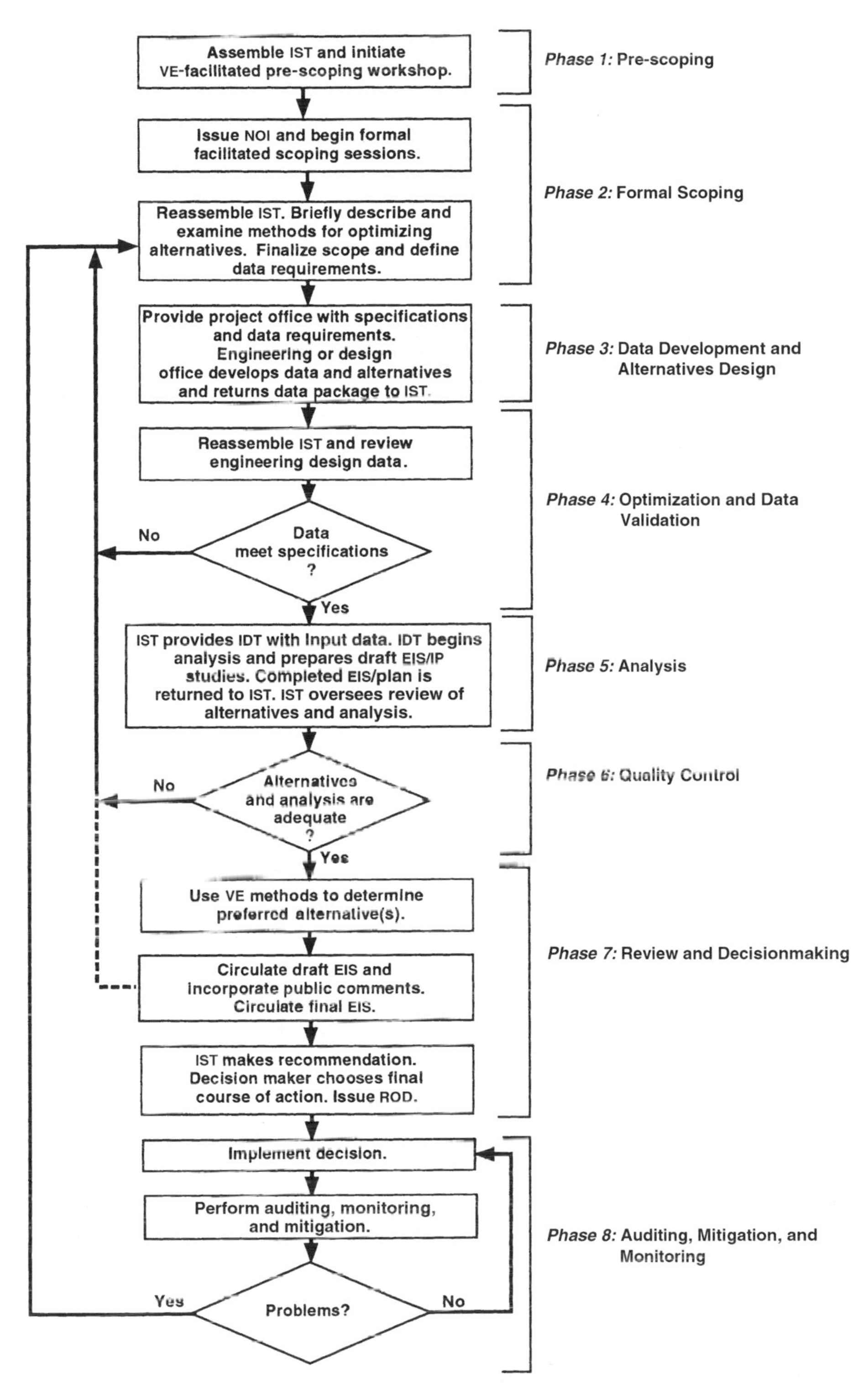

Figure 5.3 Overview of the eight-phase procedural process for implementing the TFP planning process

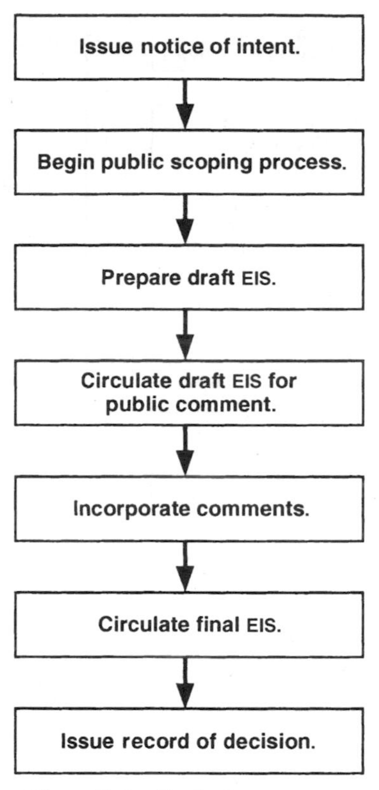

Figure 5.4 Typical EIS process

The following approach is divided into eight discrete phases for simplification purposes only. In reality, the process is iterative and various tasks described will not always be performed in the exact manner or sequence indicated. The actual process will depend on many factors, not the least of which are the agency's historical culture, complexity of the planning process, particular circumstances, and technical constraints. It is left to the reader to determine how details described in the integrated interdisciplinary planning effort, the integrated alternatives assessment phase, and the interdisciplinary analysis phase (sections 2.2, 2.3, and 2.4) are best combined with the TFP approach.

Phase 1: Pre-Scoping

Once a need for action is identified, a pre-scoping phase (Figure 5.5) is initiated. Information gathered during this step provides a coherent basis for explaining the need to take action and the potential alternatives. This phase can enhance the agency's credibility during the public scoping phase that follows. Specifically, the pre-scoping phase attempts to identify potential alter-

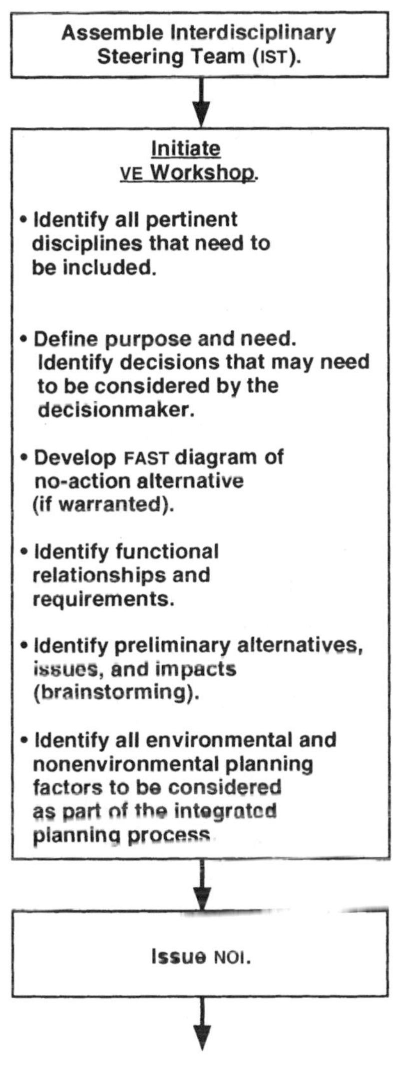

Figure 5.5 Phase 1: Pre-scoping

natives and impacts, resource and infrastructure requirements, related regulatory requirements, and potential controversy that might be encountered during the formal scoping process. The reader is referred to chapter 1 for additional details regarding specific activities that may be conducted as part of the pre-scoping stage. A description of this phase follows.

Initiating a Facilitated Workshop Using VE Tools

Once the IST is assembled, a facilitator leads a pre-scoping workshop consisting of selected IST members and other professionals representing a range of expertise and experience. The facilitated workshop, using VE tools and principles, can be conducted before publishing the EIS notice of intent (NOI). The workshop is intended to quickly identify preliminary planning factors and other information that need to be considered early in the planning process. This information is augmented during the subsequent public scoping phase and supports the IST during the formal scoping process that follows. The interdisciplinary nature of the workshop helps ensure that all pertinent planning factors are correctly identified and considered. The facilitator keeps the session on track. Guided by an experienced facilitator, this effort can be intensely focused, and a surprising amount of work can be accomplished in a short period. Specific objectives of this workshop are described as follows.

IDENTIFYING OTHER ENTITIES AND PLANNING REQUIREMENTS. One of the principal tasks of the workshop is to ensure that all pertinent entities and organizations have been identified and contacted.

An effort must also be conducted to ensure that all applicable planning requirements have been identified and will be appropriately integrated under the NEPA umbrella. An effort should also be mounted to identify preliminary environmental and regulatory drivers that may affect the course of the planning or later project implementation. Variations of this step are common to both VE and SE.

As described in chapters 1 and 2, the definition of need can have profound implications on alternatives identified for later analysis. Correctly defining the need at this step can substantially improve the effectiveness of agency planning and decisionmaking that follows. A facilitated VE workshop is a particularly effective means for reaching consensus regarding the purpose and need for taking action.

INVESTIGATING THE NO-ACTION ALTERNATIVE AND FUNCTIONAL REQUIREMENTS. As an SE principle, a list of functional requirements and constraints that the final course of action must meet is prepared (see chapter 2). This list is also useful in defining NEPA's purpose for taking action and in selection of a final course of action. The workshop participants might also be tasked with the goal of preparing a functional analysis system technique (FAST) diagram on activities currently being performed as part of the no-action alternative. This may be a useful tool for identifying or challenging various functions and actions as currently conducted (e.g., continuing operations baseline). A FAST diagram may also be useful in investigating the proposal; resource and infrastructure requirements and other planning considerations may also be evaluated as appropriate.

In some instances, a FAST diagram provides a useful tool for understanding the current baseline, identifying and investigating functional requirements and relationships, challenging preconceived assumptions and ideas, and identifying alternatives that might otherwise go unnoticed. The exercise might contribute valuable information, although, in other cases, little may be gained. For this reason, one must exercise prudence in determining the appropriate use and application (if any) of a FAST diagram. A description of the preparation and application of a FAST diagram is beyond the scope of this text. For more information, contact a facilitator trained in VE.

IDENTIFYING PRELIMINARY ALTERNATIVES AND ISSUES. Promoting an atmosphere that encourages innovative thinking is a cornerstone of the TFP strategy. This step emphasizes thinking outside the box. Brainstorming, a technique commonly used in VE, is a proven methodology for achieving this goal. During this phase, the group participants are challenged to consider every conceivable approach for identifying alternatives that meet the agency's underlying need. The facilitator maintains an open and nonhostile atmosphere where prevailing assumptions, mindsets, and paradigms are openly challenged in an effort to identify potential courses of action that might otherwise be overlooked.

If existing information is sufficient, the group next examines each of the alternatives in an effort to identify potentially significant impacts or issues that could be encountered. The interdisciplinary format provides an ideal atmosphere for this objective. Identification of potential impacts at this early stage alerts the agency to issues that might later be confronted. Consistent with the TFP philosophy, a review also is conducted to identify non–environmental-planning considerations and factors that need to be investigated to provide information sufficient to support a reasoned decision.

COMPLETING THE PRE-SCOPING PHASE. Once the facilitated workshop is completed, the stage is set for preparing and issuing the NOI, which completes the pre-scoping phase.

Phase 2: Formal Scoping

Once the NOI is issued, the agency is ready to begin the formal scoping phase, depicted in Figure 5.6. As described in Phase 1, modern management principles, such as VE, are used to enhance the formal scoping process.

Using Value Engineering Techniques to Conduct Public Scoping Meetings

In addition to the general public scoping process, some agencies pursue alternative public involvement efforts, such as conducting focus group meetings specifically directed at seeking in-depth comments from special-interest groups or parties, particularly those possessing a high degree of technical insight. Facilitated workshops in conjunction with selected VE principles offer

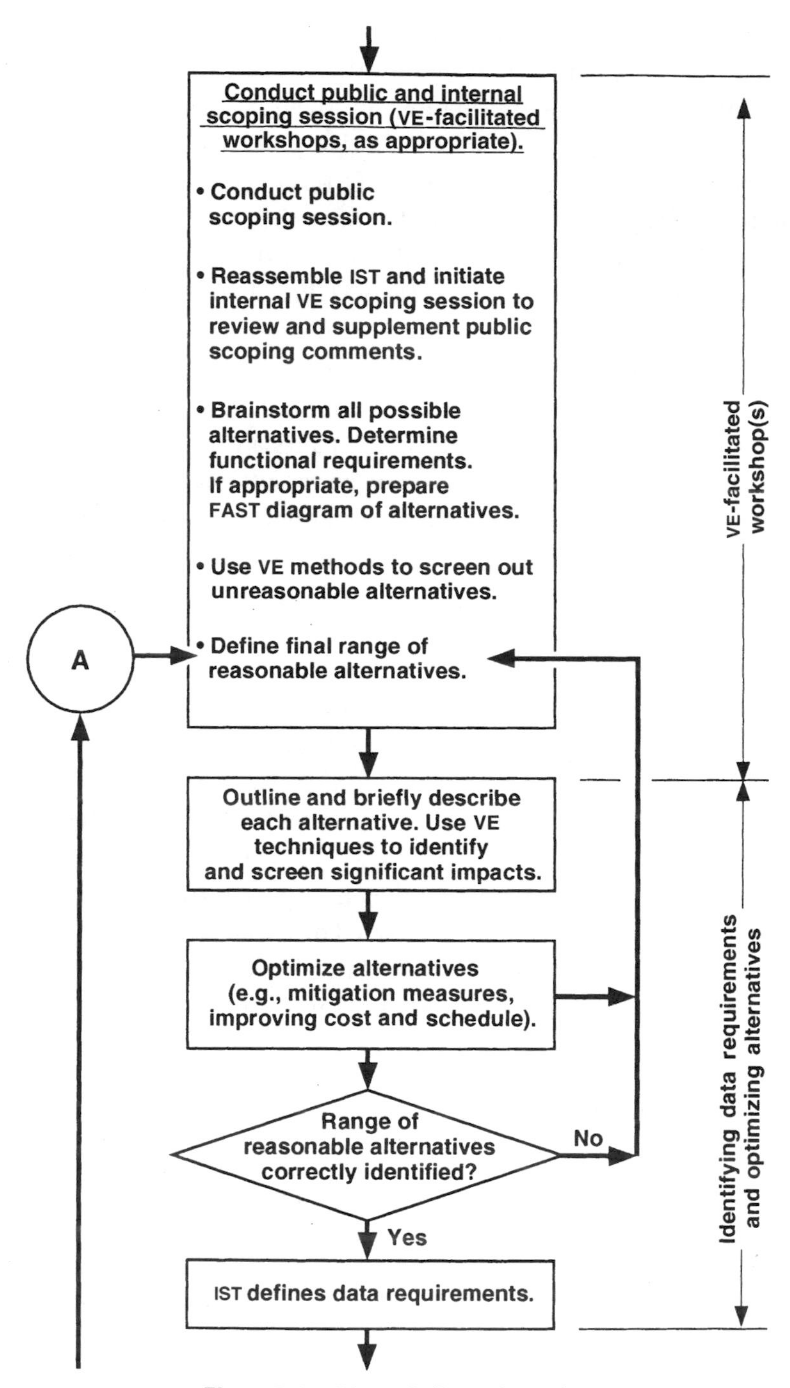

Figure 5.6 Phase 2: Formal scoping

a useful framework for managing and enhancing the effectiveness of public scoping meetings. Properly conducted, a publicly facilitated workshop can promote an open atmosphere where prevailing assumptions and mindsets are challenged in an effort to identify innovative and imaginative alternatives that might otherwise go unnoticed. This approach also has the advantage of ensuring that all participants have an opportunity to provide feedback and voice opinions. Specific VE techniques, such as preparing problem and issue statements and group brainstorming, can be employed to identify alternatives and issues.

As described in chapter 2, one approach is to divide participants into small groups based on the interests and concerns voiced by the participants. The facilitator then assigns each group a specific issue(s) to examine.

Conducting an Internal VE Workshop to Supplement Public Scoping

On completing the public scoping sessions, the IST is reassembled. Where appropriate (e.g., voluminous or complex scoping comments have been received), a facilitated workshop can be conducted internally to review and respond to public comments. VE management techniques, such as matrix methods, can be used to screen, combine, and group scoping comments into categories. Once comments are categorized, the IST/IDT reviews, evaluates, and prioritizes the scoping issues.

In concert with this effort, an internal scoping session, similar to that conducted during the pre-scoping phase, can be performed to supplement the public session. Specifically, this session can be used to finalize unresolved issues, provide further clarification, and investigate potential alternatives and issues in greater depth. The internal scoping session may add value because agency specialists are often in a position to explore technical issues in greater detail than might be possible in a general public setting.

Unlike public scoping, the internal scoping session is not necessarily restricted to consideration of environmental issues. Consistent with the TFP strategy, all pertinent planning factors should be captured during this effort. Public scoping comments are combined with results of the internal session to produce a final scope for the comprehensive planning analysis. Results of the internal scoping session (relevant to environmental concerns) are released to the public. Specific efforts performed during this internal workshop are described in the following sections.

CONSIDERING ALL POSSIBLE ALTERNATIVES. A brainstorming effort is used to explore potential courses of action that may have been overlooked. Emphasis is placed on promoting an atmosphere, encouraging free exchange of ideas and thinking outside of the box.

To promote an objective analysis, the TFP strategy intentionally avoids identification of a proposed action. Instead, a proposal consisting of a set of reasonable alternatives is identified and defined for later analysis. Forces tending to influence an analysis in favor of a particular direction are therefore minimized. Ultimately, this approach provides greater opportunity to actually identify the alternative best meeting all relevant planning requirements and decision-making factors.

SCREENING AND DEFINING RANGE OF ALTERNATIVES. The list of alternatives generated during the brainstorming session might be much larger than the range that can be reasonably analyzed. Where appropriate, VE techniques, such as matrix methods, are used to systematically dismiss, combine, sort, and rank potential alternatives in an effort to screen or reduce this list to a manageable range of reasonable alternatives for later analysis. Prior to beginning this task, specific elimination criteria for distinguishing reasonable from unreasonable alternatives should be established to provide a systematic basis for making such determinations.

For example, one approach (consisting of two distinct stages) may provide a useful methodology for screening important attributes of each alternative. During the first stage, group consensus is used to identify and segregate alternatives that are clearly unreasonable and can therefore be eliminated from further consideration. Next, the facilitator leads the group through an exercise designed to identify alternatives that are similar. Where appropriate, similar or duplicate alternatives are combined.

Outlining and Describing Alternatives

After consensus is reached, a brief document is prepared describing each of the alternatives. This document provides only enough detail so that engineers and analysts understand the scope and specifications of the alternatives.

Based on this document, engineers and analysts are tasked with responsibility for developing (i.e., describing) the alternatives to the point where detail is sufficient to support an adequate analysis. This is a principle intrinsic to SE. A preliminary matrix is prepared of potential regulatory requirements versus alternatives that will be analyzed. The regulatory matrix provides a heads-up so that the agency is not surprised by unanticipated requirements.

IDENTIFYING AND SCREENING SIGNIFICANT IMPACTS. Because environmental effects are a function of the specific alternative under consideration, identification of significant impacts is deferred until after the IST reaches consensus regarding the range of reasonable alternatives for detailed analysis. This task is performed by the IDT and may or may not involve the IST members.

The IDT might also use a facilitated workshop to manage this effort. To adequately reflect all potentially affected resources, the IDT workshop must include specialists representing a cross section of all potentially affected resources and disciplines. Brainstorming or other techniques may be employed in reviewing each of the alternatives to identify potential impacts and issues for later analysis. An interdisciplinary workshop allows experts representing all affected disciplines to interface, advancing the understanding of interrelationships among environmental resources, impacts, safety issues, and technical constraints. Environmental disturbances and impacts that might otherwise go unnoticed are more likely to be identified.

For example, a forester might recognize that a proposal to construct a tree farm on a hill would also require construction of an irrigation system that would result in infiltration of water into the soil column. This information alerts a geologist, who understands that infiltration of this water has the potential to instigate landslides along an interbedded clay layer that underlies the hill slope. A hydrologist is now forewarned that landslide debris could be transported into a stream meandering near the base of the hillside, affecting water quality several miles downstream. Finally, a biologist becomes aware that a spawning site for an endangered species could be affected by construction of a tree farm on a hill several miles upstream. Lacking this interdisciplinary communication, such relationships can go unrecognized.

Once these impacts and issues are identified, data management methods can be used to screen and group potentially significant impacts.

Optimizing Alternatives

Next, a technique basic to TQM and SE is applied in an effort to optimize the alternatives prior to detailed analysis. The IST/IDT reviews each alternative, searching for approaches or methods that can optimize the alternatives (e.g., performance specifications, cost, schedule, environmental considerations, and mitigation measures). Emphasis is placed on identifying methods for minimizing environmental degradation, cost, and schedule delays.

If appropriate, a FAST diagram might be constructed for each of the identified alternatives. The diagram can be used to challenge prevailing ideas, question the actual need for conducting individual activities within a given alternative, and search for more cost-effective and environmentally sound approaches.

Verifying the Range of Alternatives

A review is performed to ensure that the alternatives selected for analysis properly reflect the range of reasonable alternatives as required by NEPA.

Accordingly, the loop shown on the right side of Figure 5.6 indicates this step might be iterative. For example, the optimization phase might inadvertently limit the range of reasonable alternatives identified in the earlier scoping process. Thus, a final check is made to ensure the analyzed alternatives still encompass the range of all reasonable alternatives.

Defining Data Requirements

Once the range of alternatives slated for analysis is briefly defined, optimized, and verified, an effort is initiated to define specific data requirements necessary for performing the integrated NEPA planning analysis. Data requirements could involve parameters such as emissions, effluents, estimates of ground disturbance, and natural resource requirements. Nonenvironmental data requirements might include factors such as cost-benefit data, reliability studies, performance projections, and time schedules.

Accurate identification of data requirements is a critical task. In some cases, entities might have a vested interest in generating data, even if the need for such information is questionable; for example, an entity may be paid or receive funding for generating additional data that is of little or no use in the analysis. In other cases, necessary data may be overlooked or poorly defined requirements may lead to confusion or inaccurate data. An important feature distinguishing the TFP process from many other approaches is that data requirements necessary for the subsequent analysis are defined by an independent body—the IST, in collaboration with the IDT. This step is crucial in instilling an objective analysis, which minimizes problems discussed earlier. In reality, the IDT may define the data requirements while the IST acts as an oversight body that reviews and formally approves the set of data requirements.

The IST, using input from numerous sources (e.g., public scoping, analysts, engineers, design office), provides the project proponent with a specific definition of the alternatives that must be developed (i.e., designed and described) and related data that must be generated to support the analysis to follow (see Figure 5.2). Because these specifications are identified in an open, interdisciplinary forum, data are much more likely to be generated on schedule without wasted effort. Moreover, engineering and technical groups are not left with the problem of having to second-guess the data requirements. This step is also a principle inherent in a well-orchestrated SE process.

The scoping phase is now complete. Data requirements, along with a brief but concise description of the alternatives, are presented to the project proponent (engineering or design office) for detailed development. This office is responsible for preparing a detailed description of the alternatives and generating technical input data necessary for conducting the integrated planning analysis.

Phase 3: Data Development and Alternatives Design

Incorporating principles from SE, the TFP process now enters the data development and alternatives design phase depicted by the three boxes in Figure 5.7. As appropriate, an SE methodology provides a structured basis for developing designs, coordinating technical efforts, and ensuring quality of the data for the analysis phase to follow. Specialists skilled in SE may be employed to oversee this step. Chapter 2 proposes a modified SE approach to developing alternatives that is tightly integrated with the NEPA planning process.

Based on the specifications and data requirements provided by the IST, a technical group consisting of engineers and technical specialists develops descriptions of the alternatives and generates technical data sufficient to support the integrated analysis to be conducted by the IDT. Depending on the nature of the potential actions, this step must be tailored to meet specific circumstances. On completion, this data package is transmitted to the IST, according to a specified schedule, for review and validation.

Phase 4: Optimization and Data Validation

As a TQM monitoring measure, the optimization and data validation phase infuses accountability into the planning process by verifying that the data

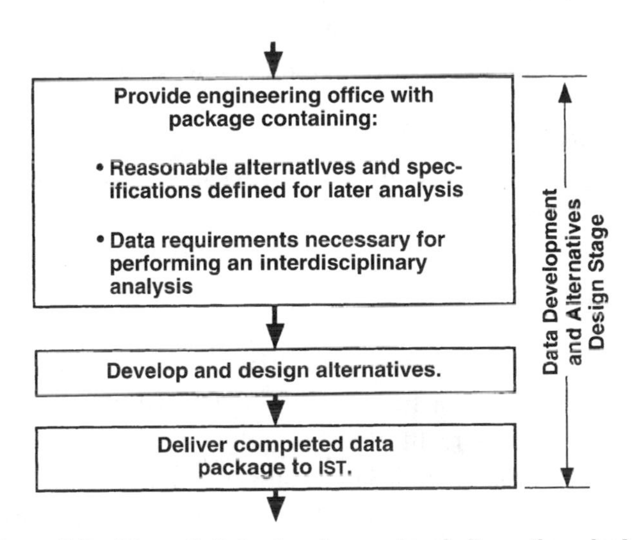

Figure 5.7 Phase 3: Data development and alternatives design

meet original specifications, no more and no less. Thus, the technical group developing the data and alternatives is accountable for ensuring data accuracy and that only specifically requested data are generated. This step is designed to control costs, improve quality and performance, and ensure that all reasonable alternatives identified by the IST are given equitable consideration.

On receiving the completed data package from the project proponent (see Figures 5.2 and 5.7), the IST/IDT reviews the data package to ensure that it is complete, accurate, and meets original specifications (Figure 5.8). The principal validation and optimization issues that are specifically addressed include:

- Do alternatives meet functional requirements?
- Are reasonable alternatives adequately described?
- Do additional methods exist for reducing cost or schedule?
- Is the environmental impact data adequate?
- Is the nonenvironmental impact data adequate?
- Are potential mitigation measures adequately addressed?

The details of this phase are represented by the seven diamonds in Figure 5.8. If the data package is large, a facilitated workshop could be convened to review it. The facilitated workshop approach ensures that alternatives are optimized and that the agency has the best courses of action available from which to choose. If the data are judged inadequate, the process is reiterated; this step is depicted by the arrow shown looping back to the box titled "Define final range of reasonable alterative" in Figure 5.6. In reality, professional judgment is exercised in determining the appropriate bounds of this loop.

Phase 5: Analysis

After the data are validated, analysts begin the analysis phase, represented together with the quality control stage in Figure 5.9. Preparation of the EIS document and other integrated planning studies are performed during this phase. This step is also depicted in Figures 5.2 and 5.3.

Conducting an Integrated Interdisciplinary Analysis

As discussed earlier, the actual environmental analysis is performed by the IDT. The IST provides the IDT with the alternatives data package that was developed by the project proponent. Related planning studies, such as risk and cost-benefit analyses, are coordinated and integrated with the environmental analysis. Results of these studies might either be incorporated directly into the

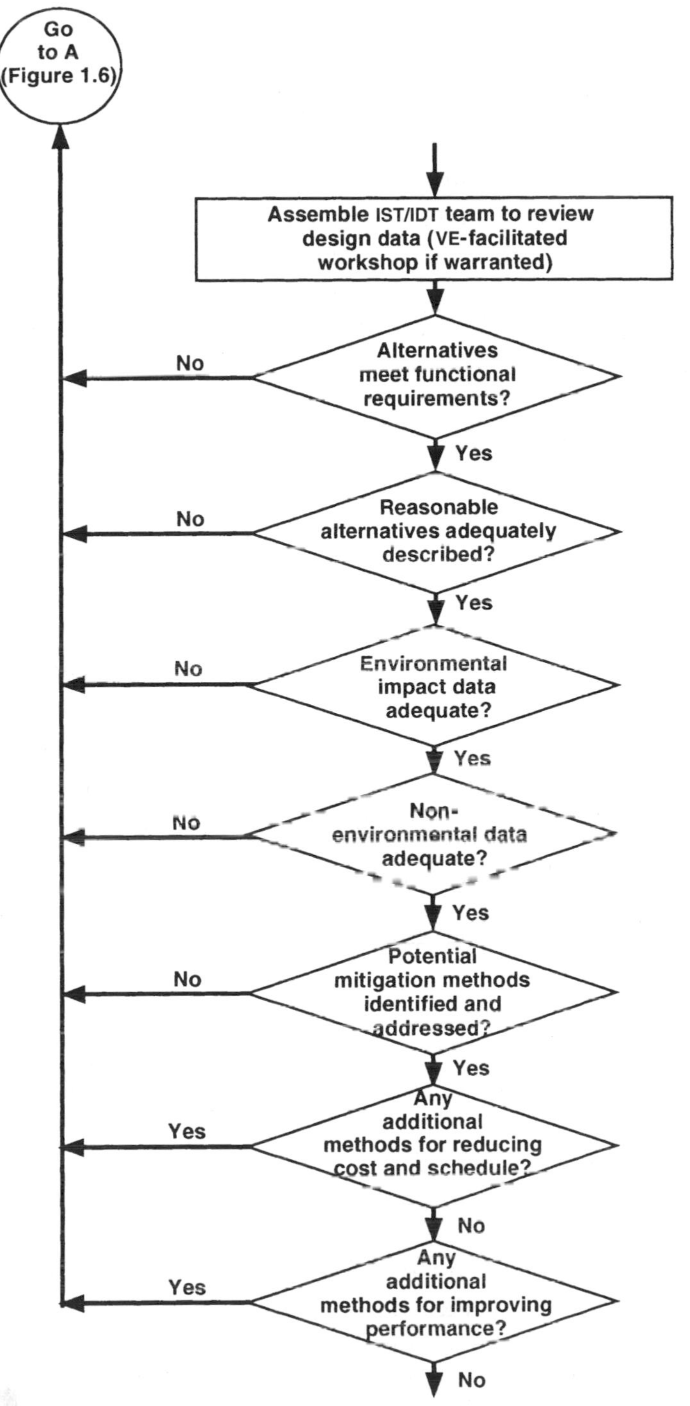

Figure 5.8 Phase 4: Optimization and data validation

draft EIS document or prepared as separate but tightly coupled studies to accompany the EIS through the agency's decisionmaking process.[8] A monitoring plan should be prepared for any circumstances considered controversial or where the effectiveness of the mitigation measures is questionable.

Emphasis is also placed on identifying inconsistencies with other regulatory requirements and planning efforts and on addressing potential infrastructure, funding, scheduling, and resource requirements. Ultimately, the goal is to produce a comprehensive and integrated study that provides decisionmakers with all pertinent planning factors necessary to reach an informed decision. A systematic process is outlined in chapter 2 for preparing an integrated interdisciplinary analysis.

Phase 6: Quality Control

The IDT returns the completed analysis to the IST for independent review (see Figure 5.2). As a TQM measure, the IST is responsible for ensuring that both the draft and the final EIS, as well as other related studies, meet specific planning requirements (Figure 5.9). Specifically, the IST ensures that the analyzed alternatives support potential decisions that need to be made (including functional requirements), are accurately described, and adequately evaluate impacts.

This goal is represented by the four diamonds shown in Figure 5.9. In addition to ensuring that the document meets all regulatory requirements, a review is conducted to ensure that the draft EIS meets the agency's underlying need and properly addresses decisions that may eventually need to be considered and made. Specific issues that are reviewed include:

- Do the alternatives bound the range of potential decisions that might be made?
- Do the alternatives meet functional requirements?
- Are the alternatives reasonable and adequately described?
- Are the impacts, mitigation measures, and other planning factors adequately described?

The IST is responsible for verifying that the alternatives analysis is consistent with the scope originally agreed upon. If the alternatives are not adequately described, the process is reiterated. This step is depicted by the upward-pointing arrow on the left side of Figure 5.9, which is shown looping back to the box labeled "Define final range of reasonable alternatives" in Figure 5.6. In reality, professional judgment is exercised in defining the appropriate bounds of this loop.

Similarly, if the impacts or other planning factors are not adequately evaluated, the analysis is reiterated. This step is depicted by the arrow on the right side of Figure 5.9, shown looping back to the box labeled "IDT begins analysis." Again, professional judgment must be exercised in determining the bounds of this loop.

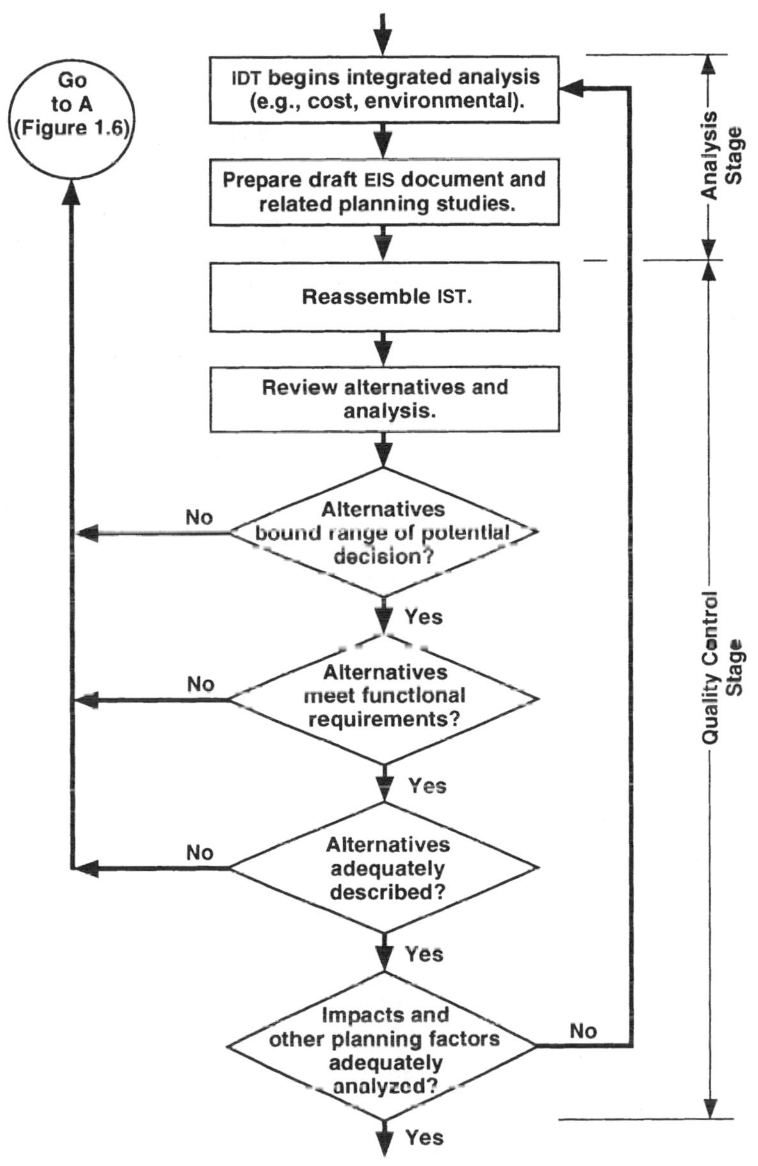

Figure 5.9 Phases 5 and 6: Analysis and quality control

Phase 7: Review and Decisionmaking

The review and decisionmaking stage completes the TFP process. This step is represented by Figure 5.10.

Using VE Tools to Determine the Preferred Alternative

The NEPA regulations require that agencies are to identify their preferred alternative(s), if one or more exists, within the draft EIS; in any event, the preferred alternative must be identified in the final EIS unless another law prohibits the expression of such a preference (§502.14[e]). The preferred alternative can be based on myriad factors, including environmental, economic, and technical considerations.[9] The agency also must identify the environmentally preferable alternative(s) in the ROD.[10]

As discussed previously, to promote a more objective analysis, the TFP approach intentionally avoids identifying a proposed action. Consistent with this objective, identification of the agency's preferred alternative is intentionally delayed until this phase in the planning process. As appropriate, a short facilitated session may be performed to identify the agency's preferred and the environmentally preferable alternatives. VE techniques provide effective tools for objectively comparing, discriminating among, and prioritizing the relative merits of the analyzed alternatives in an effort to identify the alternative(s) that best meets all planning and environmental considerations. Pertinent decisionmaking factors, such as environmental impacts, schedules, cost, and performance factors, are considered and balanced against one another.

Methods such as the *nominal grouping technique* can be used to rank and prioritize the alternatives. For example, with respect to the agency's preferred alternative, each member of the IST/IDT is instructed by the facilitator to identify one or more alternatives believed to be optimum from the standpoint of factors such as economics, performance, and environmental quality. Participants rank their selection of alternatives in terms of their relative merits (highest to lowest). Special techniques are then used to assimilate each member's selection in an effort to achieve a group consensus.

Circulating the EIS

The draft EIS (and integrated planning documents) are circulated for public review and comment. If warranted, selected members of the IST/IDT may be reassembled and VE tools can be used to screen, combine, and group public review comments. The team determines or assigns action items for the disposition of appropriate comment responses. The final comments and responses are incorporated into the EIS. As a TQM measure, a formal review is performed to ensure that all comments are appropriately addressed and incorporated.

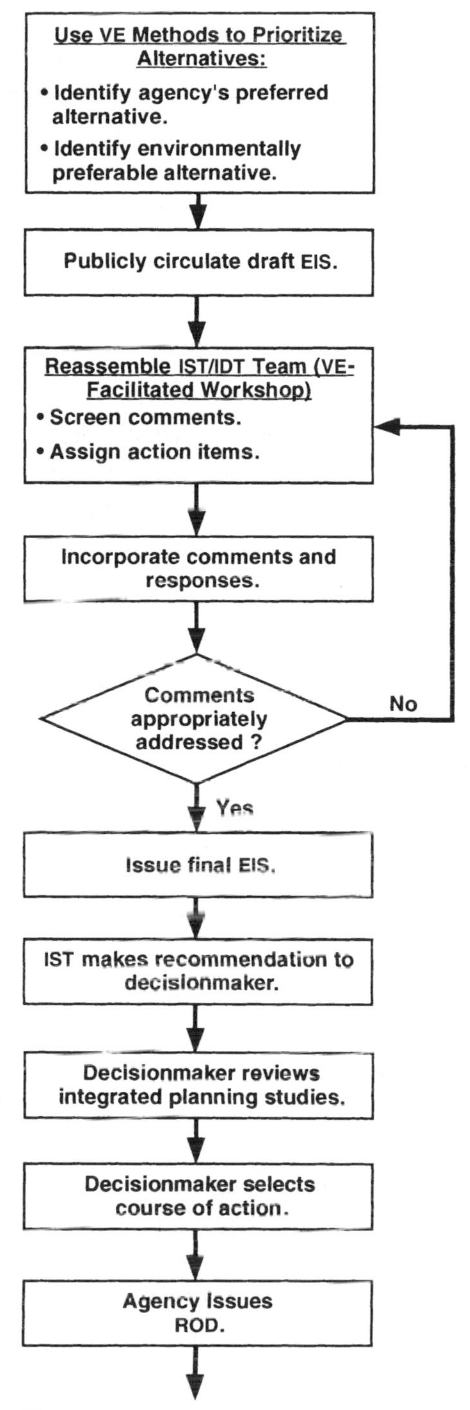

Figure 5.10 Phase 7: Review and decisionmaking

On completing this review, the final EIS is issued. Normally, the IST makes a recommendation to the decisionmaker regarding an appropriate course of action (see Figures 5.2 and 5.10). The decisionmaker reviews the integrated plan and makes a final decision regarding the cause of action to be taken. Once the final decision is made, the ROD is issued to publicly record the agency's final decision. As promoted in this book, the IST is dissolved and implementation of the selected alternative is turned over to the project proponent or project engineering office.

Phase 8: Auditing, Mitigation, and Monitoring

The eighth and final phase involves implementing the final decision, auditing, mitigating and monitoring. An independent audit should be conducted periodically to ensure that the project or action is being implemented in accordance with the agency's final decision and other applicable plans or requirements.

Mitigation and postmonitoring efforts are critical to the goal of environmental protection. The auditing procedure should review the results of the monitoring program to ensure that impacts are in accordance with projections cited in the EIS.

As indicated by the loop on the left side of Figure 5.11, problems should be referred to the appropriate officials. The loop on the right side of Figure 5.11 indicates that this step is ongoing and iterative in nature. This step can also be integrated with an environmental management system, as described in chapter 4.

Conclusion

Concinnity is essential to the success of a comprehensive and holistic planning process. The TFP strategy offers a systematic approach to achieving concinnity by unifying the disjointed planning processes currently followed by many federal agencies.

The TFP strategy offers clear advantages over other approaches in that it enhances the ability to determine an optimum course of action; provides decisionmakers with a more open and objective analysis; increases the understanding and diagnosis of interrelationships and potential conflicts; enhances the identification and analysis of trade-offs; reduces redundancies, inconsistencies, and miscommunications; promotes identification of problems that might otherwise go unrecognized; and enhances the understanding of infrastructure and resource requirements. When implemented using skilled professional judgment, such advantages substantially improve the entire federal planning process while reducing scheduling and cost constraints.

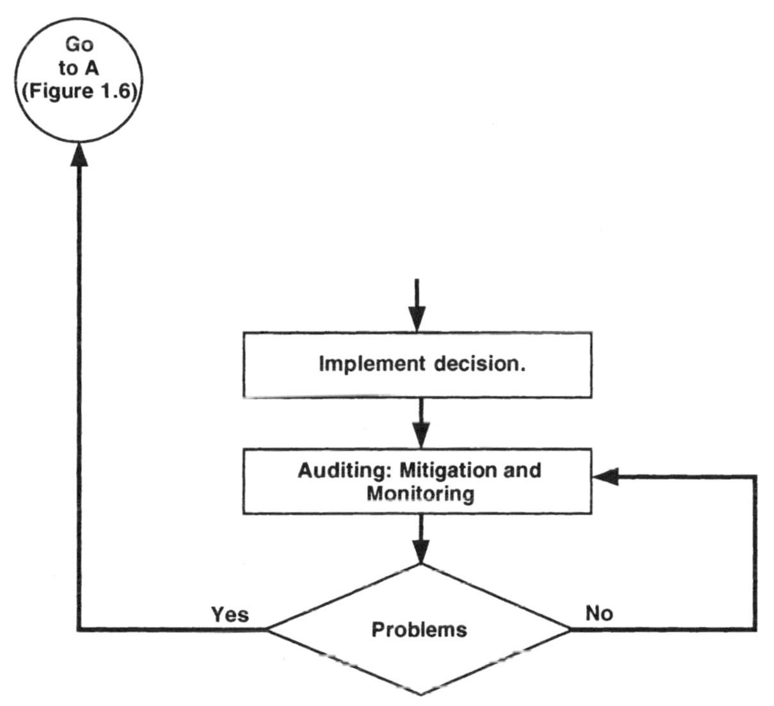

Figure 5.11 Phase 8: Implementing the final decision, auditing, mitigation, and monitoring

References

1. C. H. Eccleston, "Reengineering the Federal Planning Process: A Total Federal Planning Strategy: Integrating NEPA with Modern Management Tools," *Federal Facilities Environmental Journal* (Spring 1998).
2. 40 CFR 1501.2
3. C. H. Eccleston, *The NEPA Planning Process: A Comprehensive Guide with Emphasis on Efficiency*, (New York: John Wiley & Sons, 1999).
4. CEQ, Regulations for Implementing the Procedural Provisions of the National Environmental Policy Act, 40 Code of Federal Regulations (CFR) 1500–1508.
5. 40 CFR 1502.2 (c).
6. 40 CFR 1501.2.
7. 40 CFR 1502.14 (a).
8. 40 CFR 1506.4.
9. 40 CFR 1502.14(e).
10. 40 CFR 1505.2 (b).

APPENDIXES

APPENDIX A

Environmental Impact Statement Checklists

A set of checklists is provided in Tables A.1 through A.21 to assist the reader in preparing and reviewing an environmental impact statement (EIS) for adequacy.[1] Not all questions apply in all circumstances. Professional experience must be exercised in responding to each question. The reader should note that these lists address many but not necessarily all of the regulatory requirements that an EIS must meet. Each proposal has specific nuances that must be addressed individually. Therefore, these checklists should not be relied on as the sole method for ensuring quality and compliance with regulatory requirements. Prudence should be exercised in applying the checklists as:

- No checklist can be prepared that is universally applicable to all circumstances.
- No checklist can guarantee that the EIS will be adequate or in full compliance with NEPA and other related regulations, guidance, or laws.

A price list and copies of the checklists can be obtained at a nominal fee by contacting The NEPA and Environmental Strategies Company at ecclestonc@aol.com.

Abbreviation/Acronyms Used in the EIS Checklists

CEQ	President's Council on Environmental Quality
CERCLA	Comprehensive Environmental Response, Compensation, and Liability Act
CFR	United States Code of Federal Regulations
EA	Environmental Assessment
EIS	Environmental Impact Statement
EO	Executive Order
FONSI	Finding of No Significant Impact
FR	Federal Register
N/A	not applicable

[1] Modified from Environmental Impact Statement Checklist, U.S. Department of Energy, 1997.

NEPA	National Environmental Policy Act
NMFS	United States National Marine Fisheries Service
RCRA	Resource Conservation and Recovery Act
SHPO	State Historic Preservation Officer
US	United States
U.S.C.	United States Code
USFWS	United States Fish and Wildlife Service

Table A.1 Cover sheet

	Yes	No	N/A	EIS Page Number	Adequacy Evaluation and Comments
1.1 Does the cover sheet include: • A list of responsible agencies, including the lead agency and any cooperating agencies?					
• The title of the proposal and its location (state[s], other jurisdictions)?					
• The name(s), address(s), and telephone number(s) of a person (or persons) to contact for further information (on the general NEPA process or on the specific EIS)?					
• The EIS designation as draft, final, or supplemental?					
• A one-paragraph abstract of the EIS?					
• The date (for a draft EIS) by which comments must be received? [40 CFR 1502.11]					
1.2 Is the cover sheet one page in length? [40 CFR 1502.11]					

Table A.2 Summary

	Yes	No	N/A	EIS Page Number	Adequacy Evaluation and Comments
2.1 Does the summary describe the underlying purpose and need for agency action? • The proposed action?					
• Each of the alternatives?					
• The preferred alternative, if any?					
• The principal environmental issues analyzed and the results?					
2.2. Does the summary highlight key differences among the alternatives?					
2.3 Does the summary stress:					
• The major conclusions?					
• Areas of controversy (including issues raised by agencies and the public)?					
• The issues to be resolved (including the choice among alternatives)? [40 CFR 1502.12]					
2.4 Are the discussions in the summary consistent with the EIS text and appendices?					
2.5 Does the summary adequately and accurately summarize the EIS? [40 CFR 1502.12]					

Table A.3 Purpose and need for action

	Yes	No	N/A	EIS Page Number	Adequacy Evaluation and Comments
3.1 Does the EIS specify the underlying purpose and need to which the agency is responding in proposing the alternatives including the proposed action? [40 CFR 1502.13]					
3.2 Does the statement of purpose and need relate to the broad requirement or desire for action and not to the need for one specific proposal or the need for the EIS?					
3.3 Does the statement of purpose and need adequately explain the problem or opportunity to which the agency is responding?					
3.4 Is the statement of purpose and need written so that it: (a) does not inappropriately narrow the range of reasonable alternatives, and (b) is not too broadly defined as to make the number of alternatives virtually limitless?					

Table A.4 Description of the proposed action and alternatives

	Yes	No	N/A	EIS Page Number	Adequacy Evaluation and Comments
4.1 Does the EIS clearly describe the proposed action and alternatives?					
4.2 Is the proposed action described in terms of the actions to be taken (even a private action that has been federalized or enabled by funding)?					
4.3 Does the proposed action exclude elements that are more appropriate to the statement of purpose and need?					
4.4 Does the EIS identify the range of reasonable alternatives that satisfy the agency's purpose and need?					
4.5 Does the EIS "rigorously explore and objectively evaluate" all reasonable alternatives that encompass the range to considered by the decisionmaker? [40 CFR 1502.14(a)]					
4.6a For a draft EIS, does the document indicate whether a preferred alternative(s) exist(s) and, if so, is it identified? [40 CFR 1502.14(e)]					
4.6b For a final EIS, is the preferred alternative identified? [40 CFR 1502.14(e)]					
4.7 Does the EIS include the no-action alternative? [40 CFR 1502.14(d)]					
4.8 Is the no-action alternative described in sufficient detail so that its scope is clear and potential impacts can be identified?					
4.9 Does the no-action alternative include a discussion of the legal ramifications of taking no action, if appropriate?					
4.10 As appropriate, does the EIS identify and analyze reasonable technology, transportation, and siting alternatives, including those that could occur off-site?					
4.11 Does the EIS include reasonable alternatives outside the agency's jurisdiction? [40 CFR 1502.14(c)]					

(continues)

Table A.4 *Continued*

	Yes	No	N/A	EIS Page Number	Adequacy Evaluation and Comments
4.12 For alternatives that were eliminated from detailed study (including those that appear obvious), does the EIS fully and objectively explain why they were found unreasonable? [40 CFR 1502.14(a)]					
4.13 For each alternative analyzed in detail (including the no-action alternative), is the depth of analysis approximately the same, allowing reviewers to evaluate their comparative merits? [40 CFR 1502.14(b)]					
4.14 Are the proposed action and alternatives described in sufficient detail so that potential impacts can be identified?					
4.15 Are all phases of the proposed action and alternatives described (e.g., construction, operation, and postoperation and decommissioning)?					
4.16 Are environmental releases associated with the proposed action and alternatives quantified, including both the rates and durations?					
4.17 As appropriate, are mitigation measures included in the description of the proposed action and alternatives? [40 CFR 1502.14(f)]					
4.18 Are cost-effective waste minimization and pollution prevention activities included in the description of the proposed action and alternatives?					
4.19 As appropriate, are environmentally and economically beneficial landscape practices included in the description of the proposed action and alternatives?					
4.20 Are the descriptions of the proposed action and alternatives written broadly enough to encompass future modifications?					

Table A.4 *Continued*

	Yes	No	N/A	EIS Page Number	Adequacy Evaluation and Comments
4.21 Does the proposed action comply with CEQ regulations for interim actions? [40 CFR 1506.1]					
4.22 Does the EIS take into account relationships between the proposed action and other actions to be taken by the agency in order to avoid improper segmentation?					

Table A.5 Description of the affected environment

	Yes	No	N/A	EIS Page Number	Adequacy Evaluation and Comments
5.1 Does the EIS succinctly describe the environment of the area(s) to be affected or created by the proposed action and alternatives? [40 CFR 1502.15]					
5.2 Does the EIS identify either the presence or absence of the following within the area potentially affected by the proposed action and alternatives: • Floodplains? [EO 11988; 10 CFR 1022]					
• Wetlands? [EO 11990; 10 CFR 1022; 40 CFR 1508.27(b)(3)]					
• Threatened, endangered, or candidate species and/or their critical habitat, and other special status (e.g., state-listed) species? [16 U.S.C. 1531; 40 CFR 1508.27(b)(9)]					
• Prime or unique farmland? [7 U.S.C. 4201; 7 CFR 658; 40 CFR 1508.27(b)(3)]					
• State or national parks, forests, conservation areas, or other areas of recreational, ecological, scenic, or aesthetic importance? [40 CFR 1508.27(b)(3)]					
• Wild and scenic rivers? [16 U.S.C. 1271; 40 CFR 1508.27(b)(3)]					
• Natural resources (e.g., timber, range, soils, minerals, fish, migratory birds, wildlife, water bodies, aquifers)? [40 CFR 1508.8]					
• Property of historic, archaeological, or architectural significance (including sites on or eligible for the National Registry of Natural Landmarks)? [EO 11593; 16 U.S.C. 470; 36 CFR 800; 40 CFR 1508.27(b)(3) and (8)]					
• Native Americans' concerns? [EO 13007; 25 U.S.C. 3001; 16 U.S.C. 470; 42 U.S.C. 1996]					

Table A.5 ***Continued***

	Yes	No	N/A	EIS Page Number	Adequacy Evaluation and Comments
• Minority and low-income populations (including a description of their use and consumption of environmental resources)? [EO 12898]					
5.3 Does the description of the affected environment provide the necessary information to support the impact analysis, including cumulative impact analysis? [40 CFR 1502.15]					
5.4 Are the descriptions of the affected environment substantially consistent with current baseline studies (e.g., descriptions of plant communities, wildlife habitat, and cultural resources)?					
5.5 Is the discussion appropriately limited to information that is directly related to the scope of the proposed action and alternatives? [40 CFR 1502.15]					
5.6 Is the extent of each component of the affected environment appropriately described with respect to potential impacts)?					
5.7 Does the EIS concentrate on important issues, avoiding useless bulk and verbose descriptions of the affected environment? [40 CFR 1502.15]					

Table A.6 Environmental effects

	Yes	No	N/A	EIS Page Number	Adequacy Evaluation and Comments
6.1 Does the EIS adequately identify the direct and the indirect impacts of the proposed action and alternatives and discuss their significance? [40 CFR 1502.16(a) and (b)]					
6.2 Does the EIS adequately analyze both short-term and long-term effects?					
6.3 Does the EIS analyze both beneficial and adverse impacts? [40 CFR 1508.27(b)(1)]					
6.4 Does the EIS discuss reasonably foreseeable impacts of cumulative actions with regard to both the proposed action and alternatives? [40 CFR 1508.25(a)(2)]					
6.5 Does the EIS discuss the potential direct, indirect, and cumulative effects to the following, as identified in Question 5.2? • Floodplains? [EO 11988; 10 CFR 1022]					
• Wetlands? [EO 11990; 10 CFR 1022; 40 CFR 1508.27(b)(3)]					
• Threatened, endangered, or candidate species and/or their critical habitat, and other special status (e.g., state-listed) species? [16 U.S.C. 1531; 40 CFR 1508.27(b)(9)]					
• Prime or unique farmland? [7 U.S.C. 4201; 7 CFR 658; 40 CFR 1508.27(b)(3)]					
• State or national parks, forests, conservation areas, or other areas of recreational, ecological, scenic, or aesthetic importance? [40 CFR 1508.27(b)(3)]					
• Wild and scenic rivers? [16 U.S.C. 1271; 40 CFR 1508.27(b)(3)]					
• Natural resources (e.g., timber, range, soils, minerals, fish, migratory birds, wildlife, water bodies, aquifers)? [40 CFR 1508.8]					

Table A.6 ***Continued***

	Yes	No	N/A	EIS Page Number	Adequacy Evaluation and Comments
• Property of historic, archaeological, or architectural significance (including sites on or eligible for the National Registry of Natural Landmarks)? [EO 11593; 16 U.S.C. 470; 36 CFR 800; 40 CFR 1508.27(b)(3) and (8)]					
• Native Americans' concerns? [EO 13007; 25 U.S.C. 3001; 16 U.S.C. 470; 42 U.S.C. 1996]					
• Minority and low-income populations, to the extent that such effects are disproportionately high and adverse? [EO 12898]					
6.6 Does the EIS discuss: • Possible conflicts with land use plans, policies, or controls? [40 CFR 1502.16(c)					
• Energy requirements and conservation potential of various alternatives and mitigation measures? [40 CFR 1502.16(e)]					
• Natural or depletable resource requirements and conservation potential of the proposed action and alternatives? [40 CFR 1502.16(f)]					
• Urban quality, historic, and cultural resources, and the design of the built environment, including the reuse and conservation potential of the proposed action and alternatives? [40 CFR 1502.16(g)]					
• The means to mitigate adverse impacts? [40 CFR 1502.16(h)]					
6.7 Does the EIS discuss: • Unavoidable, adverse environmental effects?					
• The relationship between short-term uses of the environment and long-term productivity?					
• Irreversible or irretrievable commitments of resources? [40 CFR 1502.16]					

(continues)

Table A.6 ***Continued***

	Yes	No	N/A	EIS Page Number	Adequacy Evaluation and Comments
6.8 Do the discussions of environmental impacts include (as appropriate): • Human health effects?					
• Effects of accidents?					
• Transportation effects?					
6.9 Does the EIS discuss the potential effects of released pollutants rather than just identifying the releases?					
6.10 Does the EIS avoid presenting a description of severe impacts (e.g., from accidents) without also describing the likelihood or probability of such impacts occurring?					
6.11 Are the methodologies used for impact assessment generally accepted and recognized in the scientific community? [40 CFR 1502.22 and 1504.24]					
6.12 Does the EIS quantify environmental impacts where practical?					
6.13 Are impacts analyzed using a sliding-scale approach, as appropriate—that is, proportional to their potential significance?					
6.14 Does the EIS avoid presenting bounding impact estimates that obscure differences among alternatives?					
6.15 Are sufficient data and references presented to allow validation of analysis methods and results?					
6.16a If information related to significant adverse effects is incomplete or unavailable, does the EIS state that such information is lacking?					
6.16b If this information is essential to a choice among alternatives and the cost of obtaining it is not exorbitant, is the information included?					

Table A.6 ***Continued***

	Yes	No	N/A	EIS Page Number	Adequacy Evaluation and Comments
6.16c If this information cannot be obtained, does the EIS include: (1) a statement that the information is incomplete or unavailable, (2) the relevance of the information in evaluating significant effects, (3) a summary of credible scientific evidence, and (4) an evaluation based on theoretical approaches? [40 CFR 1502.22]					
6.17 As appropriate, does the EIS identify important sources of uncertainty in the analyses and conclusions?					

Table A.7 Overall considerations and incorporation of NEPA Values

	Yes	No	N/A	EIS Page Number	Adequacy Evaluation and Comments
7.1 Does the EIS identify all reasonably foreseeable impacts? [40 CFR 1508.8]					
7.2 Do the conclusions regarding potential impacts follow from the information and analyses presented in the EIS?					
7.3 Does the EIS avoid the implication that compliance with regulatory requirements demonstrates the absence of environmental effects?					
7.4 To the extent possible, does the EIS assess reasonable alternatives and identify measures to restore and enhance the environment and avoid or minimize potential adverse effects? [40 CFR 1500.2(f)]					
7.5 Does the EIS identify best-management practices associated with the proposed action or with mitigation measures that would help avoid or minimize environmental disturbance, emissions, and other adverse effects?					
7.6 Does the EIS avoid (including the appearance) justifying decisions that have already been made? [40 CFR 1502.5]					
7.7 Are all assumptions conservative, and are the analyses and methodologies generally accepted and recognized by the scientific community? [40 CFR 1502.22 and 1502.24]					
7.8 Does the EIS indicate that the agency "has taken a hard look at environmental consequences"? [*Kleppe v. Sierra Club,* 427 US 390, 410 (1976)]					
7.9 Does the EIS present the potential environmental effects of the proposal and the alternatives in comparative form, sharply defining the issues and providing a clear basis for choice? [40 CFR 1502.14]					

Table A.8 Format, general document quality, user friendliness

	Yes	No	N/A	EIS Page Number	Adequacy Evaluation and Comments
8.1 Is the EIS written precisely and concisely, using plain language and defining technical terms that must be used?					
8.2 Is information in tables and figures consistent with information in the text and appendices?					
8.3 As appropriate, is the metric system of units used (with English units in parentheses)?					
8.4 Are the units consistent throughout the document?					
8.5 Are technical terms defined, using plain language?					
8.6 If scientific notation is used, is an explanation provided?					
8.7 If regulatory terms are used, are they consistent with their regulatory definitions?					
8.8 Does the EIS use conditional language (e.g., "would" rather than "will") in describing the proposed action and alternatives and their potential consequences?					
8.9 Are graphics and other visual aids used whenever possible to simplify the EIS?					
8.10 Are abbreviations and acronyms defined the first time they are used?					
8.11 Is the use of abbreviations and acronyms minimized to the extent practical?					
8.12 Does the EIS make appropriate use of appendixes (e.g., for material prepared in connection with the EIS and related environmental reviews, substantiating material, official communications, and descriptions of methodologies)? [40 CFR 1502.18 and 1502.24]					

(continues)

Table A.8 ***Continued***

	Yes	No	N/A	EIS Page Number	Adequacy Evaluation and Comments
8.13 Do the appendixes support the content and conclusions contained in the main body of the EIS?					
8.14 Is there a discussion of the relationship between this EIS and related NEPA documents?					
8.15 Is the issue date (month and year of approval) on the cover?					

Table A.9 Other regulatory requirements

	Yes	No	N/A	EIS Page Number	Adequacy Evaluation and Comments
9.1 Unless there is a compelling reason to do otherwise, does the EIS include a: • Table of contents?					
• Index?					
• List of agencies, organizations, and persons to whom copies of the EIS were sent? [40 CFR 1502.19]					
9.2 Does the EIS identify all federal permits, licenses, and other entitlements that must be obtained in implementing the proposal? [40 CFR 1502.25(b)]					
9.3 Does the EIS identify methodologies used in the analyses, include references to sources relied upon for conclusions, supporting material, and methodologies? [40 CFR 1502.24]					
9.4 If a cost-benefit analysis is prepared, is it incorporated by reference or appended to the EIS? [40 CFR 1502.23]					
9.5 If this EIS adopts, in whole or in part, a NEPA document prepared by another federal agency, has the agency independently evaluated this information? [40 CFR 1506.3]					
9.6 Does the EIS appropriately use incorporation by reference? • Is the information up to date?					
• Is the information summarized in the EIS?					
• Are cited references publicly available? [40 CFR 1502.21]					
9.7 Does the EIS contain a list of preparers and their qualifications? [40 CFR 1502.17]					
9.8 If an EIS contractor was used, was a disclosure statement prepared? [40 CFR 1506.5(c)]					
9.9 If the EIS was prepared by a contractor, is the agency's name listed as the preparer on the title page of the EIS and has the agency evaluated all information and accepted responsibility for the contents? [40 CFR 1506.5]					

Table A.10 Procedural considerations

	Yes	No	N/A	EIS Page Number	Adequacy Evaluation and Comments
10.1 If appropriate, did the agency notify the host state and host tribe, and other affected states and tribes, of the determination to prepare the EIS?					
10.2 Did the agency publish a notice of intent in the *Federal Register,* allowing reasonable time for public comment? [40 CFR 1501.7]					
10.3 Is a floodplain/wetlands assessment required and, if so, has a notice been published in the *Federal Register?*					
10.4 In addition to EPA's notice of availability, has the agency otherwise publicized the availability of the draft EIS, focusing on potentially interested or affected persons? [40 CFR 1506.6]					
10.5 Has the agency actively sought the participation of low-income and minority communities in the preparation and review of the EIS? [EO 12898; Effective Public Participation Guidance, p. 11]					
10.6 Is the EIS administrative record being maintained contemporaneously, and does it provide evidence that the agency considered all relevant issues?					
10.7 To the fullest extent possible, have other environmental review and consultation requirements been integrated with NEPA requirements? [40 CFR 1502.25]					

Table A.11 Draft EIS considerations

	Yes	No	N/A	EIS Page Number	Adequacy Evaluation and Comments
11.1 Has the agency considered scoping comments from other agencies and the public?					
11.2 Does the draft EIS demonstrate that the agency considered possible connected actions, cumulative actions, and similar actions? [40 CFR 1508.25(a)]					
11.3 If the draft EIS identifies a preferred alternative(s), does the document present the criteria and selection process? [40 CFR 1502.14(e)]					
11.4a Does the draft EIS demonstrate adequate consultation with appropriate agencies to ensure compliance with sensitive resource laws and regulations?					
11.4b Does the document contain a list of agencies and persons consulted?					
11.4c Are letters of consultation (e.g., SHPO, USFWS) appended? [40 CFR 1502.25]					

Table A.12 Final EIS considerations

	Yes	No	N/A	EIS Page Number	Adequacy Evaluation and Comments
12.1 Does the final EIS discuss, at appropriate points, responsible opposing views not adequately addressed in the draft EIS and indicate the agency's responses to the issues raised? [40 CFR 1502.9(b)]					
12.2a Is the preferred alternative identified? [40 CFR 1502.14(e)]					
12.2b Does the document present the criteria and selection process for the preferred alternative?					
12.3 Does the final EIS demonstrate, through appropriate responses, that all substantive comments from other agencies, organizations, and the public were objectively considered, both individually and cumulatively (i.e., by modifying the alternatives, developing new alternatives, modifying and improving the analyses, making factual corrections, or explaining why the comments do not warrant agency response)? [40 CFR 1503.4]					
12.4 Are all substantive comments (or summaries thereof) and the agency's responses included with the final EIS? [40 CFR 1503.4(b)]					
12.5 Are changes to the draft EIS clearly marked or otherwise identified in the final EIS?					
12.6 Is the final EIS suitable for filing with EPA—that is, does it: • Have a new cover sheet?					
• Include comments and responses?					
• Include revisions or supplements to the draft? [40 CFR 1503.4 and 1506.9]					

Table A.13 Water resources and water quality

	Yes	No	N/A	EIS Page Number	Adequacy Evaluation and Comments
13.1 Does the EIS discuss potential effects of the proposed action and alternatives: • On surface water quantity under normal operations?					
• Under accident conditions?					
• On surface water quality under normal operations?					
• Under accident conditions?					
13.2 Does the EIS assess the effect of the proposed action and alternatives on the quantity, quality, location, and timing of stormwater runoff (e.g., will new impervious surfaces create a need for stormwater management or pollution controls)?					
13.3 Does the EIS state whether or not the proposed action or alternatives would require a stormwater discharge permit?					
13.4 Does the EIS evaluate whether the proposed action or alternatives would be subject to: • Water quality or effluent standards?					
• National Primary Drinking Water Regulations?					
• National Secondary Drinking Water Regulations?					
13.5 Does the EIS state whether or not the proposed action and alternatives would involve: • Work in, under, over, or having an effect on navigable waters of the United States?					
• Discharge of dredged or fill material into waters of the United States?					
• Deposit of fill material or an excavation that alters or modifies the course, location, condition, or capacity of navigable waters of the United States?					

(continues)

Table A.13 *Continued*

	Yes	No	N/A	EIS Page Number	Adequacy Evaluation and Comments
• Obtaining a Rivers and Harbors Act (Section 10) permit or a Clean Water Act (Section 402 or Section 404) permit?					
• Obtaining a determination under the Coastal Zone Management Act? If so, is such a determination included in the draft EIS?					
13.6 Does the EIS discuss potential effects of the proposed action and alternatives: • On groundwater quantity under normal operations?					
• Under accident conditions?					
• On groundwater quality under normal operations?					
• Under accident conditions?					
13.7 Does the EIS consider whether the proposed action or alternatives may affect municipal or private drinking water supplies?					
13.8 Does the EIS evaluate the incremental effect of effluents associated with the proposed action and alternatives in terms of cumulative water quality conditions?					
13.9 If the proposed action may involve a floodplain, does the document discuss alternative actions to avoid or minimize impacts and preserve floodplain values?					

Table A.14 Geology and soils

	Yes	No	N/A	EIS Page Number	Adequacy Evaluation and Comments
14.1 Does the EIS describe and quantify the land area proposed to be altered, excavated, or otherwise disturbed?					
14.2 Is the description of the disturbed area consistent with other sections (e.g., land use, habitat area)?					
14.3 Are issues related to seismicity sufficiently characterized, quantified, and analyzed?					
14.4 If the action involves disturbance of surface soils, are appropriate best-management practices (e.g., erosion control measures) discussed?					
14.5 Have soil stability and suitability been adequately discussed?					
14.6 Does the EIS consider whether the proposed action may disturb or cause releases of preexisting contaminants or hazardous substances in the soil?					

Table A.15 Air quality

	Yes	No	N/A	EIS Page Number	Adequacy Evaluation and Comments
15.1 Does the EIS discuss potential effects of the proposed action on ambient air quality: • Under normal operations?					
• Under accident conditions?					
15.2 Are potential emissions quantified to the extent practicable (amount and rate of release)?					
15.3 Does the EIS evaluate potential effects to human health and the environment from exposure to radioactive emissions?					
15.4 Does the EIS evaluate potential effects to human health and the environment from exposure to hazardous chemical emissions?					
15.5 When applicable, does the EIS evaluate whether the proposed action and alternatives would: • Be in compliance with the National Ambient Air Quality Standards?					
• Conform to the state implementation plan?					
• Potentially affect any area designated as Class I under the Clean Air Act?					
• Be subject to National Emissions Standards for Hazardous Air Pollutants?					
• Be subject to emissions limitations in an Air Quality Control Region?					
15.6 Does the EIS evaluate the incremental effect of emissions associated with the proposed action and alternatives in terms of cumulative air quality?					

Table A.16 Wildlife and habitat

	Yes	No	N/A	EIS Page Number	Adequacy Evaluation and Comments
16.1 If the EIS identifies potential effects of the proposed action and alternatives on threatened or endangered species and/or critical habitat, has consultation with the USFWS or other applicable agencies been concluded?					
16.2 Does the EIS discuss candidate species?					
16.3 Are state listed species identified and, if so, are results of state consultation documented?					
16.4 Are potential effects (including cumulative effects) analyzed for species other than threatened and endangered species and for habitats other than critical habitat (e.g., fish and wildlife)?					
16.5 Does the EIS analyze impacts on the biodiversity of the affected ecosystem, including genetic diversity and species diversity?					
16.6 Are habitat types identified and estimates provided by type for the amount of habitat lost or adversely affected?					
16.7 Does the EIS consider measures to protect, restore, and enhance wildlife and habitat?					

Table A.17 Human health effects

	Yes	No	N/A	EIS Page Number	Adequacy Evaluation and Comments
17.1 Have the following potentially affected populations been identified? • Involved workers?					
• Noninvolved workers?					
• The public?					
• Minority and low-income communities (as appropriate)? [EO 12898]					
17.2 Does the EIS establish the period of exposure (e.g., 30 years, 70 years) for exposed workers and the public?					
17.3 Does the EIS identify all potential routes of exposure?					
17.4 When providing quantitative estimates of impacts, does the EIS use current dose-to-risk conversion factors adopted by cognizant health and environmental agencies?					
17.5 When providing quantitative estimates of health effects due to radiation exposure, are collective effects expressed in estimated numbers of fatal cancers or cancer incidences?					
17.6 Are maximum individual effects expressed as the estimated maximum probability of a fatality or cancer incidence for an individual?					
17.7 Does the EIS describe assumptions used in the health effects calculations?					
17.8 As appropriate, does the EIS analyze radiological impacts under normal operation conditions for: • Involved workers: –Population dose and corresponding latent cancer fatalities?					
–Maximum individual dose and corresponding cancer risk?					
• Noninvolved workers: –Population dose and corresponding latent cancer fatalities?					
–Maximum individual dose and corresponding cancer risk?					

Table A.17 ***Continued***

	Yes	No	N/A	EIS Page Number	Adequacy Evaluation and Comments
17.9 Does the EIS identify a reasonable spectrum of potential accident scenarios that could occur over the life of the action, including the maximum reasonably foreseeable accident?					
17.10 Does the EIS identify failure scenarios from both natural events (e.g., tornados, earthquakes) and from human error (e.g., forklift accident)?					
17.11 As appropriate, does the EIS analyze radiological impacts under accident conditions for: • Involved workers: –Population dose and corresponding latent cancer fatalities?					
–Maximum individual dose and corresponding cancer risk?					
• Noninvolved workers: –Population dose and corresponding latent cancer fatalities?					
–Maximum individual dose and corresponding cancer risk?					
• Public: –Population dose and corresponding latent cancer fatalities?					
–Maximum individual dose and corresponding cancer risk?					
17.12 Does the EIS discuss toxic and carcinogenic health effects from exposure to hazardous chemicals: • For involved workers?					
• For noninvolved workers?					
• For the public?					
• Under routine operations?					
• Under accident conditions?					
17.13 Does the EIS adequately consider physical safety issues for involved and noninvolved workers?					

Table A.18 Transportation

	Yes	No	N/A	EIS Page Number	Adequacy Evaluation and Comments
18.1 If transportation of hazardous or radioactive waste or materials would be involved or if transportation is a major factor, are the potential effects analyzed (to a site, on site, and from a site)?					
18.2 Does the EIS analyze all reasonably foreseeable transportation links (e.g., overland transport, port transfer, marine transport, global commons)? [E.O. 12114]					
18.3 Does the EIS avoid relying exclusively on statements that transportation will be in accordance with all applicable state and federal regulations and requirements?					
18.4 Does the EIS discuss routine and reasonably foreseeable transportation accidents?					
18.5 Are the estimation methods used for assessing impacts of transportation among those generally accepted and recognized within the scientific community?					
18.6 Does the EIS discuss the annual, total, and cumulative impacts of all transportation actions to the extent that such transportation can be estimated on specific routes?					
18.7 Have transportation analyses adequately considered potential disproportionately high and adverse impacts to minority and low-income populations? [E.O. 12898]					

Table A.19 Waste management and waste minimization

	Yes	No	N/A	EIS Page Number	Adequacy Evaluation and Comments
19.1 Are pollution prevention and waste minimization practices applied in the proposed action and alternatives (e.g., is pollution prevented or reduced at the source when feasible; would waste products be recycled when feasible; are byproducts that cannot be prevented or recycled treated in and environmentally safe manner when feasible; is disposal only used as a last resort)?					
19.2 If waste would be generated, does the EIS examine the human health effects and environmental impacts of managing that waste, including waste generated during facility decontamination or decommissioning?					
19.3 Are waste materials characterized by type and estimated quantity where possible?					
19.4 Does the EIS identify RCRA/CERCLA issues related to the proposed action and alternatives?					
19.5 Does the EIS establish whether the proposal would be in compliance with federal or state laws and guidelines affecting the generation, transportation, treatment, storage, and disposal of hazardous and other waste?					

Table A.20 Socioeconomic considerations

	Yes	No	N/A	EIS Page Number	Adequacy Evaluation and Comments
20.1 Does the EIS consider potential direct, indirect, and cumulative effects on: • Land use patterns?					
• Consistency with applicable land use plans, including site-comprehensive plans, and special designation lands (e.g., farmlands, parks, wildlife, conservation areas)?					
• Compatibility of nearby uses?					
20.2 Does the EIS consider possible changes in the local population due to the proposed action?					
20.3 Does the EIS consider potential economic impacts, such as effects on jobs and housing?					
20.4 Does the EIS consider potential effects on public water and wastewater services, stormwater management, community services, and utilities?					
20.5 Does the EIS evaluate potential noise effects of the proposed action and the application of community noise level standards?					
20.6 Does the EIS state whether the proposal could result in a disproportionately large adverse impact to minority or low-income populations? [EO 12898]					

Table A.21 Cultural resources

	Yes	No	N/A	EIS Page Number	Adequacy Evaluation and Comments
21.1 Was the State Historic Preservation Officer (SHPO) consulted?					
21.2 Was a cultural resources survey conducted for both archaeological and historical resources (while maintaining confidentiality by not disclosing locations for sensitive sites)?					
21.3 Does the EIS discuss potential access conflicts and other adverse impacts to Native American sacred sites (while maintaining confidentiality by not disclosing locations)? [EO 13007]					
21.4 Does the EIS include a provision for mitigation in the event unanticipated archaeological materials (e.g., sites or artifacts) are encountered?					
21.5 Does the EIS address consistency of the proposal with any applicable or proposed cultural resources management plan?					

Appendix B

Reprint of the CEQ NEPA Regulations

TABLE OF CONTENTS

PART 1500—PURPOSE, POLICY, AND MANDATE

AUTHORITY: NEPA, the Environmental Quality Improvement Act of 1970, as amended (42 U.S.C. 4371 *et seq.*), sec. 309 of the Clean Air Act, as amended (42 U.S.C. 7609) and E.O. 11514, Mar. 5, 1970, as amended by E.O. 11991, May 24, 1977).

SOURCE: 43 FR 55990, Nov. 28, 1978, unless otherwise noted.

§ 1500.1 Purpose.

(a) The National Environmental Policy Act (NEPA) is our basic national charter for protection of the environment. It establishes policy, sets goals (section 101), and provides means (section 102) for carrying out the policy. Section 102(2) contains "action-forcing" provisions to make sure that federal agencies act according to the letter and spirit of the Act. The regulations that follow implement section 102(2). Their purpose is to tell federal agencies what they must do to comply with the procedures and achieve the goals of the Act. The President, the federal agencies, and the courts share responsibility for enforcing the Act so as to achieve the substantive requirements of section 101.

(b) NEPA procedures must insure that environmental information is available to public officials and citizens before decisions are made and before actions are taken. The information must be of high quality. Accurate scientific analysis, expert agency comments, and public scrutiny are essential to implementing NEPA. Most important, NEPA documents must concentrate on the issues that are truly significant to the action in question, rather than amassing needless detail.

(c) Ultimately, of course, it is not better documents but better decisions that count. NEPA's purpose is not to generate paperwork—even excellent paperwork—but to foster excellent action. The NEPA process is intended to help public officials make decisions that are based on understanding of environmental consequences, and take actions that protect, restore, and enhance the environment. These regulations provide the direction to achieve this purpose.

§ 1500.2 Policy.

Federal agencies shall to the fullest extent possible:

(a) Interpret and administer the policies, regulations, and public laws of the United States in accordance with the policies set forth in the Act and in these regulations.

(b) Implement procedures to make the NEPA process more useful to decisionmakers and the public; to reduce paperwork and the accumulation of extraneous background data; and to emphasize real environmental issues and alternatives. Environmental impact statements shall be concise, clear, and to the point, and shall be supported by evidence that agencies have made the necessary environmental analyses.

(c) Integrate the requirements of NEPA with other planning and environmental review procedures required by law or by agency practice so that all such procedures run concurrently rather than consecutively.

(d) Encourage and facilitate public involvement in decisions which affect the quality of the human environment.

(e) Use the NEPA process to identify and assess the reasonable alternatives to proposed actions that will avoid or minimize adverse effects of these actions upon the quality of the human environment.

(f) Use all practicable means, consistent with the requirements of the Act and other essential considerations of national policy, to restore and enhance the quality of the human environment and avoid or minimize any possible adverse effects of their actions upon the quality of the human environment.

§ 1500.3 Mandate.

Parts 1500 through 1508 of this title provide regulations applicable to and binding on all Federal agencies for implementing the procedural provisions of the National Environmental Policy Act of 1969, as amended (Pub. L. 91-190, 42 U.S.C. 4321 et seq.) (NEPA or the Act) except where compliance would be inconsistent with other statutory requirements. These regulations are issued pursuant to NEPA, the Environmental Quality Improvement Act of 1970, as amended (42 U.S.C. 4371 et seq.) section 309 of the Clean Air Act, as amended (42 U.S.C. 7609) and Executive Order 11514, Protection and Enhancement of Environmental Quality (March 5, 1970, as amended by Executive Order 11991, May 24, 1977). These regulations, unlike the predecessor guidelines, are not confined to sec. 102(2)(C) (environmental impact statements). The regulations apply to the whole of section 102(2). The provisions of the Act and of these regulations must be read together as a whole in order to comply with the spirit and letter of the law. It is the Council's intention that judicial review of agency compliance with these regulations not occur before an agency has filed the final environmental impact statement, or has made a final finding of no significant impact (when such a finding will result in action affecting the environment), or takes action that will result in irreparable injury. Furthermore, it is the Council's intention that any trivial violation of these regulations not give rise to any independent cause of action.

§ 1500.4 Reducing paperwork.

Agencies shall reduce excessive paperwork by:

(a) Reducing the length of environmental impact statements (§ 1502.2(c)), by means such as setting appropriate page limits (§§ 1501.7(b)(1) and 1502.7).

(b) Preparing analytic rather than encyclopedic environmental impact statements (§ 1502.2(a)).

(c) Discussing only briefly issues other than significant ones (§ 1502.2(b)).

(d) Writing environmental impact statements in plain language (§ 1502.8).

(e) Following a clear format for environmental impact statements (§ 1502.10).

(f) Emphasizing the portions of the environmental impact statement that are useful to decisionmakers and the public (§§ 1502.14 and 1502.15) and reducing emphasis on background material (§ 1502.16).

(g) Using the scoping process, not only to identify significant environmental issues deserving of study, but also to deemphasize insignificant issues, narrowing the scope of the environmental impact statement process accordingly (§ 1501.7).

(h) Summarizing the environmental impact statement (§ 1502.12) and circulating the summary instead of the entire environmental impact statement if the latter is unusually long (§ 1502.19).

(i) Using program, policy, or plan environmental impact statements and tiering from statements of broad scope to those of narrower scope, to eliminate repetitive discussions of the same issues (§§ 1502.4 and 1502.20).

(j) Incorporating by reference (§ 1502.21).

(k) Integrating NEPA requirements with other environmental review and consultation requirements (§ 1502.25).

(l) Requiring comments to be as specific as possible (§ 1503.3).

(m) Attaching and circulating only changes to the draft environmental impact statement, rather than rewriting and circulating the entire statement when changes are minor (§ 1503.4(c)).

(n) Eliminating duplication with State and local procedures, by providing for joint preparation (§ 1506.2), and with other Federal procedures, by providing that an agency may adopt appropriate environmental documents prepared by another agency (§ 1506.3).

(o) Combining environmental documents with other documents (§ 1506.4).

(p) Using categorical exclusions to define categories of actions which do not individually or cumulatively have a significant effect on the human environment and which are therefore exempt from requirements to prepare an environmental impact statement (§ 1508.4).

(q) Using a finding of no significant impact when an action not otherwise excluded will not have a significant effect on the human environment and is therefore exempt from requirements to prepare an environmental impact statement (§ 1508.13).

[43 FR 55990, Nov. 29, 1978; 44 FR 873, Jan. 3, 1979]

§ 1500.5 Reducing delay.

Agencies shall reduce delay by:

(a) Integrating the NEPA process into early planning (§ 1501.2).

(b) Emphasizing interagency cooperation before the environmental impact statement is prepared, rather than submission of adversary comments on a completed document (§ 1501.6).

(c) Insuring the swift and fair resolution of lead agency disputes (§ 1501.5).

(d) Using the scoping process for an early identification of what are and what are not the real issues (§ 1501.7).

(e) Establishing appropriate time limits for the environmental impact statement process (§§ 1501.7(b)(2) and 1501.8).

(f) Preparing environmental impact statements early in the process (§ 1502.5).

(g) Integrating NEPA requirements with other environmental review and consultation requirements (§ 1502.25).

(h) Eliminating duplication with State and local procedures by providing for joint preparation (§ 1506.2) and with other Federal procedures by providing that an agency may adopt appropriate environmental documents prepared by another agency (§ 1506.3).

(i) Combining environmental documents with other documents (§ 1506.4).

(j) Using accelerated procedures for proposals for legislation (§ 1506.8).

(k) Using categorical exclusions to define categories of actions which do not individually or cumulatively have a significant effect on the human environment (§ 1508.4) and which are therefore exempt from requirements to prepare an environmental impact statement.

(l) Using a finding of no significant impact when an action not otherwise excluded will not have a significant effect on the human environment (§ 1508.13) and is therefore exempt from requirements to prepare an environmental impact statement.

§ 1500.6 Agency authority.

Each agency shall interpret the provisions of the Act as a supplement to its existing authority and as a mandate to view traditional policies and missions in the light of the Act's national environmental objectives. Agencies shall review their policies, procedures, and regulations accordingly and revise them as necessary to insure full compliance with the purposes and provisions of the Act. The phrase "to the fullest extent possible" in section 102 means that each agency of the Federal Government shall comply with that section unless existing law applicable to the agency's operations expressly prohibits or makes compliance impossible.

PART 1501—NEPA AND AGENCY PLANNING

Sec.

AUTHORITY: NEPA, the Environmental Quality Improvement Act of 1970, as amended (42 U.S.C. 4371 *et seq.*), sec. 309 of the Clean Air Act, as amended (42 U.S.C. 7609, and E.O. 11514 (Mar. 5, 1970, as amended by E.O. 11991, May 24, 1977).

SOURCE: 43 FR 55992, Nov. 29, 1978, unless otherwise noted.

§ 1501.1 Purpose.

The purposes of this part include:

(a) Integrating the NEPA process into early planning to insure appropriate consideration of NEPA's policies and to eliminate delay.

(b) Emphasizing cooperative consultation among agencies before the environmental impact statement is prepared rather than submission of adversary comments on a completed document.

(c) Providing for the swift and fair resolution of lead agency disputes.

(d) Identifying at an early stage the significant environmental issues deserving of study and deemphasizing insignificant issues, narrowing the scope of the environmental impact statement accordingly.

(e) Providing a mechanism for putting appropriate time limits on the environmental impact statement process.

§ 1501.2 Apply NEPA early in the process.

Agencies shall integrate the NEPA process with other planning at the earliest possible time to insure that planning and decisions reflect environmental values, to avoid delays later in the process, and to head off potential conflicts. Each agency shall:

(a) Comply with the mandate of section 102(2)(A) to "utilize a systematic, interdisciplinary approach which will insure the integrated use of the natural and social sciences and the environmental design arts in planning and in decisionmaking which may have an impact on man's environment," as specified by § 1507.2.

(b) Identify environmental effects and values in adequate detail so they can be compared to economic and technical analyses. Environmental documents and appropriate analyses shall be circulated and reviewed at the same time as other planning documents.

(c) Study, develop, and describe appropriate alternatives to recommended courses of action in any proposal which involves unresolved conflicts concerning alternative uses of available resources as provided by section 102(2)(E) of the Act.

(d) Provide for cases where actions are planned by private applicants or other non-Federal entities before Federal involvement so that:

(1) Policies or designated staff are available to advise potential applicants of studies or other information foreseeably required for later Federal action.

(2) The Federal agency consults early with appropriate State and local agencies and Indian tribes and with interested private persons and organizations when its own involvement is reasonably foreseeable.

(3) The Federal agency commences its NEPA process at the earliest possible time.

§ 1501.3 When to prepare an environmental assessment.

(a) Agencies shall prepare an environmental assessment (§ 1508.9) when necessary under the procedures adopted by individual agencies to supplement these regulations as described in § 1507.3. An assessment is not necessary if the agency has decided to prepare an environmental impact statement.

(b) Agencies may prepare an environmental assessment on any action at any time in order to assist agency planning and decisionmaking.

§ 1501.4 Whether to prepare an environmental impact statement.

In determining whether to prepare an environmental impact statement the Federal agency shall:

(a) Determine under its procedures supplementing these regulations (described in § 1507.3) whether the proposal is one which:

(1) Normally requires an environmental impact statement, or

(2) Normally does not require either an environmental impact statement or an environmental assessment (categorical exclusion).

(b) If the proposed action is not covered by paragraph (a) of this section, prepare an environmental assessment (§ 1508.9). The agency shall involve environmental agencies, applicants, and the public, to the extent practicable, in preparing assessments required by § 1508.9(a)(1).

(c) Based on the environmental assessment make its determination whether to prepare an environmental impact statement.

(d) Commence the scoping process (§ 1501.7), if the agency will prepare an environmental impact statement.

(e) Prepare a finding of no significant impact (§ 1508.13), if the agency determines on the basis of the environmental assessment not to prepare a statement.

(1) The agency shall make the finding of no significant impact available

to the affected public as specified in § 1506.6.

(2) In certain limited circumstances, which the agency may cover in its procedures under § 1507.3, the agency shall make the finding of no significant impact available for public review (including State and areawide clearinghouses) for 30 days before the agency makes its final determination whether to prepare an environmental impact statement and before the action may begin. The circumstances are:

(i) The proposed action is, or is closely similar to, one which normally requires the preparation of an environmental impact statement under the procedures adopted by the agency pursuant to § 1507.3, or

(ii) The nature of the proposed action is one without precedent.

§ 1501.5 Lead agencies.

(a) A lead agency shall supervise the preparation of an environmental impact statement if more than one Federal agency either:

(1) Proposes or is involved in the same action; or

(2) Is involved in a group of actions directly related to each other because of their functional interdependence or geographical proximity.

(b) Federal, State, or local agencies, including at least one Federal agency, may act as joint lead agencies to prepare an environmental impact statement (§ 1506.2).

(c) If an action falls within the provisions of paragraph (a) of this section the potential lead agencies shall determine by letter or memorandum which agency shall be the lead agency and which shall be cooperating agencies. The agencies shall resolve the lead agency question so as not to cause delay. If there is disagreement among the agencies, the following factors (which are listed in order of descending importance) shall determine lead agency designation:

(1) Magnitude of agency's involvement.

(2) Project approval/disapproval authority.

(3) Expertise concerning the action's environmental effects.

(4) Duration of agency's involvement.

(5) Sequence of agency's involvement.

(d) Any Federal agency, or any State or local agency or private person substantially affected by the absence of lead agency designation, may make a written request to the potential lead agencies that a lead agency be designated.

(e) If Federal agencies are unable to agree on which agency will be the lead agency or if the procedure described in paragraph (c) of this section has not resulted within 45 days in a lead agency designation, any of the agencies or persons concerned may file a request with the Council asking it to determine which Federal agency shall be the lead agency.

A copy of the request shall be transmitted to each potential lead agency. The request shall consist of:

(1) A precise description of the nature and extent of the proposed action.

(2) A detailed statement of why each potential lead agency should or should not be the lead agency under the criteria specified in paragraph (c) of this section.

(f) A response may be filed by any potential lead agency concerned within 20 days after a request is filed with the Council. The Council shall determine as soon as possible but not later than 20 days after receiving the request and all responses to it which Federal agency shall be the lead agency and which other Federal agencies shall be cooperating agencies.

[43 FR 55992, Nov. 29, 1978; 44 FR 873, Jan. 3, 1979]

§ 1501.6 Cooperating agencies.

The purpose of this section is to emphasize agency cooperation early in the NEPA process. Upon request of the lead agency, any other Federal agency which has jurisdiction by law shall be a cooperating agency. In addition any other Federal agency which has special expertise with respect to any environmental issue, which should be addressed in the statement may be a cooperating agency upon request of the lead agency. An agency may request the lead agency to designate it a cooperating agency.

(a) The lead agency shall:

(1) Request the participation of each cooperating agency in the NEPA process at the earliest possible time.

(2) Use the environmental analysis and proposals of cooperating agencies with jurisdiction by law or special expertise, to the maximum extent possible consistent with its responsibility as lead agency.

(3) Meet with a cooperating agency at the latter's request.

(b) Each cooperating agency shall:

(1) Participate in the NEPA process at the earliest possible time.

(2) Participate in the scoping process (described below in § 1501.7).

(3) Assume on request of the lead agency responsibility for developing information and preparing environmental analyses including portions of the environmental impact statement concerning which the cooperating agency has special expertise.

(4) Make available staff support at the lead agency's request to enhance the latter's interdisciplinary capability.

(5) Normally use its own funds. The lead agency shall, to the extent available funds permit, fund those major activities or analyses it requests from cooperating agencies. Potential lead agencies shall include such funding requirements in their budget requests.

(c) A cooperating agency may in response to a lead agency's request for assistance in preparing the environmental impact statement (described in paragraph (b) (3), (4), or (5) of this section) reply that other program commitments preclude any involvement or the degree of involvement requested in the action that is the subject of the environmental impact statement. A copy of this reply shall be submitted to the Council.

§ 1501.7 Scoping.

There shall be an early and open process for determining the scope of issues to be addressed and for identifying the significant issues related to a proposed action. This process shall be termed scoping. As soon as practicable after its decision to prepare an environmental impact statement and before the scoping process the lead agency shall publish a notice of intent (§ 1508.22) in the FEDERAL REGISTER except as provided in § 1507.3(e).

(a) As part of the scoping process the lead agency shall:

(1) Invite the participation of affected Federal, State, and local agencies, any affected Indian tribe, the proponent of the action, and other interested persons (including those who might not be in accord with the action on environmental grounds), unless there is a limited exception under § 1507.3(c). An agency may give notice in accordance with § 1506.6.

(2) Determine the scope (§ 1508.25) and the significant issues to be analyzed in depth in the environmental impact statement.

(3) Identify and eliminate from detailed study the issues which are not significant or which have been covered by prior environmental review (§ 1506.3), narrowing the discussion of these issues in the statement to a brief presentation of why they will not have a significant effect on the human environment or providing a reference to their coverage elsewhere.

(4) Allocate assignments for preparation of the environmental impact statement among the lead and cooperating agencies, with the lead agency retaining responsibility for the statement.

(5) Indicate any public environmental assessments and other environmental impact statements which are being or will be prepared that are related to but are not part of the scope of the impact statement under consideration.

(6) Identify other environmental review and consultation requirements so the lead and cooperating agencies may prepare other required analyses and studies concurrently with, and integrated with, the environmental impact statement as provided in § 1502.25.

(7) Indicate the relationship between the timing of the preparation of environmental analyses and the agency's tentative planning and decisionmaking schedule.

(b) As part of the scoping process the lead agency may:

(1) Set page limits on environmental documents (§ 1502.7).

(2) Set time limits (§ 1501.8).

(3) Adopt procedures under § 1507.3 to combine its environmental assessment process with its scoping process.

(4) Hold an early scoping meeting or meetings which may be integrated with any other early planning meeting the agency has. Such a scoping meeting will often be appropriate when the impacts of a particular action are confined to specific sites.

(c) An agency shall revise the determinations made under paragraphs (a) and (b) of this section if substantial changes are made later in the proposed action, or if significant new circumstances or information arise which bear on the proposal or its impacts.

§ 1501.8 Time limits.

Although the Council has decided that prescribed universal time limits for the entire NEPA process are too inflexible, Federal agencies are encouraged to set time limits appropriate to individual actions (consistent with the time intervals required by § 1506.10). When multiple agencies are involved the reference to agency below means lead agency.

(a) The agency shall set time limits if an applicant for the proposed action requests them: *Provided,* That the limits are consistent with the purposes of NEPA and other essential considerations of national policy.

(b) The agency may:

(1) Consider the following factors in determining time limits:

(i) Potential for environmental harm.

(ii) Size of the proposed action.

(iii) State of the art of analytic techniques.

(iv) Degree of public need for the proposed action, including the consequences of delay.

(v) Number of persons and agencies affected.

(vi) Degree to which relevant information is known and if not known the time required for obtaining it.

(vii) Degree to which the action is controversial.

(viii) Other time limits imposed on the agency by law, regulations, or executive order.

(2) Set overall time limits or limits for each constituent part of the NEPA process, which may include:

(i) Decision on whether to prepare an environmental impact statement (if not already decided).

(ii) Determination of the scope of the environmental impact statement.

(iii) Preparation of the draft environmental impact statement.

(iv) Review of any comments on the draft environmental impact statement from the public and agencies.

(v) Preparation of the final environmental impact statement.

(vi) Review of any comments on the final environmental impact statement.

(vii) Decision on the action based in part on the environmental impact statement.

(3) Designate a person (such as the project manager or a person in the agency's office with NEPA responsibilities) to expedite the NEPA process.

(c) State or local agencies or members of the public may request a Federal Agency to set time limits.

PART 1502—ENVIRONMENTAL IMPACT STATEMENT

Sec.
1502.23 Cost-benefit analysis.
1502.24 Methodology and scientific accuracy.
1502.25 Environmental review and consultation requirements.

AUTHORITY: NEPA, the Environmental Quality Improvement Act of 1970, as amended (42 U.S.C. 4371 *et seq.*), sec. 309 of the Clean Air Act, as amended (42 U.S.C. 7609), and E.O. 11514 (Mar. 5, 1970, as amended by E.O. 11991, May 24, 1977).

SOURCE: 43 FR 55994, Nov. 29, 1978, unless otherwise noted.

§ 1502.1 Purpose.

The primary purpose of an environmental impact statement is to serve as an action-forcing device to insure that the policies and goals defined in the Act are infused into the ongoing programs and actions of the Federal Government. It shall provide full and fair discussion of significant environmental impacts and shall inform decisionmakers and the public of the reasonable alternatives which would avoid or minimize adverse impacts or enhance the quality of the human environment. Agencies shall focus on significant environmental issues and alternatives and shall reduce paperwork and the accumulation of extraneous background data. Statements shall be concise, clear, and to the point, and shall be supported by evidence that the agency has made the necessary environmental analyses. An environmental impact statement is more than a disclosure document. It shall be used by Federal officials in conjunction with other relevant material to plan actions and make decisions.

§ 1502.2 Implementation.

To achieve the purposes set forth in § 1502.1 agencies shall prepare environmental impact statements in the following manner:

(a) Environmental impact statements shall be analytic rather than encyclopedic.

(b) Impacts shall be discussed in proportion to their significance. There shall be only brief discussion of other than significant issues. As in a finding of no significant impact, there should be only enough discussion to show why more study is not warranted.

(c) Environmental impact statements shall be kept concise and shall be no longer than absolutely necessary to comply with NEPA and with these regulations. Length should vary first with potential environmental problems and then with project size.

(d) Environmental impact statements shall state how alternatives considered in it and decisions based on it will or will not achieve the requirements of sections 101 and 102(1) of the Act and other environmental laws and policies.

(e) The range of alternatives discussed in environmental impact statements shall encompass those to be considered by the ultimate agency decisionmaker.

(f) Agencies shall not commit resources prejudicing selection of alternatives before making a final decision (§ 1506.1).

(g) Environmental impact statements shall serve as the means of assessing the environmental impact of proposed agency actions, rather than justifying decisions already made.

§ 1502.3 Statutory requirements for statements.

As required by sec. 102(2)(C) of NEPA environmental impact statements (§ 1508.11) are to be included in every recommendation or report.

On proposals (§ 1508.23).

For legislation and (§ 1508.17).

Other major Federal actions (§ 1508.18).

Significantly (§ 1508.27).

Affecting (§§ 1508.3, 1508.8).

The quality of the human environment (§ 1508.14).

§ 1502.4 Major Federal actions requiring the preparation of environmental impact statements.

(a) Agencies shall make sure the proposal which is the subject of an environmental impact statement is properly defined. Agencies shall use the criteria for scope (§ 1508.25) to determine which proposal(s) shall be the subject of a particular statement. Proposals or parts of proposals which are related to each other closely enough to be, in effect, a single course of action shall be evaluated in a single impact statement.

(b) Environmental impact statements may be prepared, and are sometimes required, for broad Federal actions such as the adoption of new agency programs or regulations (§ 1508.18). Agencies shall prepare statements on broad actions so that they are relevant to policy and are timed to coincide with meaningful points in agency planning and decisionmaking.

(c) When preparing statements on broad actions (including proposals by more than one agency), agencies may find it useful to evaluate the proposal(s) in one of the following ways:

(1) Geographically, including actions occurring in the same general location, such as body of water, region, or metropolitan area.

(2) Generically, including actions which have relevant similarities, such as common timing, impacts, alternatives, methods of implementation, media, or subject matter.

(3) By stage of technological development including federal or federally assisted research, development or demonstration programs for new technologies which, if applied, could significantly affect the quality of the human environment. Statements shall be prepared on such programs and shall be available before the program has reached a stage of investment or commitment to implementation likely to determine subsequent development or restrict later alternatives.

(d) Agencies shall as appropriate employ scoping (§ 1501.7), tiering (§ 1502.20), and other methods listed in §§ 1500.4 and 1500.5 to relate broad and narrow actions and to avoid duplication and delay.

§ 1502.5 Timing.

An agency shall commence preparation of an environmental impact statement as close as possible to the time the agency is developing or is presented with a proposal (§ 1508.23) so that preparation can be completed in time for the final statement to be included in any recommendation or report on the proposal. The statement shall be prepared early enough so that it can serve practically as an important contribution to the decisionmaking process and will not be used to rationalize or justify decisions already made (§§ 1500.2(c), 1501.2, and 1502.2). For instance:

(a) For projects directly undertaken by Federal agencies the environmental impact statement shall be prepared at the feasibility analysis (go-no go) stage and may be supplemented at a later stage if necessary.

(b) For applications to the agency appropriate environmental assessments or statements shall be commenced no later than immediately after the application is received. Federal agencies are encouraged to begin preparation of such assessments or statements earlier, preferably jointly with applicable State or local agencies.

(c) For adjudication, the final environmental impact statement shall normally precede the final staff recommendation and that portion of the public hearing related to the impact study. In appropriate circumstances the statement may follow preliminary hearings designed to gather information for use in the statements.

(d) For informal rulemaking the draft environmental impact statement shall normally accompany the proposed rule.

§ 1502.6 Interdisciplinary preparation.

Environmental impact statements shall be prepared using an inter-disciplinary approach which will insure the integrated use of the natural and social sciences and the environmental design arts (section 102(2)(A) of the Act). The disciplines of the preparers shall be appropriate to the scope and issues identified in the scoping process (§ 1501.7).

§ 1502.7 Page limits.

The text of final environmental impact statements (e.g., paragraphs (d) through (g) of § 1502.10) shall normally be less than 150 pages and for proposals of unusual scope or complexity shall normally be less than 300 pages.

§ 1502.8 Writing.

Environmental impact statements shall be written in plain language and may use appropriate graphics so that decisionmakers and the public can readily understand them. Agencies should employ writers of clear prose or editors to write, review, or edit statements, which will be based upon the analysis and supporting data from the natural and social sciences and the environmental design arts.

§ 1502.9 Draft, final, and supplemental statements.

Except for proposals for legislation as provided in § 1506.8 environmental impact statements shall be prepared in two stages and may be supplemented.

(a) Draft environmental impact statements shall be prepared in accordance with the scope decided upon in the scoping process. The lead agency shall work with the cooperating agencies and shall obtain comments as required in Part 1503 of this chapter. The draft statement must fulfill and satisfy to the fullest extent possible the requirements established for final statements in section 102(2)(C) of the Act. If a draft statement is so inadequate as to preclude meaningful analysis, the agency shall prepare and circulate a revised draft of the appropriate portion. The agency shall make every effort to disclose and discuss at appropriate points in the draft statement all major points of view on the environmental impacts of the alternatives including the proposed action.

(b) Final environmental impact statements shall respond to comments as required in Part 1503 of this chapter. The agency shall discuss at appropriate points in the final statement any responsible opposing view which was not adequately discussed in the draft statement and shall indicate the agency's response to the issues raised.

(c) Agencies:

(1) Shall prepare supplements to either draft or final environmental impact statements if:

(i) The agency makes substantial changes in the proposed action that are relevant to environmental concerns; or

(ii) There are significant new circumstances or information relevant to environmental concerns and bearing on the proposed action or its impacts.

(2) May also prepare supplements when the agency determines that the purposes of the Act will be furthered by doing so.

(3) Shall adopt procedures for introducing a supplement into its formal administrative record, if such a record exists.

(4) Shall prepare, circulate, and file a supplement to a statement in the same fashion (exclusive of scoping) as a draft and final statement unless alternative procedures are approved by the Council.

§ 1502.10 Recommended format.

Agencies shall use a format for environmental impact statements which will encourage good analysis and clear presentation of the alternatives including the proposed action. The following standard format for environmental impact statements should be followed unless the agency determines that there is a compelling reason to do otherwise:

(a) Cover sheet.

(b) Summary.

(c) Table of contents.

(d) Purpose of and need for action.

(e) Alternatives including proposed action (sections 102(2)(C)(iii) and 102(2)(E) of the Act).

(f) Affected environment.

(g) Environmental consequences (especially sections 102(2)(C) (i), (ii), (iv), and (v) of the Act).

(h) List of preparers.

(i) List of Agencies, Organizations, and persons to whom copies of the statement are sent.

(j) Index.

(k) Appendices (if any).

If a different format is used, it shall include paragraphs (a), (b), (c), (h), (i), and (j), of this section and shall include the substance of paragraphs (d), (e), (f), (g), and (k) of this section, as further described in §§ 1502.11 through 1502.18, in any appropriate format.

§ 1502.11 Cover sheet.

The cover sheet shall not exceed one page. It shall include:

(a) A list of the responsible agencies including the lead agency and any cooperating agencies.

(b) The title of the proposed action that is the subject of the statement (and if appropriate the titles of related cooperating agency actions), together with the State(s) and county(ies) (or other jurisdiction if applicable) where the action is located.

(c) The name, address, and telephone number of the person at the agency who can supply further information.

(d) A designation of the statement as a draft, final, or draft or final supplement.

(e) A one paragraph abstract of the statement.

(f) The date by which comments must be received (computed in cooperation with EPA under § 1506.10).

The information required by this section may be entered on Standard Form 424 (in items 4, 6, 7, 10, and 18).

§ 1502.12 Summary.

Each environmental impact statement shall contain a summary which adequately and accurately summarizes the statement. The summary shall stress the major conclusions, areas of controversy (including issues raised by agencies and the public), and the issues to be resolved (including the choice among alternatives). The summary will normally not exceed 15 pages.

§ 1502.13 Purpose and need.

The statement shall briefly specify the underlying purpose and need to which the agency is responding in proposing the alternatives including the proposed action.

§ 1502.14 Alternatives including the proposed action.

This section is the heart of the environmental impact statement. Based on the information and analysis presented in the sections on the Affected Environment (§ 1502.15) and the Environmental Consequences (§ 1502.16), it should present the environmental impacts of the proposal and the alternatives in comparative form, thus sharply defining the issues and providing a clear basis for choice among options by the decisionmaker and the public. In this section agencies shall:

(a) Rigorously explore and objectively evaluate all reasonable alternatives, and for alternatives which were eliminated from detailed study, briefly discuss the reasons for their having been eliminated.

(b) Devote substantial treatment to each alternative considered in detail including the proposed action so that reviewers may evaluate their comparative merits.

(c) Include reasonable alternatives not within the jurisdiction of the lead agency.

(d) Include the alternative of no action.

(e) Identify the agency's preferred alternative or alternatives, if one or more exists, in the draft statement and identify such alternative in the final statement unless another law prohibits the expression of such a preference.

(f) Include appropriate mitigation measures not already included in the proposed action or alternatives.

§ 1502.15 Affected environment.

The environmental impact statement shall succinctly describe the environment of the area(s) to be affected or created by the alternatives under consideration. The descriptions shall be no longer than is necessary to understand the effects of the alternatives. Data and analyses in a statement shall be commensurate with the importance of the impact, with less important material summarized, consolidated, or simply referenced. Agencies shall avoid useless bulk in statements and shall concentrate effort and attention on important issues. Verbose descriptions of the affected environment are themselves no measure of the adequacy of an environmental impact statement.

§ 1502.16 Environmental consequences.

This section forms the scientific and analytic basis for the comparisons under § 1502.14. It shall consolidate the discussions of those elements required by sections 102(2)(C) (i), (ii), (iv), and (v) of NEPA which are within the scope of the statement and as much of section 102(2)(C)(iii) as is necessary to support the comparisons. The discussion will include the environmental impacts of the alternatives including the proposed action, any adverse environmental effects which cannot be avoided should the proposal be implemented, the relationship between short-term uses of man's environment and the maintenance and enhancement of long-term productivity, and any irreversible or irretrievable commitments of resources which would be involved in the proposal should it be implemented. This section should not duplicate discussions in § 1502.14. It shall include discussions of:

(a) Direct effects and their significance (§ 1508.8).

(b) Indirect effects and their significance (§ 1508.8).

(c) Possible conflicts between the proposed action and the objectives of Federal, regional, State, and local (and in the case of a reservation, Indian tribe) land use plans, policies and controls for the area concerned. (See § 1506.2(d).)

(d) The environmental effects of alternatives including the proposed action. The comparisons under § 1502.14 will be based on this discussion.

(e) Energy requirements and conservation potential of various alternatives and mitigation measures.

(f) Natural or depletable resource requirements and conservation potential of various alternatives and mitigation measures.

(g) Urban quality, historic and cultural resources, and the design of the built environment, including the reuse and conservation potential of various alternatives and mitigation measures.

(h) Means to mitigate adverse environmental impacts (if not fully covered under § 1502.14(f)).

[43 FR 55994, Nov. 29, 1978; 44 FR 873, Jan. 3, 1979]

§ 1502.17 List of preparers.

The environmental impact statement shall list the names, together with their qualifications (expertise, experience, professional disciplines), of the persons who were primarily responsible for preparing the environmental impact statement or significant background papers, including basic components of the statement (§§ 1502.6 and 1502.8). Where possible the persons who are responsible for a particular analysis, including analyses in background papers, shall be identified. Normally the list will not exceed two pages.

§ 1502.18 Appendix.

If an agency prepares an appendix to an environmental impact statement the appendix shall:

(a) Consist of material prepared in connection with an environmental impact statement (as distinct from material which is not so prepared and which is incorporated by reference (§ 1502.21)).

(b) Normally consist of material which substantiates any analysis fundamental to the impact statement.

(c) Normally be analytic and relevant to the decision to be made.

(d) Be circulated with the environmental impact statement or be readily available on request.

§ 1502.19 Circulation of the environmental impact statement.

Agencies shall circulate the entire draft and final environmental impact statements except for certain appendices as provided in § 1502.18(d) and unchanged statements as provided in § 1503.4(c). However, if the statement is unusually long, the agency may circulate the summary instead, except that the entire statement shall be furnished to:

(a) Any Federal agency which has jurisdiction by law or special expertise with respect to any environmental impact involved and any appropriate Federal, State or local agency authorized to develop and enforce environmental standards.

(b) The applicant, if any.

(c) Any person, organization, or agency requesting the entire environmental impact statement.

(d) In the case of a final environmental impact statement any person, organization, or agency which submitted substantive comments on the draft.

If the agency circulates the summary and thereafter receives a timely request for the entire statement and for additional time to comment, the time for that requestor only shall be extended by at least 15 days beyond the minimum period.

§ 1502.20 Tiering.

Agencies are encouraged to tier their environmental impact statements to eliminate repetitive discussions of the same issues and to focus on the actual issues ripe for decision at each level of environmental review (§ 1508.28). Whenever a broad environmental impact statement has been prepared (such as a program or policy statement) and a subsequent statement or environmental assessment is then prepared on an action included within the entire program or policy (such as a site specific action) the subsequent statement or environmental assessment need only summarize the issues discussed in the broader statement and incorporate discussions from the broader statement by reference and shall concentrate on the issues specific to the subsequent action. The subsequent document shall state where the earlier document is available. Tiering may also be appropriate for different stages of actions. (Section 1508.28).

§ 1502.21 Incorporation by reference.

Agencies shall incorporate material into an environmental impact statement by reference when the effect will be to cut down on bulk without impeding agency and public review of the action. The incorporated material shall be cited in the statement and its content briefly described. No material may be incorporated by reference unless it is reasonably available for inspection by potentially interested persons within the time allowed for comment. Material based on proprietary data which is itself not available for review and comment shall not be incorporated by reference.

§ 1502.22 Incomplete or unavailable information.

When an agency is evaluating reasonably foreseeable significant adverse effects on the human environment in an environmental impact statement and there is incomplete or unavailable information, the agency shall always make clear that such information is lacking.

(a) If the incomplete information relevant to reasonably foreseeable significant adverse impacts is essential to a reasoned choice among alternatives and the overall costs of obtaining it are not exorbitant, the agency shall include the information in the environmental impact statement.

(b) If the information relevant to reasonably foreseeable significant adverse impacts cannot be obtained because the overall costs of obtaining it are exorbitant or the means to obtain it are not known, the agency shall include within the environmental impact statement: (1) A statement that such information is incomplete or unavailable; (2) a statement of the relevance of the incomplete or unavailable information to evaluating reasonably foreseeable significant adverse impacts on the human environment; (3) a summary of existing credible scientific evidence which is relevant to evaluating the reasonably foreseeable significant adverse impacts on the human environment, and (4) the agency's evaluation of such impacts based upon theoretical approaches or research methods generally accepted in the scientific community. For the purposes of this section, "reasonably foreseeable" includes impacts which have catastrophic consequences, even if their probability of occurrence is low, provided that the analysis of the impacts is supported by credible scientific evidence, is not based on pure conjecture, and is within the rule of reason.

(c) The amended regulation will be applicable to all environmental impact statements for which a Notice of Intent (40 CFR 1508.22) is published in the Federal Register on or after May 27, 1986. For environmental impact statements in progress, agencies may choose to comply with the requirements of either the original or amended regulation.

[51 FR 15625, Apr. 25, 1986]

§ 1502.23 Cost-benefit analysis.

If a cost-benefit analysis relevant to the choice among environmentally different alternatives is being considered for the proposed action, it shall be incorporated by reference or appended to the statement as an aid in evaluating the environmental consequences. To assess the adequacy of compliance with section 102(2)(B) of the Act the statement shall, when a cost-benefit analysis is prepared, discuss the relationship between that analysis and any analyses of unquantified environmental impacts, values, and amenities. For purposes of complying with the Act, the weighing of the merits and drawbacks of the various alternatives need not be displayed in a monetary cost-benefit analysis and should not be when there are important qualitative considerations. In any event, an environmental impact statement should at least indicate those considerations, including factors not related to environmental quality, which are likely to be relevant and important to a decision.

§ 1502.24 Methodology and scientific accuracy.

Agencies shall insure the professional integrity, including scientific integrity, of the discussions and analyses in environmental impact statements. They shall identify any methodologies used and shall make explicit reference by footnote to the scientific and other sources relied upon for conclusions in the statement. An agency may place discussion of methodology in an appendix.

§ 1502.25 Environmental review and consultation requirements.

(a) To the fullest extent possible, agencies shall prepare draft environmental impact statements concurrently with and integrated with environmental impact analyses and related surveys and studies required by the Fish and Wildlife Coordination Act (16 U.S.C. 661 et seq.), the National Historic Preservation Act of 1966 (16 U.S.C. 470 et seq.), the Endangered Species Act of 1973 (16 U.S.C. 1531 et seq.), and other environmental review laws and executive orders.

(b) The draft environmental impact statement shall list all Federal permits, licenses, and other entitlements which must be obtained in implementing the proposal. If it is uncertain whether a Federal permit, license, or other entitlement is necessary, the draft environmental impact statement shall so indicate.

PART 1503—COMMENTING

AUTHORITY: NEPA, the Environmental Quality Improvement Act of 1970, as amended (42 U.S.C. 4371 *et seq.*), sec. 309 of the Clean Air Act, as amended (42 U.S.C. 7609), and E.O. 11514 (Mar. 5, 1970, as amended by E.O. 11991, May 24, 1977).

SOURCE: 43 FR 55997, Nov. 29, 1978, unless otherwise noted.

§ 1503.1 Inviting comments.

(a) After preparing a draft environmental impact statement and before preparing a final environmental impact statement the agency shall:

(1) Obtain the comments of any Federal agency which has jurisdiction by law or special expertise with respect to any environmental impact involved or which is authorized to develop and enforce environmental standards.

(2) Request the comments of:

(i) Appropriate State and local agencies which are authorized to develop and enforce environmental standards;

(ii) Indian tribes, when the effects may be on a reservation; and

(iii) Any agency which has requested that it receive statements on actions of the kind proposed.

Under Executive Order No. 12372, the Office of Management and Budget, through its system of clearinghouses, provides a means of securing the views of State and local environmental agencies. The clearinghouses may be used, by mutual agreement of the lead agency and the clearinghouse, for securing

State and local reviews of the draft environmental impact statements.

(3) Request comments from the applicant, if any.

(4) Request comments from the public, affirmatively soliciting comments from those persons or organizations who may be interested or affected.

(b) An agency may request comments on a final environmental impact statement before the decision is finally made. In any case other agencies or persons may make comments before the final decision unless a different time is provided under § 1506.10.

§ 1503.2 Duty to comment.

Federal agencies with jurisdiction by law or special expertise with respect to any environmental impact involved and agencies which are authorized to develop and enforce environmental standards shall comment on statements within their jurisdiction, expertise, or authority. Agencies shall comment within the time period specified for comment in § 1506.10. A Federal agency may reply that it has no comment. If a cooperating agency is satisfied that its views are adequately reflected in the environmental impact statement, it should reply that it has no comment.

§ 1503.3 Specificity of comments.

(a) Comments on an environmental impact statement or on a proposed action shall be as specific as possible and may address either the adequacy of the statement or the merits of the alternatives discussed or both.

(b) When a commenting agency criticizes a lead agency's predictive methodology, the commenting agency should describe the alternative methodology which it prefers and why.

(c) A cooperating agency shall specify in its comments whether it needs additional information to fulfill other applicable environmental reviews or consultation requirements and what information it needs. In particular, it shall specify any additional information it needs to comment adequately on the draft statement's analysis of significant site-specific effects associated with the granting or approving by that cooperating agency of necessary Federal permits, licenses, or entitlements.

(d) When a cooperating agency with jurisdiction by law objects to or expresses reservations about the proposal on grounds of environmental impacts, the agency expressing the objection or reservation shall specify the mitigation measures it considers necessary to allow the agency to grant or approve applicable permit, license, or related requirements or concurrences.

§ 1503.4 Response to comments.

(a) An agency preparing a final environmental impact statement shall assess and consider comments both individually and collectively, and shall respond by one or more of the means listed below, stating its response in the final statement. Possible responses are to:

(1) Modify alternatives including the proposed action.

(2) Develop and evaluate alternatives not previously given serious consideration by the agency.

(3) Supplement, improve, or modify its analyses.

(4) Make factual corrections.

(5) Explain why the comments do not warrant further agency response, citing the sources, authorities, or reasons which support the agency's position and, if appropriate, indicate those circumstances which would trigger agency reappraisal or further response.

(b) All substantive comments received on the draft statement (or summaries thereof where the response has been exceptionally voluminous), should be attached to the final statement whether or not the comment is thought to merit individual discussion by the agency in the text of the statement.

(c) If changes in response to comments are minor and are confined to the responses described in paragraphs (a) (4) and (5) of this section, agencies may write them on errata sheets and attach them to the statement instead of rewriting the draft statement. In such cases only the comments, the responses, and the changes and not the final statement need be circulated (§ 1502.19). The entire document with a new cover sheet shall be filed as the final statement (§ 1506.9).

PART 1504—PREDECISION REFERRALS TO THE COUNCIL OF PROPOSED FEDERAL ACTIONS DETERMINED TO BE ENVIRONMENTALLY UNSATISFACTORY

AUTHORITY: NEPA, the Environmental Quality Improvement Act of 1970, as amended (42 U.S.C. 4371 *et seq.*), sec. 309 of the Clean Air Act, as amended (42 U.S.C. 7609), and E.O. 11514 (Mar. 5, 1970, as amended by E.O. 11991, May 24, 1977).

SOURCE: 43 FR 55998, Nov. 29, 1978, unless otherwise noted.

§ 1504.1 Purpose.

(a) This part establishes procedures for referring to the Council Federal interagency disagreements concerning proposed major Federal actions that might cause unsatisfactory environmental effects. It provides means for early resolution of such disagreements.

(b) Under section 309 of the Clean Air Act (42 U.S.C. 7609), the Administrator of the Environmental Protection Agency is directed to review and comment publicly on the environmental impacts of Federal activities, including actions for which environmental impact statements are prepared. If after this review the Administrator determines that the matter is "unsatisfactory from the standpoint of public health or welfare or environmental quality," section 309 directs that the matter be referred to the Council (hereafter "environmental referrals").

(c) Under section 102(2)(C) of the Act other Federal agencies may make similar reviews of environmental impact statements, including judgments on the acceptability of anticipated environmental impacts. These reviews must be made available to the President, the Council and the public.

§ 1504.2 Criteria for referral.

Environmental referrals should be made to the Council only after concerted, timely (as early as possible in the process), but unsuccessful attempts to resolve differences with the lead agency. In determining what environmental objections to the matter are appropriate to refer to the Council, an agency should weigh potential adverse environmental impacts, considering:

(a) Possible violation of national environmental standards or policies.

(b) Severity.

(c) Geographical scope.

(d) Duration.

(e) Importance as precedents.

(f) Availability of environmentally preferable alternatives.

§ 1504.3 Procedure for referrals and response.

(a) A Federal agency making the referral to the Council shall:

(1) Advise the lead agency at the earliest possible time that it intends to refer a matter to the Council unless a satisfactory agreement is reached.

(2) Include such advice in the referring agency's comments on the draft environmental impact statement, except when the statement does not contain adequate information to permit an assessment of the matter's environmental acceptability.

(3) Identify any essential information that is lacking and request that it be made available at the earliest possible time.

(4) Send copies of such advice to the Council.

(b) The referring agency shall deliver its referral to the Council not later than twenty-five (25) days after the final environmental impact statement has been made available to the Environmental Protection Agency, commenting agencies, and the public. Except when an extension of this period has been granted by the lead agency, the Council will not accept a referral after that date.

(c) The referral shall consist of:

(1) A copy of the letter signed by the head of the referring agency and delivered to the lead agency informing the lead agency of the referral and the reasons for it, and requesting that no action be taken to implement the matter until the Council acts upon the

referral. The letter shall include a copy of the statement referred to in (c)(2) of this section.

(2) A statement supported by factual evidence leading to the conclusion that the matter is unsatisfactory from the standpoint of public health or welfare or environmental quality. The statement shall:

(i) Identify any material facts in controversy and incorporate (by reference if appropriate) agreed upon facts,

(ii) Identify any existing environmental requirements or policies which would be violated by the matter,

(iii) Present the reasons why the referring agency believes the matter is environmentally unsatisfactory,

(iv) Contain a finding by the agency whether the issue raised is of national importance because of the threat to national environmental resources or policies or for some other reason,

(v) Review the steps taken by the referring agency to bring its concerns to the attention of the lead agency at the earliest possible time, and

(vi) Give the referring agency's recommendations as to what mitigation alternative, further study, or other course of action (including abandonment of the matter) are necessary to remedy the situation.

(d) Not later than twenty-five (25) days after the referral to the Council the lead agency may deliver a response to the Council, and the referring agency. If the lead agency requests more time and gives assurance that the matter will not go forward in the interim, the Council may grant an extension. The response shall:

(1) Address fully the issues raised in the referral.

(2) Be supported by evidence.

(3) Give the lead agency's response to the referring agency's recommendations.

(e) Interested persons (including the applicant) may deliver their views in writing to the Council. Views in support of the referral should be delivered not later than the referral. Views in support of the response shall be delivered not later than the response.

(f) Not later than twenty-five (25) days after receipt of both the referral and any response or upon being informed that there will be no response (unless the lead agency agrees to a longer time), the Council may take one or more of the following actions:

(1) Conclude that the process of referral and response has successfully resolved the problem.

(2) Initiate discussions with the agencies with the objective of mediation with referring and lead agencies.

(3) Hold public meetings or hearings to obtain additional views and information.

(4) Determine that the issue is not one of national importance and request the referring and lead agencies to pursue their decision process.

(5) Determine that the issue should be further negotiated by the referring and lead agencies and is not appropriate for Council consideration until one or more heads of agencies report to the Council that the agencies' disagreements are irreconcilable.

(6) Publish its findings and recommendations (including where appropriate a finding that the submitted evidence does not support the position of an agency).

(7) When appropriate, submit the referral and the response together with the Council's recommendation to the President for action.

(g) The Council shall take no longer than 60 days to complete the actions specified in paragraph (f) (2), (3), or (5) of this section.

(h) When the referral involves an action required by statute to be determined on the record after opportunity for agency hearing, the referral shall be conducted in a manner consistent with 5 U.S.C. 557(d) (Administrative Procedure Act).

[43 FR 55998, Nov. 29, 1978; 44 FR 873, Jan. 3, 1979]

PART 1505—NEPA AND AGENCY DECISIONMAKING

Sec.

AUTHORITY: NEPA, the Environmental Quality Improvement Act of 1970, as amended (42 U.S.C. 4371 *et seq.*), sec. 309 of the Clean Air Act, as amended (42 U.S.C.

7609), and E.O. 11514 (Mar. 5, 1970, as amended by E.O. 11991, May 24, 1977).

SOURCE: 43 FR 55999, Nov. 29, 1978, unless otherwise noted.

§ 1505.1 Agency decisionmaking procedures.

Agencies shall adopt procedures (§ 1507.3) to ensure that decisions are made in accordance with the policies and purposes of the Act. Such procedures shall include but not be limited to:

(a) Implementing procedures under section 102(2) to achieve the requirements of sections 101 and 102(1).

(b) Designating the major decision points for the agency's principal programs likely to have a significant effect on the human environment and assuring that the NEPA process corresponds with them.

(c) Requiring that relevant environmental documents, comments, and responses be part of the record in formal rulemaking or adjudicatory proceedings.

(d) Requiring that relevant environmental documents, comments, and responses accompany the proposal through existing agency review processes so that agency officials use the statement in making decisions.

(e) Requiring that the alternatives considered by the decisionmaker are encompassed by the range of alternatives discussed in the relevant environmental documents and that the decisionmaker consider the alternatives described in the environmental impact statement. If another decision document accompanies the relevant environmental documents to the decisionmaker, agencies are encouraged to make available to the public before the decision is made any part of that document that relates to the comparison of alternatives.

§ 1505.2 Record of decision in cases requiring environmental impact statements.

At the time of its decision (§ 1506.10) or, if appropriate, its recommendation to Congress, each agency shall prepare a concise public record of decision. The record, which may be integrated into any other record prepared by the agency, shall:

(a) State what the decision was.

(b) Identify all alternatives considered by the agency in reaching its decision, specifying the alternative or alternatives which were considered to be environmentally preferable. An agency may discuss preferences among alternatives based on relevant factors including economic and technical considerations and agency statutory missions. An agency shall identify and discuss all such factors including any essential considerations of national policy which were balanced by the agency in making its decision and state how those considerations entered into its decision.

(c) State whether all practicable means to avoid or minimize environmental harm from the alternative selected have been adopted, and if not, why they were not. A monitoring and enforcement program shall be adopted and summarized where applicable for any mitigation.

§ 1505.3 Implementing the decision.

Agencies may provide for monitoring to assure that their decisions are carried out and should do so in important cases. Mitigation (§ 1505.2(c)) and other conditions established in the environmental impact statement or during its review and committed as part of the decision shall be implemented by the lead agency or other appropriate consenting agency. The lead agency shall:

(a) Include appropriate conditions in grants, permits or other approvals.

(b) Condition funding of actions on mitigation.

(c) Upon request, inform cooperating or commenting agencies on progress in carrying out mitigation measures which they have proposed and which were adopted by the agency making the decision.

(d) Upon request, make available to the public the results of relevant monitoring.

PART 1506—OTHER REQUIREMENTS OF NEPA

AUTHORITY: NEPA, the Environmental Quality Improvement Act of 1970, as amended (42 U.S.C. 4371 *et seq.*), sec. 309 of the Clean Air Act, as amended (42 U.S.C. 7609), and E.O. 11514 (Mar. 5, 1970, as amended by E.O. 11991, May 24, 1977).

SOURCE: 43 FR 56000, Nov. 29, 1978, unless otherwise noted.

§ 1506.1 Limitations on actions during NEPA process.

(a) Until an agency issues a record of decision as provided in § 1505.2 (except as provided in paragraph (c) of this section), no action concerning the proposal shall be taken which would:

(1) Have an adverse environmental impact; or

(2) Limit the choice of reasonable alternatives.

(b) If any agency is considering an application from a non-Federal entity, and is aware that the applicant is about to take an action within the agency's jurisdiction that would meet either of the criteria in paragraph (a) of this section, then the agency shall promptly notify the applicant that the agency will take appropriate action to insure that the objectives and procedures of NEPA are achieved.

(c) While work on a required program environmental impact statement is in progress and the action is not covered by an existing program statement, agencies shall not undertake in the interim any major Federal action covered by the program which may significantly affect the quality of the human environment unless such action:

(1) Is justified independently of the program;

(2) Is itself accompanied by an adequate environmental impact statement; and

(3) Will not prejudice the ultimate decision on the program. Interim action prejudices the ultimate decision on the program when it tends to determine subsequent development or limit alternatives.

(d) This section does not preclude development by applicants of plans or designs or performance of other work necessary to support an application for Federal, State or local permits or assistance. Nothing in this section shall preclude Rural Electrification Administration approval of minimal expenditures not affecting the environment (*e.g.* long leadtime equipment and purchase options) made by nongovernmental entities seeking loan guarantees from the Administration.

§ 1506.2 Elimination of duplication with State and local procedures.

(a) Agencies authorized by law to cooperate with State agencies of statewide jurisdiction pursuant to section 102(2)(D) of the Act may do so.

(b) Agencies shall cooperate with State and local agencies to the fullest extent possible to reduce duplication between NEPA and State and local requirements, unless the agencies are specifically barred from doing so by some other law. Except for cases covered by paragraph (a) of this section, such cooperation shall to the fullest extent possible include:

(1) Joint planning processes.

(2) Joint environmental research and studies.

(3) Joint public hearings (except where otherwise provided by statute).

(4) Joint environmental assessments.

(c) Agencies shall cooperate with State and local agencies to the fullest extent possible to reduce duplication between NEPA and comparable State and local requirements, unless the agencies are specifically barred from doing so by some other law. Except for cases covered by paragraph (a) of this section, such cooperation shall to the fullest extent possible include joint environmental impact statements. In such cases one or more Federal agencies and one or more State or local agencies shall be joint lead agencies. Where State laws or local ordinances have environmental impact statement requirements in addition to but not in conflict with those in NEPA, Federal agencies shall cooperate in fulfilling these requirements as well as those of Federal laws so that one document will comply with all applicable laws.

(d) To better integrate environmental impact statements into State or local planning processes, statements shall discuss any inconsistency of a proposed action with any approved State or local plan and laws (whether or not federally sanctioned). Where an inconsistency exists, the statement should describe the extent to which the agency would reconcile its proposed action with the plan or law.

§ 1506.3 Adoption.

(a) An agency may adopt a Federal draft or final environmental impact statement or portion thereof provided that the statement or portion thereof meets the standards for an adequate statement under these regulations.

(b) If the actions covered by the original environmental impact statement and the proposed action are substantially the same, the agency adopting another agency's statement is not required to recirculate it except as a final statement. Otherwise the adopting agency shall treat the statement as a draft and recirculate it (except as provided in paragraph (c) of this section).

(c) A cooperating agency may adopt without recirculating the environmental impact statement of a lead agency when, after an independent review of the statement, the cooperating agency concludes that its comments and suggestions have been satisfied.

(d) When an agency adopts a statement which is not final within the agency that prepared it, or when the action it assesses is the subject of a referral under Part 1504, or when the statement's adequacy is the subject of a judicial action which is not final, the agency shall so specify.

§ 1506.4 Combining documents.

Any environmental document in compliance with NEPA may be combined with any other agency document to reduce duplication and paperwork.

§ 1506.5 Agency responsibility.

(a) *Information.* If an agency requires an applicant to submit environmental information for possible use by the agency in preparing an environmental impact statement, then the agency should assist the applicant by outlining the types of information required. The agency shall independently evaluate the information submitted and shall be responsible for its accuracy. If the agency chooses to use the information submitted by the applicant in the environmental impact statement, either directly or by reference, then the names of the persons responsible for the independent evaluation shall be included in the list of preparers (§ 1502.17). It is the intent of this paragraph that acceptable work not be redone, but that it be verified by the agency.

(b) *Environmental assessments.* If an agency permits an applicant to prepare an environmental assessment, the agency, besides fulfilling the requirements of paragraph (a) of this section, shall make its own evaluation of the environmental issues and take responsibility for the scope and content of the environmental assessment.

(c) *Environmental impact statements.* Except as provided in §§ 1506.2 and 1506.3 any environmental impact statement prepared pursuant to the requirements of NEPA shall be prepared directly by or by a contractor selected by the lead agency or where appropriate under § 1501.6(b), a cooperating agency. It is the intent of these regulations that the contractor be chosen solely by the lead agency, or by the lead agency in cooperation with cooperating agencies, or where appropriate by a cooperating agency to avoid any conflict of interest. Contractors shall execute a disclosure statement prepared by the lead agency, or where appropriate the cooperating agency, specifying that they have no financial or other interest in the outcome of the project. If the document is prepared by contract, the responsible Federal official shall furnish guidance and participate in the preparation and shall independently evaluate the statement prior to its approval and take responsibility for its scope and contents. Nothing in this section is intended to prohibit any agency from requesting any person to submit information to it or to prohibit any person from submitting information to any agency.

§ 1506.6 Public involvement.

Agencies shall:

(a) Make diligent efforts to involve the public in preparing and implementing their NEPA procedures.

(b) Provide public notice of NEPA-related hearings, public meetings, and the availability of environmental documents so as to inform those persons and agencies who may be interested or affected.

(1) In all cases the agency shall mail notice to those who have requested it on an individual action.

(2) In the case of an action with effects of national concern notice shall include publication in the FEDERAL REGISTER and notice by mail to national organizations reasonably expected to be interested in the matter and may include listing in the *102 Monitor*. An agency engaged in rulemaking may provide notice by mail to national organizations who have requested that notice regularly be provided. Agencies shall maintain a list of such organizations.

(3) In the case of an action with effects primarily of local concern the notice may include:

(i) Notice to State and areawide clearinghouses pursuant to EO 12372, the Intergovernmental Review Process.

(ii) Notice to Indian tribes when effects may occur on reservations.

(iii) Following the affected State's public notice procedures for comparable actions.

(iv) Publication in local newspapers (in papers of general circulation rather than legal papers).

(v) Notice through other local media.

(vi) Notice to potentially interested community organizations including small business associations.

(vii) Publication in newsletters that may be expected to reach potentially interested persons.

(viii) Direct mailing to owners and occupants of nearby or affected property.

(ix) Posting of notice on and off site in the area where the action is to be located.

(c) Hold or sponsor public hearings or public meetings whenever appropriate or in accordance with statutory requirements applicable to the agency. Criteria shall include whether there is:

(1) Substantial environmental controversy concerning the proposed action or substantial interest in holding the hearing.

(2) A request for a hearing by another agency with jurisdiction over the action supported by reasons why a hearing will be helpful. If a draft environmental impact statement is to be considered at a public hearing, the agency should make the statement available to the public at least 15 days in advance (unless the purpose of the hearing is to provide information for the draft environmental impact statement).

(d) Solicit appropriate information from the public.

(e) Explain in its procedures where interested persons can get information or status reports on environmental impact statements and other elements of the NEPA process.

(f) Make environmental impact statements, the comments received, and any underlying documents available to the public pursuant to the provisions of the Freedom of Information Act (5 U.S.C. 552), without regard to the exclusion for interagency memoranda where such memoranda transmit comments of Federal agencies on the environmental impact of the proposed action. Materials to be made available to the public shall be provided to the public without charge to the extent practicable, or at a fee which is not more than the actual costs of reproducing copies required to be sent to other Federal agencies, including the Council.

§ 1506.7 Further guidance.

The Council may provide further guidance concerning NEPA and its procedures including:

(a) A handbook which the Council may supplement from time to time, which shall in plain language provide guidance and instructions concerning the application of NEPA and these regulations.

(b) Publication of the Council's Memoranda to Heads of Agencies.

(c) In conjunction with the Environmental Protection Agency and the publication of the 102 Monitor, notice of:

(1) Research activities;

(2) Meetings and conferences related to NEPA; and

(3) Successful and innovative procedures used by agencies to implement NEPA.

§ 1506.8 Proposals for legislation.

(a) The NEPA process for proposals for legislation (§ 1508.17) significantly affecting the quality of the human environment shall be integrated with the legislative process of the Congress. A legislative environmental impact statement is the detailed statement required by law to be included in a recommendation or report on a legislative proposal to Congress. A legislative environmental impact statement shall be considered part of the formal transmittal of a legislative proposal to Congress; however, it may be transmitted to Congress up to 30 days later in order to allow time for completion of an accurate statement which can serve as the basis for public and Congressional debate. The statement must be available in time for Congressional hearings and deliberations.

(b) Preparation of a legislative environmental impact statement shall conform to the requirements of these regulations except as follows:

(1) There need not be a scoping process.

(2) The legislative statement shall be prepared in the same manner as a draft statement, but shall be considered the "detailed statement" required by statute; *Provided,* That when any of the following conditions exist both the draft and final environmental impact statement on the legislative proposal shall be prepared and circulated as provided by §§ 1503.1 and 1506.10.

(i) A Congressional Committee with jurisdiction over the proposal has a rule requiring both draft and final environmental impact statements.

(ii) The proposal results from a study process required by statute (such as those required by the Wild and Scenic Rivers Act (16 U.S.C. 1271 et seq.) and the Wilderness Act (16 U.S.C. 1131 et seq.)).

(iii) Legislative approval is sought for Federal or federally assisted construction or other projects which the agency recommends be located at specific geographic locations. For proposals requiring an environmental impact statement for the acquisition of space by the General Services Administration, a draft statement shall accompany the Prospectus or the 11(b) Report of Building Project Surveys to the Congress, and a final statement shall be completed before site acquisition.

(iv) The agency decides to prepare draft and final statements.

(c) Comments on the legislative statement shall be given to the lead agency which shall forward them along with its own responses to the Congressional committees with jurisdiction.

§ 1506.9 Filing requirements.

Environmental impact statements together with comments and responses shall be filed with the Environmental Protection Agency, attention Office of Federal Activities (A-104), 401 M Street SW., Washington, D.C. 20460. Statements shall be filed with EPA no earlier than they are also transmitted to commenting agencies and made available to the public. EPA shall deliver one copy of each statement to the Council, which shall satisfy the requirement of availability to the President. EPA may issue guidelines to agencies to implement its responsibilities under this section and § 1506.10.

§ 1506.10 Timing of agency action.

(a) The Environmental Protection Agency shall publish a notice in the FEDERAL REGISTER each week of the environmental impact statements filed during the preceding week. The minimum time periods set forth in this section shall be calculated from the date of publication of this notice.

(b) No decision on the proposed action shall be made or recorded under § 1505.2 by a Federal agency until the later of the following dates:

(1) Ninety (90) days after publication of the notice described above in paragraph (a) of this section for a draft environmental impact statement.

(2) Thirty (30) days after publication of the notice described above in paragraph (a) of this section for a final environmental impact statement.

An exception to the rules on timing may be made in the case of an agency decision which is subject to a formal internal appeal. Some agencies have a formally established appeal process which allows other agencies or the public to take appeals on a decision and make their views known, after publication of the final environmental impact statement. In such cases, where a real opportunity exists to alter the decision, the decision may be made and recorded at the same time the environmental impact statement is published. This means that the period for appeal of the decision and the 30-day period prescribed in paragraph (b)(2) of this section may run concurrently. In such cases the environmental impact statement shall explain the timing and the public's right of appeal. An agency engaged in rule-making under the Administrative Procedure Act or other statute for the purpose of protecting the public health or safety, may waive the time period in paragraph (b)(2) of this section and publish a decision on the final rule simultaneously with publication of the notice of the availability of the final environmental impact statement as described in paragraph (a) of this section.

(c) If the final environmental impact statement is filed within ninety (90) days after a draft environmental impact statement is filed with the Environmental Protection Agency, the minimum thirty (30) day period and the minimum ninety (90) day period may run concurrently. However, subject to paragraph (d) of this section agencies shall allow not less than 45 days for comments on draft statements.

(d) The lead agency may extend prescribed periods. The Environmental Protection Agency may upon a showing by the lead agency of compelling reasons of national policy reduce the prescribed periods and may upon a showing by any other Federal agency of compelling reasons of national policy also extend prescribed periods, but only after consultation with the lead agency. (Also see § 1507.3(d).) Failure to file timely comments shall not be a sufficient reason for extending a period. If the lead agency does not concur with the extension of time, EPA may not extend it for more than 30 days. When the Environmental Protection Agency reduces or extends any period of time it shall notify the Council.

[43 FR 56000, Nov. 29, 1978; 44 FR 874, Jan. 3, 1979]

§ 1506.11 Emergencies.

Where emergency circumstances make it necessary to take an action with significant environmental impact without observing the provisions of these regulations, the Federal agency taking the action should consult with the Council about alternative arrangements. Agencies and the Council will limit such arrangements to actions necessary to control the immediate impacts of the emergency. Other actions remain subject to NEPA review.

§ 1506.12 Effective date.

The effective date of these regulations is July 30, 1979, except that for agencies that administer programs that qualify under section 102(2)(D) of the Act or under sec. 104(h) of the Housing and Community Development Act of 1974 an additional four months shall be allowed for the State or local agencies to adopt their implementing procedures.

(a) These regulations shall apply to the fullest extent practicable to ongoing activities and environmental documents begun before the effective date. These regulations do not apply to an environmental impact statement or supplement if the draft statement was filed before the effective date of these regulations. No completed environmental documents need be redone by reasons of these regulations. Until these regulations are applicable, the Council's guidelines published in the FEDERAL REGISTER of August 1, 1973, shall continue to be applicable. In cases where these regulations are applicable the guidelines are superseded. However, nothing shall prevent an agency from proceeding under these regulations at an earlier time.

(b) NEPA shall continue to be applicable to actions begun before January 1, 1970, to the fullest extent possible.

PART 1507—AGENCY COMPLIANCE

AUTHORITY: NEPA, the Environmental Quality Improvement Act of 1970, as amended (42 U.S.C. 4371 *et seq.*), sec. 309 of the Clean Air Act, as amended (42 U.S.C. 7609), and E.O. 11514 (Mar. 5, 1970, as amended by E.O. 11991, May 24, 1977).

SOURCE: 43 FR 56002, Nov. 29, 1978, unless otherwise noted.

§ 1507.1 Compliance.

All agencies of the Federal Government shall comply with these regulations. It is the intent of these regulations to allow each agency flexibility in adapting its implementing procedures authorized by § 1507.3 to the requirements of other applicable laws.

§ 1507.2 Agency capability to comply.

Each agency shall be capable (in terms of personnel and other resources) of complying with the requirements enumerated below. Such compliance may include use of other's resources, but the using agency shall itself have sufficient capability to evaluate what others do for it. Agencies shall:

(a) Fulfill the requirements of section 102(2)(A) of the Act to utilize a systematic, interdisciplinary approach which will insure the integrated use of the natural and social sciences and the environmental design arts in planning and in decisionmaking which may have an impact on the human environment. Agencies shall designate a person to be responsible for overall review of agency NEPA compliance.

(b) Identify methods and procedures required by section 102(2)(B) to insure that presently unquantified environmental amenities and values may be given appropriate consideration.

(c) Prepare adequate environmental impact statements pursuant to section 102(2)(C) and comment on statements in the areas where the agency has jurisdiction by law or special expertise or is authorized to develop and enforce environmental standards.

(d) Study, develop, and describe alternatives to recommended courses of action in any proposal which involves unresolved conflicts concerning alternative uses of available resources. This requirement of section 102(2)(E) extends to all such proposals, not just the more limited scope of section 102(2)(C)(iii) where the discussion of alternatives is confined to impact statements.

(e) Comply with the requirements of section 102(2)(H) that the agency initiate and utilize ecological information in the planning and development of resource-oriented projects.

(f) Fulfill the requirements of sections 102(2)(F), 102(2)(G), and 102(2)(I), of the Act and of Executive Order 11514, Protection and Enhancement of Environmental Quality, Sec. 2.

§ 1507.3 Agency procedures.

(a) Not later than eight months after publication of these regulations as finally adopted in the FEDERAL REGISTER, or five months after the establishment of an agency, whichever shall come later, each agency shall as necessary adopt procedures to supplement these regulations. When the agency is a department, major subunits are encouraged (with the consent of the department) to adopt their own procedures. Such procedures shall not paraphrase these regulations. They shall confine themselves to implementing procedures. Each agency shall consult with the Council while developing its procedures and before publishing them in the FEDERAL REGISTER for comment. Agencies with similar programs should consult with each other and the Council to coordinate their procedures, especially for programs requesting similar information from applicants. The procedures shall be adopted only after an opportunity for public review and after review by the Council for conformity with the Act and these regulations. The Council shall complete its review within 30 days. Once in effect they shall be filed with the Council and made readily available to the public. Agencies are encouraged to publish explanatory guidance for these regulations and their own procedures. Agencies shall continue to review their policies and procedures and in consultation with the Council to revise them as necessary to ensure full compliance with the purposes and provisions of the Act.

(b) Agency procedures shall comply with these regulations except where compliance would be inconsistent with statutory requirements and shall include:

(1) Those procedures required by §§ 1501.2(d), 1502.9(c)(3), 1505.1, 1506.6(e), and 1508.4.

(2) Specific criteria for and identification of those typical classes of action:

(i) Which normally do require environmental impact statements.

(ii) Which normally do not require either an environmental impact statement or an environmental assessment (categorical exclusions (§ 1508.4)).

(iii) Which normally require environmental assessments but not necessarily environmental impact statements.

(c) Agency procedures may include specific criteria for providing limited exceptions to the provisions of these regulations for classified proposals. They are proposed actions which are specifically authorized under criteria established by an Executive Order or statute to be kept secret in the interest of national defense or foreign policy and are in fact properly classified pursuant to such Executive Order or statute. Environmental assessments and environmental impact statements which address classified proposals may be safeguarded and restricted from public dissemination in accordance with agencies' own regulations applicable to classified information. These documents may be organized so that classified portions can be included as annexes, in order that the unclassified portions can be made available to the public.

(d) Agency procedures may provide for periods of time other than those presented in § 1506.10 when necessary to comply with other specific statutory requirements.

(e) Agency procedures may provide that where there is a lengthy period between the agency's decision to prepare an environmental impact statement and the time of actual preparation, the notice of intent required by § 1501.7 may be published at a reasonable time in advance of preparation of the draft statement.

PART 1508—TERMINOLOGY AND INDEX

Sec.

AUTHORITY: NEPA, the Environmental Quality Improvement Act of 1970, as amended (42 U.S.C. 4371 *et seq.*), sec. 309 of the Clean Air Act, as amended (42 U.S.C. 7609), and E.O. 11514 (Mar. 5, 1970, as amended by E.O. 11991, May 24, 1977).

SOURCE: 43 FR 56003, Nov. 29, 1978, unless otherwise noted.

§ 1508.1 Terminology.

The terminology of this part shall be uniform throughout the Federal Government.

§ 1508.2 Act.

"Act" means the National Environmental Policy Act, as amended (42 U.S.C. 4321, et seq.) which is also referred to as "NEPA."

§ 1508.3 Affecting.

"Affecting" means will or may have an effect on.

§ 1508.4 Categorical exclusion.

"Categorical exclusion" means a category of actions which do not individually or cumulatively have a significant effect on the human environment and which have been found to have no such effect in procedures adopted by a Federal agency in implementation of these regulations (§ 1507.3) and for which, therefore, neither an environmental assessment nor an environmental impact statement is required. An agency may decide in its procedures or otherwise, to prepare environmental assessments for the reasons stated in § 1508.9 even though it is not required to do so. Any procedures under this section shall provide for extraordinary circumstances in which a normally excluded action may have a significant environmental effect.

§ 1508.5 Cooperating agency.

"Cooperating agency" means any Federal agency other than a lead agency which has jurisdiction by law or special expertise with respect to any environmental impact involved in a proposal (or a reasonable alternative) for legislation or other major Federal action significantly affecting the quality of the human environment. The selection and responsibilities of a cooperating agency are described in § 1501.6. A State or local agency of similar qualifications or, when the effects are on a reservation, an Indian Tribe, may by agreement with the lead agency become a cooperating agency.

§ 1508.6 Council.

"Council" means the Council on Environmental Quality established by Title II of the Act.

§ 1508.7 Cumulative impact.

"Cumulative impact" is the impact on the environment which results from the incremental impact of the action when added to other past, present, and reasonably foreseeable future actions regardless of what agency (Federal or non-Federal) or person undertakes such other actions. Cumulative impacts can result from individually minor but collectively significant actions taking place over a period of time.

§ 1508.8 Effects.

"Effects" include:

(a) Direct effects, which are caused by the action and occur at the same time and place.

(b) Indirect effects, which are caused by the action and are later in time or farther removed in distance, but are still reasonably foreseeable. Indirect effects may include growth inducing effects and other effects related to induced changes in the pattern of land use, population density or growth rate, and related effects on air and water and other natural systems, including ecosystems.

Effects and impacts as used in these regulations are synonymous. Effects includes ecological (such as the effects on natural resources and on the components, structures, and functioning of affected ecosystems), aesthetic, historic, cultural, economic, social, or health, whether direct, indirect, or cumulative. Effects may also include those resulting from actions which may have both beneficial and detrimental effects, even if on balance the agency believes that the effect will be beneficial.

§ 1508.9 Environmental assessment.

"Environmental assessment":

(a) Means a concise public document for which a Federal agency is responsible that serves to:

(1) Briefly provide sufficient evidence and analysis for determining

whether to prepare an environmental impact statement or a finding of no significant impact.

(2) Aid an agency's compliance with the Act when no environmental impact statement is necessary.

(3) Facilitate preparation of a statement when one is necessary.

(b) Shall include brief discussions of the need for the proposal, of alternatives as required by section 102(2)(E), of the environmental impacts of the proposed action and alternatives, and a listing of agencies and persons consulted.

§ 1508.10 Environmental document.

"Environmental document" includes the documents specified in § 1508.9 (environmental assessment), § 1508.11 (environmental impact statement), § 1508.13 (finding of no significant impact), and § 1508.22 (notice of intent).

§ 1508.11 Environmental impact statement.

"Environmental impact statement" means a detailed written statement as required by section 102(2)(C) of the Act.

§ 1508.12 Federal agency.

"Federal agency" means all agencies of the Federal Government. It does not mean the Congress, the Judiciary, or the President, including the performance of staff functions for the President in his Executive Office. It also includes for purposes of these regulations States and units of general local government and Indian tribes assuming NEPA responsibilities under section 104(h) of the Housing and Community Development Act of 1974.

§ 1508.13 Finding of no significant impact.

"Finding of no significant impact" means a document by a Federal agency briefly presenting the reasons why an action, not otherwise excluded (§ 1508.4), will not have a significant effect on the human environment and for which an environmental impact statement therefore will not be prepared. It shall include the environmental assessment or a summary of it and shall note any other environmental documents related to it (§ 1501.7(a)(5)). If the assessment is included, the finding need not repeat any of the discussion in the assessment but may incorporate it by reference.

§ 1508.14 Human environment.

"Human environment" shall be interpreted comprehensively to include the natural and physical environment and the relationship of people with that environment. (See the definition of "effects" (§ 1508.8).) This means that economic or social effects are not intended by themselves to require preparation of an environmental impact statement. When an environmental impact statement is prepared and economic or social and natural or physical environmental effects are interrelated, then the environmental impact statement will discuss all of these effects on the human environment.

§ 1508.15 Jurisdiction by law.

"Jurisdiction by law" means agency authority to approve, veto, or finance all or part of the proposal.

§ 1508.16 Lead agency.

"Lead agency" means the agency or agencies preparing or having taken primary responsibility for preparing the environmental impact statement.

§ 1508.17 Legislation.

"Legislation" includes a bill or legislative proposal to Congress developed by or with the significant cooperation and support of a Federal agency, but does not include requests for appropriations. The test for significant cooperation is whether the proposal is in fact predominantly that of the agency rather than another source. Drafting does not by itself constitute significant cooperation. Proposals for legislation include requests for ratification of treaties. Only the agency which has primary responsibility for the subject matter involved will prepare a legislative environmental impact statement.

§ 1508.18 Major Federal action.

"Major Federal action" includes actions with effects that may be major and which are potentially subject to

Federal control and responsibility. Major reinforces but does not have a meaning independent of significantly (§ 1508.27). Actions include the circumstance where the responsible officials fail to act and that failure to act is reviewable by courts or administrative tribunals under the Administrative Procedure Act or other applicable law as agency action.

(a) Actions include new and continuing activities, including projects and programs entirely or partly financed, assisted, conducted, regulated, or approved by federal agencies; new or revised agency rules, regulations, plans, policies, or procedures; and legislative proposals (§§ 1506.8, 1508.17). Actions do not include funding assistance solely in the form of general revenue sharing funds, distributed under the State and Local Fiscal Assistance Act of 1972, 31 U.S.C. 1221 et seq., with no Federal agency control over the subsequent use of such funds. Actions do not include bringing judicial or administrative civil or criminal enforcement actions.

(b) Federal actions tend to fall within one of the following categories:

(1) Adoption of official policy, such as rules, regulations, and interpretations adopted pursuant to the Administrative Procedure Act, 5 U.S.C. 551 et seq.; treaties and international conventions or agreements; formal documents establishing an agency's policies which will result in or substantially alter agency programs.

(2) Adoption of formal plans, such as official documents prepared or approved by federal agencies which guide or prescribe alternative uses of federal resources, upon which future agency actions will be based.

(3) Adoption of programs, such as a group of concerted actions to implement a specific policy or plan; systematic and connected agency decisions allocating agency resources to implement a specific statutory program or executive directive.

(4) Approval of specific projects, such as construction or management activities located in a defined geographic area. Projects include actions approved by permit or other regulatory decision as well as federal and federally assisted activities.

§ 1508.19 Matter.

"Matter" includes for purposes of Part 1504:

(a) With respect to the Environmental Protection Agency, any proposed legislation, project, action or regulation as those terms are used in section 309(a) of the Clean Air Act (42 U.S.C. 7609).

(b) With respect to all other agencies, any proposed major federal action to which section 102(2)(C) of NEPA applies.

§ 1508.20 Mitigation.

"Mitigation" includes:

(a) Avoiding the impact altogether by not taking a certain action or parts of an action.

(b) Minimizing impacts by limiting the degree or magnitude of the action and its implementation.

(c) Rectifying the impact by repairing, rehabilitating, or restoring the affected environment.

(d) Reducing or eliminating the impact over time by preservation and maintenance operations during the life of the action.

(e) Compensating for the impact by replacing or providing substitute resources or environments.

§ 1508.21 NEPA process.

"NEPA process" means all measures necessary for compliance with the requirements of section 2 and Title I of NEPA.

§ 1508.22 Notice of intent.

"Notice of intent" means a notice that an environmental impact statement will be prepared and considered. The notice shall briefly:

(a) Describe the proposed action and possible alternatives.

(b) Describe the agency's proposed scoping process including whether, when, and where any scoping meeting will be held.

(c) State the name and address of a person within the agency who can answer questions about the proposed action and the environmental impact statement.

§ 1508.23 Proposal.

"Proposal" exists at that stage in the development of an action when an agency subject to the Act has a goal and is actively preparing to make a decision on one or more alternative means of accomplishing that goal and the effects can be meaningfully evaluated. Preparation of an environmental impact statement on a proposal should be timed (§ 1502.5) so that the final statement may be completed in time for the statement to be included in any recommendation or report on the proposal. A proposal may exist in fact as well as by agency declaration that one exists.

§ 1508.24 Referring agency.

"Referring agency" means the federal agency which has referred any matter to the Council after a determination that the matter is unsatisfactory from the standpoint of public health or welfare or environmental quality.

§ 1508.25 Scope.

Scope consists of the range of actions, alternatives, and impacts to be considered in an environmental impact statement. The scope of an individual statement may depend on its relationships to other statements (§§1502.20 and 1508.28). To determine the scope of environmental impact statements, agencies shall consider 3 types of actions, 3 types of alternatives, and 3 types of impacts. They include:

(a) Actions (other than unconnected single actions) which may be:

(1) Connected actions, which means that they are closely related and therefore should be discussed in the same impact statement. Actions are connected if they:

(i) Automatically trigger other actions which may require environmental impact statements.

(ii) Cannot or will not proceed unless other actions are taken previously or simultaneously.

(iii) Are interdependent parts of a larger action and depend on the larger action for their justification.

(2) Cumulative actions, which when viewed with other proposed actions have cumulatively significant impacts and should therefore be discussed in the same impact statement.

(3) Similar actions, which when viewed with other reasonably foreseeable or proposed agency actions, have similarities that provide a basis for evaluating their environmental consequencies together, such as common timing or geography. An agency may wish to analyze these actions in the same impact statement. It should do so when the best way to assess adequately the combined impacts of similar actions or reasonable alternatives to such actions is to treat them in a single impact statement.

(b) Alternatives, which include: (1) No action alternative.

(2) Other reasonable courses of actions.

(3) Mitigation measures (not in the proposed action).

(c) Impacts, which may be: (1) Direct; (2) indirect; (3) cumulative.

§ 1508.26 Special expertise.

"Special expertise" means statutory responsibility, agency mission, or related program experience.

§ 1508.27 Significantly.

"Significantly" as used in NEPA requires considerations of both context and intensity:

(a) *Context.* This means that the significance of an action must be analyzed in several contexts such as society as a whole (human, national), the affected region, the affected interests, and the locality. Significance varies with the setting of the proposed action. For instance, in the case of a site-specific action, significance would usually depend upon the effects in the locale rather than in the world as a whole. Both short- and long-term effects are relevant.

(b) *Intensity.* This refers to the severity of impact. Responsible officials must bear in mind that more than one agency may make decisions about partial aspects of a major action. The following should be considered in evaluating intensity:

(1) Impacts that may be both beneficial and adverse. A significant effect may exist even if the Federal agency

believes that on balance the effect will be beneficial.

(2) The degree to which the proposed action affects public health or safety.

(3) Unique characteristics of the geographic area such as proximity to historic or cultural resources, park lands, prime farmlands, wetlands, wild and scenic rivers, or ecologically critical areas.

(4) The degree to which the effects on the quality of the human environment are likely to be highly controversial.

(5) The degree to which the possible effects on the human environment are highly uncertain or involve unique or unknown risks.

(6) The degree to which the action may establish a precedent for future actions with significant effects or represents a decision in principle about a future consideration.

(7) Whether the action is related to other actions with individually insignificant but cumulatively significant impacts. Significance exists if it is reasonable to anticipate a cumulatively significant impact on the environment. Significance cannot be avoided by terming an action temporary or by breaking it down into small component parts.

(8) The degree to which the action may adversely affect districts, sites, highways, structures, or objects listed in or eligible for listing in the National Register of Historic Places or may cause loss or destruction of significant scientific, cultural, or historical resources.

(9) The degree to which the action may adversely affect an endangered or threatened species or its habitat that has been determined to be critical under the Endangered Species Act of 1973.

(10) Whether the action threatens a violation of Federal, State, or local law or requirements imposed for the protection of the environment.

[43 FR 56003, Nov. 29, 1978; 44 FR 874, Jan. 3, 1979]

§ 1508.28 Tiering.

"Tiering" refers to the coverage of general matters in broader environmental impact statements (such as national program or policy statements) with subsequent narrower statements or environmental analyses (such as regional or basinwide program statements or ultimately site-specific statements) incorporating by reference the general discussions and concentrating solely on the issues specific to the statement subsequently prepared. Tiering is appropriate when the sequence of statements or analyses is:

(a) From a program, plan, or policy environmental impact statement to a program, plan, or policy statement or analysis of lesser scope or to a site-specific statement or analysis.

(b) From an environmental impact statement on a specific action at an early stage (such as need and site selection) to a supplement (which is preferred) or a subsequent statement or analysis at a later stage (such as environmental mitigation). Tiering in such cases is appropriate when it helps the lead agency to focus on the issues which are ripe for decision and exclude from consideration issues already decided or not yet ripe.

INDEX

INDEX—Continued

Glossary

The reader is directed to §1507 for additional information regarding the definition of NEPA terms.

Act A synonym used in the Council on Environmental Quality Regulations to refer to the National Environmental Policy Act, as amended (42 U.S.C. 4321, et seq.).

Actions The Council on Environmental Quality's NEPA Regulations define three types of actions, other than unconnected single actions, which must be taken into consideration during a NEPA analysis: (1) connected, (2) cumulative, and (3) similar.

Administrative Procedures Act A law that specifies the requirements and procedures that must be followed in issuing regulations.

Alternatives As used in the Council on Environmental Quality's NEPA Regulations, other reasonable options that would meet the need of a proposed action. There are three types of alternatives: (1) no-action alternative, (2) other reasonable courses of actions, and (3) mitigation measures (not in the proposed action).

Applicant An applicant is a nonfederal party that has filed an application with a federal agency that is subject to a NEPA review before the agency may approve the application. Such applications normally involve required federal approvals or permits that must be obtained before the applicant may proceed with a specified action.

Biodiversity The variety and variability of different organisms in the environment.

Categorical Exclusions (CATX) A class of actions, under NEPA, that do not have a significant effect, either individually or cumulatively, on the human environment, and therefore do not require preparation of an environmental assessment or environmental impact statement.

CEQ See Council on Environment.

Connected Actions As defined by the Council on Environmental Quality's NEPA Regulations, actions that are closely related and therefore should be discussed in the same impact statement. Actions are connected if they (1) automatically trigger other actions that may require environmental impact statements, (2) cannot or will not proceed unless other actions are taken previously or simultaneously, or (3) are interdependent parts of a larger action and depend on the larger action for their justification.

Context As used in the Council on Environmental Quality's NEPA Regulations, a factor that must be considered in making a determination regarding the significance of an impact. In making a determination regarding the significance of an action, the impacts must be analyzed in several contexts, such as society as a whole (human, national), the affected region, the affected interests, and the locality. Significance varies with the setting of the proposed action. For instance, in the case of a site-specific action, significance usually depends on the effects in the locale rather than in the world as a whole. Both short- and long-term effects are relevant.

Cooperating Agency A federal agency, other than a lead agency, that has jurisdiction by law or special expertise with respect to any environmental impact involved in a proposal (or a reasonable alternative) for legislation or other major federal action significantly affecting the quality of the human environment.

Council A synonym used for the Council on Environmental Quality.

Council on Environmental Quality (CEQ) The council created by Title II of NEPA to oversee the NEPA process.

Council on Environmental Quality Regulations The regulations issued by the Council on Environmental Quality (40 CFR parts 1500-1508) for implementing the procedural aspects of NEPA.

Cumulative Actions As defined by the Council on Environmental Quality's NEPA Regulations, actions that, when viewed with other proposed actions, have cumulatively significant impacts and should therefore be discussed in the same impact statement.

Cumulative Impact The impact on the environment that results from the incremental impact of an action when it is added to other past, present, and reasonably foreseeable future actions, regardless of what agency (federal or nonfederal) or person has undertaken them. This is an important concept because individually minor but collectively significant impacts can take place over a period of time.

Direct Impacts Effects caused by the action that occur at the same time and place as the action.

EA See Environmental Assessment.

Ecology The relationship of organisms to one another and within an environment.

Effects As used in the NEPA Regulations, synonymous with *impacts*. Effects may include impacts of an action on ecological resources (such as natural resources and the components, structures, and functioning of affected ecosystems) and aesthetic, historic, economic, social, health, and cultural resources. The concept of effects includes direct, indirect, and cumulative effects, and includes both beneficial and detrimental impacts. There are three types of impacts: (1) direct, (2) indirect, and (3) cumulative.

EIS See Environmental Impact Statement.

Emission A pollution discharge into the atmosphere from smokestacks, vents and other sources.

Endangered Species Organisms threatened with extinction by man-made or natural changes in the environment.

Environment See Human Environment.

Environmental Assessment (EA) A concise public document used to briefly provide sufficient evidence and analysis for determining whether to prepare an environmental impact statement or a finding of no significant impact for a proposed action. An EA may also be used to assist an agency in compliance with the NEPA act when no environmental impact statement is necessary and to facilitate preparation of an EIS when one is necessary. An EA must include brief discussions of the need for the proposal, alternatives, environmental impacts of the proposed action and alternatives, and a listing of agencies and persons consulted.

Environmental Document As defined in the Council on Environmental Quality's NEPA Regulations, includes environmental assessment, environmental impact statement, finding of no significant impact, and the notice of intent.

Environmental Impact Statement (EIS) A detailed document that must be prepared under the Council on Environmental Quality's NEPA Regulations for proposed actions that may result in a significant environmental impact.

Federal Agency As defined by the Council on Environmental Quality's NEPA Regulations, any agency of the federal government. This term does not include Congress, the judiciary, or the president, including the staff functions of the Executive Office.

Finding of No Significant Impact (FONSI) A document by a federal agency that briefly presents the reasons why an action not categorically excluded will not have a significant effect on the human environment and therefore, for which an environmental impact statement will not be required. It must include the environmental assessment or a summary of it. The FONSI must also note any other environmental documents related to it.

FONSI See Finding of No Significant Impact.

Groundwater The supply of water stored beneath the Earth's surface.

Habitat The location and surroundings where a population of plants or animals live.

Hazardous Waste A waste defined under the Resource Conservation and Recovery Act as posing a hazard to human health and the environment. To be designated as a hazardous, the waste must possess one of the following four characteristics: (1) reactivity, (2) corrosivity, (3) ignitability, (4) toxicity. A waste may also be designated hazardous if it is listed by the Environmental Protection Agency (EPA) as a hazardous waste.

Human Environment As defined by the Council on Environmental Quality's NEPA Regulations, the natural and physical environment and the relationship of people with that environment. This means that economic or social effects are not intended by themselves to require preparation of an EIS. When an EIS is prepared and economic or social and natural or physical environmental effects are interrelated, then the EIS will discuss all of these effects on the human environment.

Hydrology The science dealing with the distribution and movement of surface water and groundwater.

Impacts See Effects.

Indirect Impacts Reasonably foreseeable impacts that are caused by an action but that occur later or that are removed in distance from the action. Indirect impacts may include growth-inducing effects and other effects related to induced changes in the pattern of land use, population density, or growth rate, and related effects on air and water and other natural systems, including ecosystems.

Intensity As used in the Council on Environmental Quality's NEPA Regulations, a factor that must be considered in making a determination regarding the significance of an impact, which in turn contributes to the determination of the significance of an action. The intensity is the degree to which the impact would affect the environment.

Interim Action An action within the scope of a proposal that is the subject of an ongoing EIS, that an agency proposes to pursue before the ROD is issued, and that is permissible under 40 CFR 1506.1.

Implementation Plan (IP) A document used by many federal agencies to record the results of the EIS scoping process. The IP also provides a plan for preparing the EIS.

IP See Implementation Plan.

Jurisdiction by Law As used in NEPA, agency authority to approve, veto, or finance all or part of a proposal.

Land Use Plans With respect to NEPA, formally adopted documents for land use planning or zoning, including proposed plans that have been formally proposed by a government body and are under active consideration (See Council on Environmental Quality's *Forty Most Asked Questions,* question 23b).

Land Use Policies As used in reference to NEPA, formally adopted statements of land use policy embodied in laws or regulations as well as policies that have been formally proposed but not yet adopted (See Council on Environmental Quality's *Forty Most Asked Questions,* question 23b).

Lead Agency As used in the Council on Environmental Quality's NEPA Regulations, the agency or agencies preparing or having primary responsibility for preparing the environmental impact statement.

Legislation A bill or legislative proposal to Congress developed by or with the significant cooperation and support of a federal agency, not including requests for appropriations. The test for significant cooperation is whether the proposal is in fact predominantly that of the agency or another source. Drafting does not by itself constitute significant cooperation. Proposals for legislation include requests for ratification of treaties. Only the agency with primary responsibility for the subject matter involved prepares a legislative environmental impact statement.

Major Federal Action As used in the Council on Environmental Quality's NEPA Regulations, action with potentially major effects and that is potentially subject to federal control and responsibility. The term *major* reinforces but does not have a meaning independent of the term *significant.* Actions include the circumstance where responsible officials fail to act and that failure to act is reviewable by courts or administrative tribunals under the Administrative Procedure Act or other applicable law as agency action.

Mitigation Measure that may be taken to avoid, minimize, rectify, reduce, or compensate for the adverse impacts of an action on the environment.

Mitigation Action Plan a Document describing the plan for implementing commitments made in an EIS/ROD or EA/FONSI.

Monitoring The process of observing and measuring environmental impacts on environmental resources to verify compliance with the description of the proposed action and mitigation factors that were cited in a NEPA document.

National Environmental Policy Act (NEPA) Statute passed by Congress in 1969 establishing the basic environmental policy for protection of the environment (42 U.S.C. 4321 et seq.). It provides a systematic and interdisciplinary process that agencies are required to follow to reduce or prevent environment degradation. The Act contains action-forcing procedures that must be followed by federal agencies to ensure federal decisionmakers take environmental factors into account before making a final decision regarding a proposed action.

NEPA See National Environmental Policy Act.

NEPA Process All measures that are necessary for compliance with the requirements of section 2 and Title I of NEPA.

NEPA Review The process followed in complying with section 102(2) of NEPA.

NOA See Notice of Availability.

NOI See Notice of Intent.

Notice of Availability (NOA) A formal notice, defined at 40 CFR 1508.22 and published in the *Federal Register,* announcing the issuance and public availability of a draft or final EIS.

Notice of Intent (NOI) A formal notice, published in the *Federal Register,* announcing the issuance and public availability of a draft or final EIS.

Program For the purposes of NEPA, a sequence of connected or related actions as discussed in 40 CFR 1508.18(b)(3) and 1508.25(a).

Programmatic EA/EIS A broadly scoped EA or EIS prepared to evaluate an agency program and/or including a sequence of connected or related agency actions or projects as discussed at 40 CFR 1508.18(b)(3) and 1508.25(a).

Project For the purposes of NEPA, a specific agency effort, including actions approved by a permit or regulatory decision, federal and federally assisted activities, or similar activities, as described in 40 CFR 1508.18(b)(4).

Proposal As used in the Council on Environmental Quality's NEPA Regulations, that stage in the development of an action when an agency subject to the Act has a goal, is actively preparing to make a decision on one or more alternative means of accomplishing that goal, whose effects can be meaningfully evaluated. Preparation of an EIS on a proposal should be timed so that the final statement may be completed in time for the statement to be included in any recommendation or report on the proposal. A proposal may exist in fact as well as by agency declaration that one exists.

Public Scoping That portion of the scoping process where the public is invited to participate, as described at 40 CFR 1501.7 (a)(1) and (b)(4).

Record of Decision (ROD) A public document, prepared on completion of an EIS, that records the agency's final decision, rationale for making the decision, and commitments to monitoring and mitigation.

Referring Agency As used in the Council on Environmental Quality's NEPA Regulations, the federal agency that refers any matter to the Council after a determination that the matter is unsatisfactory from the standpoint of public health or welfare or environmental quality.

Regulations As used in this book, NEPA Regulations that were issued by the Council on Environmental Quality.

Resources With respect to NEPA, environmental resources, including all physical (e.g., geological, biological, atmospheric), socioeconomic, and other related aspects of the environment that may be potentially affected by the agency's action.

Risk As used in this text, the probability that an accident will occur multiplied by its consequences.

ROD See Record of Decision.

S-EIS See Supplemental EIS.

Scope As used in the Council on Environmental Quality's NEPA Regulations, the range of actions, alternatives, and impacts to be considered in an EIS. The scope of an individual statement may depend on its relationships to other statements. To determine the scope of an EIS, the agency must consider three types of actions, three types of alternatives, and three types of impacts.

Significance The degree to which an impact may affect the human environment. The term, as used in the Council on Environmental Quality's NEPA Regulations, requires consideration of both context and intensity of an impact.

Similar Actions As defined by the Council on Environmental Quality's NEPA Regulations, actions that, when viewed with other reasonably foreseeable or proposed agency actions, have similarities that provide a basis for evaluating their environmental consequences together, such as common timing or geography. An agency may wish to analyze these actions in the same impact statement; it should do so when the best way to assess adequately the combined impacts of similar actions or reasonable alternatives to such actions is to treat them in a single impact statement.

Special Expertise As defined by the Council on Environmental Quality's NEPA Regulations, statutory responsibility, agency mission, or related program experience.

Supplemental EIS (S-EIS) An EIS prepared to supplement an existing EIS as described in 40 CFR 1502.9(c). A supplemental EIS is prepared when a substantial change is made to the proposed action or when important new information is acquired regarding the action.

Tiering The coverage of general matters in broader environmental impact statements (such as national program or policy statements) with subsequent narrower statements or environmental analyses (such as regional or basinwide program statements or, ultimately, site-specific statements) that incorporate by reference the general discussions and concentrate solely on the issues specific to the statement subsequently prepared.

Tribal Lands The area of "Indian country," as defined in 18 U.S.C. 1151, that is under a tribe's jurisdiction.

Wetlands An area that is saturated or partially saturated. An area need be saturated only during a small potion of the year to be designated a wetlands. In order to be designated a wetlands, the area must be exhibit certain soil, hydrological, and vegetative characteristics.

About the Author

Charles H. Eccleston is the author of the successful text *The NEPA Planning Process: A Comprehensive Guide with Emphasis on Efficiency,* John Wiley & Sons (1999), and is the author of the upcoming text *How to Write Effective Environmental Assessments: A Comprehensive Guide to Effectively Planning and Complying with NEPA's EA Requirements.* He has lectured, taught, and authored over 20 publications on the National Environmental Policy Act (NEPA) and environmental impact assessment.

In nearly 20 years of diverse engineering and scientific experience, he has managed a diverse array of environmental analysis and planning efforts. As a principal environmental scientist at the U.S. Department of Energy's Hanford Site in Richland, Washington, he has developed innovative tools, techniques, and strategies for effectively integrating NEPA with sitewide planning, and other environmental processes, such as ISO-14000 and pollution prevention. In this position, he has developed numerous methodologies that have received national attention for their ability to streamline NEPA compliance while reducing project cost and delays.

Mr. Eccleston has chaired two national committees chartered with responsibility for establishing nationally accepted methods of professional practice (AMPPs) that address problems that have traditionally hindered NEPA and environmental planning. Currently, he chairs the National Association of Environmental Professionals Tools and Techniques (TNT) NEPA Practice Committee.

As a member of *Who's Who in Science and Engineering in America,* he has participated in two White House–sponsored environmental workshops held to develop approaches for improving NEPA effectiveness and for spearheading a national environmental-industrial coalition.

Prior to working at the Department of Energy's Hanford Site, he held a position with the GTE Corporation's Defense Electronics Division in Sunnyvale, California, where he contributed to the development of advanced strategic weapon systems that helped to bring an end to the Cold War. Before this, he was a senior engineer in the Advanced Design Branch (ADB) at the Texas Instruments Corporation in Dallas.

Mr. Eccleston holds a master of science degree in environmental geology/geophysics and bachelor of science degrees in environmental geology and computer science. The author is a Certified Environmental Professional (CEP). He consults on NEPA and planning problems and may be contacted by e-mail at ecclestonc@aol.com.

Index